Materials Horizons: From Nature to Nanomaterials

Series Editor

Vijay Kumar Thakur, School of Aerospace, Transport and Manufacturing, Cranfield University, Cranfield, UK

Materials are an indispensable part of human civilization since the inception of life on earth. With the passage of time, innumerable new materials have been explored as well as developed and the search for new innovative materials continues briskly. Keeping in mind the immense perspectives of various classes of materials, this series aims at providing a comprehensive collection of works across the breadth of materials research at cutting-edge interface of materials science with physics, chemistry, biology and engineering.

This series covers a galaxy of materials ranging from natural materials to nanomaterials. Some of the topics include but not limited to: biological materials, biomimetic materials, ceramics, composites, coatings, functional materials, glasses, inorganic materials, inorganic-organic hybrids, metals, membranes, magnetic materials, manufacturing of materials, nanomaterials, organic materials and pigments to name a few. The series provides most timely and comprehensive information on advanced synthesis, processing, characterization, manufacturing and applications in a broad range of interdisciplinary fields in science, engineering and technology.

This series accepts both authored and edited works, including textbooks, monographs, reference works, and professional books. The books in this series will provide a deep insight into the state-of-art of Materials Horizons and serve students, academic, government and industrial scientists involved in all aspects of materials research.

Review Process

The proposal for each volume is reviewed by the following:

1. Responsible (in-house) editor
2. One external subject expert
3. One of the editorial board members.

The chapters in each volume are individually reviewed single blind by expert reviewers and the volume editor.

Krashn Kumar Dwivedi · Piyush Uniyal ·
Akarsh Verma
Editors

Mechanics and Materials Science of Biological Materials

Experimental and Modeling Approaches

Springer

Editors
Krashn Kumar Dwivedi
Department of Mechanical Engineering
Washington University
Saint Louis, WA, USA

Piyush Uniyal
Department of Mechanical Engineering
KU Leuven
Leuven, Belgium

Akarsh Verma
Department of Mechanical Engineering
UPES (University of Petroleum and Energy Studies)
Dehradun, Uttarakhand, India

ISSN 2524-5384 ISSN 2524-5392 (electronic)
Materials Horizons: From Nature to Nanomaterials
ISBN 978-981-96-5365-2 ISBN 978-981-96-5366-9 (eBook)
https://doi.org/10.1007/978-981-96-5366-9

This Springer imprint is published by the registered company Springer Nature Singapore Pte Ltd.
The registered company address is: 152 Beach Road, #21-01/04 Gateway East, Singapore 189721, Singapore

Contents

Multiscale Assessment of Bone Fragility Beyond Bone Mineral Density

Praveer Sihota and Navin Kumar

Abstract Recent research highlights that impairments in bone quality, rather than solely bone mineral density (BMD), are a significant factor in the weakened mechanical properties of bone seen in various medical conditions, including type 1 and type 2 diabetes, chronic kidney disease, cancer, and certain metabolic bone disorders. These conditions may alter bone quality without significantly affecting BMD, leading to an elevated risk of fragility fractures. Extensive studies have revealed that the BMD T-score, a widely used quantitative metric, frequently falls short in accurately predicting fracture risk. To address this, both in vivo and ex vivo methods have been utilized to assess bone quality, with a focus on structural, mechanical, and compositional characteristics. This review examines the multiscale evaluation of bone quality, emphasizing the key findings of different techniques and their respective limitations.

Keywords Bone quality · Biomechanics · Bone fragility · Bone strength · Structural characterization · Compositional analysis

1 Introduction

Recent studies suggest that deficits in bone quality, rather than bone mineral density alone, underlie the compromised bone mechanical properties observed various bone-related diseases. A multiscale assessment of bone quality is essential to fully understand these changes and their impact on fracture risk. While several techniques are available to assess bone quality, key bone quality assessment methods, their main outcome, and their utility in studying specific aspects of bone hierarchy are presented in this chapter. This chapter aims to explore the multiscale bone quality assessment approach in both in vivo and ex vivo settings, focusing on three major categories: structural, mechanical, and compositional characterization.

P. Sihota (✉) · N. Kumar
Department of Mechanical Engineering, Indian Institute of Technology Ropar, Rupnagar, Punjab, India
e-mail: praveer.sihota.mez2021@alumni.iitrpr.ac.in

K. K. Dwivedi et al. (eds.), *Mechanics and Materials Science of Biological Materials*, Materials Horizons: From Nature to Nanomaterials,
https://doi.org/10.1007/978-981-96-5366-9_1

1.1 Bone Structural Characterization

The structural assessment of bone tissue can be divided into non-invasive in vivo methods (e.g., radiographic imaging and magnetic resonance imaging) and invasive ex vivo techniques (e.g., histomorphometry). In clinical practice, non-invasive radiographic techniques are widely used for their efficiency and speed in analyzing bone morphology and evaluating fracture risk. These techniques evaluate structural characteristics across multiple scales, ranging from macroscale features like bone shape, size, and cross-sectional attributes (including cortical thickness and diameters) to microscale parameters such as bone volume fraction, trabecular structure, cortical porosity, micro-crack dimensions, and the visualization of lacunae and vascular networks.

In recent years, various advanced imaging techniques have been developed to study bone structure. These include in vivo methods such as quantitative computed tomography (QCT), peripheral QCT (pQCT), high-resolution peripheral QCT (HR-pQCT), magnetic resonance imaging (MRI), and dual-energy X-ray absorptiometry (DXA), as well as ex vivo techniques like micro-CT. Of these, HR-pQCT, DXA, and micro-CT are widely used and are discussed in detail in this chapter.

In contrast, the invasive yet highly regarded method of bone histomorphometry offers comprehensive insights into 2D bone morphology, cellular parameters, and remodeling processes. These details cannot be captured by radiographic methods and are elaborated on in subsequent sections.

1.2 Bone Histomorphometry

Bone histomorphometry represents a pivotal quantitative technique for the examination of calcified bone biopsies, providing two-dimensional insights into the microstructural and remodeling characteristics of bone tissue. Esteemed as the gold standard, it serves as an indispensable tool in both clinical and research domains, enabling the investigation of bone disease mechanisms and the evaluation of therapeutic interventions. Typically, bone biopsies are collected from the iliac crest, an optimal site that yields samples encompassing both cortical plates and trabecular bone [1].

Histomorphometric analysis is categorized into structural and remodeling parameters. Structural parameters provide insights into bone mass and architecture, including trabecular volume (BV/TV), trabecular thickness (Tb.Th), trabecular separation (Tb.Sp), and cortical width (Ct.Wi). Remodeling parameters are divided into static and dynamic groups. Static parameters include cellular indices such as osteocyte density (N.Ot/B.Ar), empty lacunae (N.Em.Lc/B.Ar), and osteoblast or osteoclast counts per bone perimeter (N.Ob/B.Pm, N.Oc/B.Pm) [2]. Other metrics include osteoid surface (OS/BS), thickness (O.Th), volume (OV/BV), and eroded surface (ES/BS) [3, 4].

Dynamic remodeling parameters, however, necessitate prior tetracycline labeling to mark areas of active mineralization. This enables the measurement of key metrics such as mineral apposition rate, bone formation rate, mineralization lag time, and activation frequency [1, 5].

While bone histomorphometry provides unparalleled insights into bone remodeling processes, it is not without limitations. As an invasive procedure, it necessitates surgical expertise and specialized facilities, which introduces risks such as infection and delayed healing, particularly for patients with comorbidities like diabetes. Despite these limitations, the detailed and precise data derived from this method make it an indispensable tool for enhancing our knowledge of bone health and disease.

1.3 Micro-computed Tomography (μ-CT)

The μ-CT is a robust ex vivo imaging method used to examine the microstructural characteristics of trabecular and cortical bone. It operates on the principle that X-rays are attenuated as they pass through a material, with the attenuated rays captured by a detector opposite the X-ray source. The object, along with the source and detector, rotates to produce multiple 2D radiographic projections, which are processed using tomographic algorithms to create detailed 3D reconstructions of bone microarchitecture. Digital image processing enhances these reconstructions for precise visualization of structural details. To ensure accurate bone density measurements, a bone-equivalent calibration phantom, such as calcium hydroxyapatite, is scanned simultaneously [6].

The μ-CT is commonly employed to analyze bone biopsies, excised specimens, or skeletal regions in small animals, offering resolutions ranging from ~0.5 to 100 μm. The approach for tomographic data acquisition differs based on the system: ex vivo desktop μ-CT setups typically use a rotating specimen stage with a stationary X-ray source and detector, while in vivo systems for small animals employ a fixed stage with a rotating source-detector assembly. This technique is highly valued for its capacity to assess trabecular microarchitecture, detect microdamage, and visualize intricate networks such as the osteocyte-lacunae system and vascular canals. However, achieving higher resolutions often results in longer scanning times and increased radiation exposure.

Using μ-CT, several structural parameters can be quantified, including trabecular bone volume fraction (BV/TV), number (Tb.N), thickness (Tb.Th), separation (Tb.Sp), structure model index (SMI), anisotropy (DA), connectivity density (Conn.D), cortical area (Ct.Ar), cortical thickness (Ct.Th), and tissue mineral density (TMD). Notably, TMD differs from volumetric bone mineral density (vBMD) as it excludes porosity and soft tissue, focusing solely on the mineralized bone matrix [7–9].

Accurate measurements of cortical porosity require careful consideration of scan resolution, particularly when examining the vascular canal network and osteocyte-lacunar network (OLN). The vascular network, critical for bone remodeling and fluid flow, is analyzed through metrics such as total canal volume (Ca.V), canal volume density (Ca.V/Ct.TV), canal number (Ca.N), and canal density (Ca.N/Ct.TV). Similarly, the OLN is evaluated using parameters like individual lacuna volume (Lc.V), lacuna volume density (Lc.V/Ct.TV), and lacuna number density (N.Lc/Ct.TV), with advanced image processing software aiding these analyses [10].

1.4 High-Resolution Peripheral Quantitative Computed Tomography (HR-PQCT)

HR-pQCT provides a non-invasive alternative to techniques like bone histomorphometry and ex vivo μ-CT for evaluating bone microstructure. This method focuses on in vivo measurements of peripheral skeletal sites, such as the radius and tibia. The second-generation HR-pQCT system (Xtreme CT II, Scanco 128 Medical AG) offers enhanced resolution (61 μm isotropic voxel size) and improved reproducibility compared to its predecessor. It enables detailed assessments of cortical and trabecular bone properties, including volumetric bone mineral density (vBMD), trabecular thickness, and cortical porosity. Additionally, the system can evaluate regions like the proximal tibia, distal radius, hand, and knee, with an effective radiation dose of approximately 0.003 mSv, significantly lower than that of QCT [11].

HR-pQCT also supports micro-finite element analysis (μFEA), allowing the exploration of biomechanical properties such as cortical thickness, porosity, and trabecular structure. However, the technique is currently limited to peripheral sites and does not address fragility fractures in clinically significant locations such as the lumbar spine or proximal femur. Motion artifacts can further affect the accuracy of bone morphology assessments [6].

Ex vivo studies have demonstrated a strong correlation between HR-pQCT-derived measurements at the tibia and the biomechanical strength of proximal femora [12]. For instance, larger cortical pores (>100 μm) in the tibia were associated with reduced simulated hip strength [13]. These findings suggest a meaningful link between microstructural parameters of the tibia and overall bone strength in other regions.

1.5 Dual-Energy X-Ray Absorptiometry (DXA)

Dual-energy X-ray absorptiometry (DXA) is a widely used non-invasive imaging method for measuring areal bone mineral density (aBMD, g/cm^2) at critical skeletal sites, including the lumbar spine (L1–L4), proximal femur (femoral neck and total hip), and distal forearm. Its quick acquisition time and ease of use make it a preferred tool in clinical settings. The technique uses two X-ray beams of different energy levels, with soft tissue absorption subtracted to isolate the energy absorbed by bone.

Algorithms process this data, which is then calibrated against a standard radiographic density to calculate aBMD. These values are compared to reference populations to assess bone density and fracture risk [14, 15].

Unlike DXA, quantitative computed tomography (QCT) provides true volumetric BMD (vBMD, g/cm^3), whereas DXA measures the mass of bone mineral over a 2D area (aBMD, g/cm^2), which can sometimes lead to limitations in interpreting bone density data.

According to World Health Organization (WHO) guidelines, DXA results are used to diagnose osteoporosis, which is defined by an aBMD value of 2.5 or more standard deviations below the mean of a young reference population (T-score ≤ -2.5). As the gold standard for osteoporosis diagnosis, DXA is instrumental in identifying individuals at heightened fracture risk. However, in certain conditions, such as type 2 diabetes, patients may exhibit normal or even elevated aBMD despite an increased fracture risk. This discrepancy suggests that additional factors, beyond aBMD, contribute to bone fragility in such cases, highlighting the need for complementary diagnostic methods [16–20].

2 Mechanical Characterization

Mechanical properties of bone can be evaluated at various hierarchical levels, ranging from macroscale testing of whole bones to mesoscale analysis of bone tissue and microscale or nanoscale investigation of material properties. Typically, mechanical testing involves securing the bone sample in a testing apparatus, applying a controlled load or displacement, and recording the resulting deformation. These tests yield critical information about bone's fracture resistance through key properties such as stiffness, strength, toughness, and fracture toughness.

This section explores the key mechanical testing methodologies, the parameters they assess, and the length scales at which they operate. Additionally, it provides an overview of sample preparation, handling, and storage considerations to ensure accurate and reliable results.

2.1 Whole Bone Test

Macroscale mechanical testing of whole bones is inherently destructive and, therefore, limited to ex vivo studies using animal or cadaveric samples. Commonly tested bones include long bones and vertebrae, with bending and compression as typical loading modes.

2.2 Three-Point Bending Test

In this method, femoral bones are often used due to their elliptical diaphyseal cross-section, which facilitates mechanical analysis. The bone is positioned on two support points, with the loading roller placed anteriorly. A preload ensures proper contact before applying a constant displacement rate until failure. Load–displacement data is collected to calculate key parameters such as maximum load (Fmax), stiffness (measured as the slope of the linear portion of the curve), and work-to-failure (area under the curve, indicating toughness) [21–23]. However, this method assumes uniform cross-sections and homogeneous material properties, which can lead to discrepancies when comparing results to smaller-scale tests [24].

2.3 Uniaxial Compression

This test evaluates material properties like modulus, strength, and toughness in machined cortical or trabecular bone specimens [25, 26]. The bone core, typically prepared with a length-to-diameter ratio of ~1.5:1, is secured in customized steel end caps to prevent movement. Preloading (~5–10N) ensures proper contact, followed by preconditioning cycles to minimize initial nonlinearities (toe region). The final test captures load–displacement data, which is normalized by sample geometry to calculate tissue-level properties. Parameters derived include elastic modulus, yield point (using the 0.2% offset method), ultimate load, post-yield strain, and toughness [21, 25, 27, 28].

2.4 Cyclic Reference Point Indentation (cRPI)

The cRPI measures bone tissue's resistance to permanent deformation, offering direct measurements of material properties while minimizing the influence of larger-scale factors, such as bone geometry. The technique is applicable in both ex vivo and in vivo (animal studies) settings using the BioDent® device (Active Life Scientific, Santa Barbara, CA). The system employs a reference probe and a spherical test tip (~2.5 μm radius) to generate load-indentation distance curves, from which key parameters are derived:

- Indentation Distance (ID): Depth during the first cycle (μm).
- Indentation Distance Increase (IDI): Difference between the last and first cycles (μm), a critical indicator of bone fragility.

IDI and its derived parameters help distinguish between fragile and robust bone, providing valuable insights into bone's mechanical properties and complementing other evaluation methods [29–33].

2.5 Microindentation

The Osteoprobe® (Active Life Scientific, Santa Barbara, CA) is a handheld impact-based microindentation device designed to assess bone material properties. It measures energy dissipation due to a single high-velocity impact, creating a controlled crack in the bone. The procedure involves three steps: (1) applying a 10N linear preload to pass through soft tissue, (2) delivering a 30N single impact load, and (3) unloading the probe. Indentations are performed on the antero-medial tibial surface, with at least eight indentations spaced ≥1 mm apart. If the device slips or is not perpendicular, additional indentations are taken to ensure reliability.

The key outcome parameter, the bone material strength index (BMSi), is calculated as the average indentation distance into the bone normalized to that into a polymethyl methacrylate (PMMA) reference phantom, multiplied by 100 [30]. However, BMSi has not been validated against standard mechanical testing parameters such as stiffness, strength, or toughness [34]. Additionally, this method does not measure deformation geometry, limiting its ability to assess hardness.

2.6 Nanoindentation (NI)

Nanoindentation provides detailed insights into bone material properties using depth-sensing transducers. Bone samples are prepared by cutting with a low-speed diamond blade saw and embedding in epoxy, which cures in approximately two hours. The cured sample is ground and polished using abrasive papers (1200, 2000, and 4000 grit) and diamond solutions (1, 0.5, and 0.25 μm particle sizes) under water cooling. Ultrasonic cleaning with distilled water is performed between each step to ensure a pristine surface, as artifacts from poor polishing can impact results [35].

Testing is conducted with a Berkovich pyramidal tip (142.3° angle, ~150 nm tip radius) calibrated using standard fused quartz and aluminum samples [36, 37]. The system integrates an in situ scanning probe microscope for precise indent positioning, with several indents made at a fixed peak load. A load–displacement curve is generated during testing, allowing the calculation of hardness (H) and reduced modulus (Er) using the Oliver and Pharr method [38, 39]. Hardness is determined as the maximum load divided by the indentation contact area, while the reduced modulus reflects the material's resistance to elastic deformation, derived from the slope of the unloading curve.

Nanoindentation also supports modulus mapping and can be combined with compositional analysis techniques such as Raman spectroscopy. However, the method requires meticulous sample preparation, as surface irregularities must be significantly smaller than the indentation size to ensure accuracy [40].

3 Compositional Analysis

Bone composition comprises three primary components: an organic matrix (predominantly type I collagen), a hydroxyapatite-like mineral phase, and water. Changes in these components, influenced by aging or disease, can significantly impact bone's mechanical properties at the macroscale. To better understand bone quality and predict its structural integrity, it is essential to evaluate these individual elements. Compositional analysis encompasses various techniques, including gravimetric analysis, assessment of crystal size, mineral-to-matrix ratio, enzymatic and non-enzymatic collagen cross-links, and mineral distribution.

3.1 Fourier Transform Infrared Spectroscopy (FTIR)

FTIR is a widely used technique for analyzing bone composition, particularly the mineral-to-matrix ratio. Bone specimens are freeze-dried, cut into small pieces, and analyzed to detect macromolecular vibrations in collagen and hydroxyapatite. The spectra are recorded in total reflectance (ATR) mode, typically in the 4000–400 cm^{-1} range, and processed using Fourier transformation to identify characteristic frequencies. Thin bone sections can also be analyzed in transmission mode, although this is challenging for unfixed, unembedded samples.

Key compositional parameters derived from FTIR include [41, 42]: carbonate-to-phosphate ratio, mineral crystallinity, acid phosphate content, non-enzymatic cross-link ratio (NE-xLR) [43], while enzymatic cross-link ratio (E-xLR) and collagen maturity ratios are also evaluated [44].

FTIR complements Raman spectroscopy for identifying molecular bond vibrations. While FTIR excels in sensitivity and speed due to its high signal-to-noise ratio (SNR), it requires dehydrated samples, limiting its application for hydrated tissues. Conversely, Raman spectroscopy offers finer spatial resolution (~1 μm) and the ability to analyze hydrated specimens or tissues in vivo, though it suffers from lower SNR and longer scan times [45–47].

3.2 High-Performance Liquid Chromatography (HPLC)

HPLC is a method utilized to isolate and measure various components within ex vivo bone specimen. The technique operates on the principle of mass transfer, where a stationary phase (sorbent particles, 2–50 μm) interacts with a mobile phase (a solvent or solvent mixture under high pressure). Separation occurs as sample components chemically interact with the stationary phase and elute at specific rates, which are accurately detected at the column's exit. HPLC allows precise quantification of collagen cross-links, bone turnover markers, and amino acids, with sensitivity reaching pico-molar levels. However, it is a destructive method, requiring sample homogenization and specialized technical skills.

The primary outcomes of HPLC include [48, 49]:

- Enzymatic Cross-Links (E-xL): Beneficial cross-links formed by lysyl oxidase, transitioning from immature (DHLNL, HLNL) to mature forms (pyridinoline [Pyr], deoxypyridinoline [Dpyr]).
- Non-Enzymatic Cross-Links (NE-xL): Fluorescent cross-links, such as pentosidine, resulting from glycation and associated with aging and type 2 diabetes.

E-xL stabilizes by 10–15 years of age in human, contributing to mechanical strength, whereas NE-xL is linked to increased bone fragility. The ability to measure both types of cross-links provides critical insights into bone quality and its mechanical properties [50–52].

3.3 Scanning Electron Microscopy—Quantitative Backscattered Electron Imaging (qBEI)

SEM is an effective technique for analyzing bone surface morphology at micro- and sub-micrometer scales. When high-energy electrons interact with the bone surface, they generate two types of signals: backscattered and secondary electrons. Standard SEM imaging utilizes secondary electrons, which have lower kinetic energy, to create detailed images of the surface. To prevent surface charging during the process, bone samples are coated with a thin layer of platinum using ion sputtering.

Quantitative Backscattered Electron Imaging (qBEI) utilizes backscattered electrons, which result from interactions between the electron beam and the sample's atomic structure. The intensity of backscattered electrons correlates with the atomic number (Z) of the elements in the sample, a phenomenon known as Z contrast. Calcium, with its high atomic number ($Z = 20$), dominates the qBEI signal, enabling the study of bone tissue mineralization stages.

The main outcome of qBEI analysis is the bone mineral density distribution (BMDD), which provides detailed insights into local mineralization variations. Key parameters include: mean calcium weight percentage (mean Ca, wt %), peak calcium weight percentage (peak Ca, wt %), mineralization heterogeneity (width Ca, wt %) [3, 53].

BMDD values are calibrated using a calcium standard to quantify calcium content in bone specimens. A key limitation is the need for a meticulously polished surface, as irregularities can affect backscattering accuracy. To ensure proper polishing, bone samples are generally embedded in PMMA [54].

3.4 Energy Dispersive X-ray Spectroscopy (EDX)

EDX integrated with SEM for determining the elemental composition of a bone sample. In EDX, the sample is irradiated with high-intensity X-rays, producing characteristic peaks corresponding to the elements present. The concentration of elements can be quantified based on peak intensities.

EDX utilizes dehydrated specimens with thin metal coating to prevent charging. EDX is particularly useful for analyzing the calcium/phosphorus (Ca/P) ratio in bone samples, providing valuable insights into mineral composition [55, 56].

3.5 Thermogravimetric Analysis (TGA)

TGA is used to quantify the bulk mineral-to-matrix ratio in bone samples by monitoring mass changes as a function of temperature. This technique provides comprehensive data on the total mineral, organic, and water content, including carbonated minerals.

In a typical TGA procedure, a bone sample (8–12 mg) is placed in an alumina crucible and heated at a constant rate (usually 10 °C/min) in a controlled air atmosphere up to 1000 °C. The mass loss is recorded across specific temperature ranges. The bulk mineral-to-matrix ratio is determined as the ratio of the mineral content (percentage dry weight at 600 °C) to the organic mass loss (200–600 °C) [57, 58].

3.6 X-ray Diffraction (XRD)

XRD is used to analyze the structural characteristics of bone mineral, including mean crystal size. Bone samples are prepared by defatting and dehydrating, typically using ethanol, followed by grinding into a homogeneous powder. These samples are then dried and stored for analysis.

During XRD, a slow scan captures the diffraction pattern, focusing on peaks at angles $2\theta = 26°$ (002 plane, c-axis) and $2\theta = 40°$ (310 plane, ab-plane). These peaks provide data for calculating crystal dimensions [59–61]. Scherrer equation, $B(2\theta) = \lambda/L \cos\theta$, is used to estimate mean crystal size, where λ is the X-ray wavelength, θ is the Bragg angle, B is the mean crystal size, and L is the peak width at half-maximum [57–59]. CuKα radiation with a wavelength of 1.5406 Å is commonly employed, offering nanoscale insights into bone mineral structure.

3.7 Small-Angle X-ray Scattering (SAXS)

SAXS is a widely used technique for investigating materials at the nanometer scale. It measures scattered rays from an incident X-ray beam to analyze parameters such as average crystal thickness, shape, and predominant orientation [60, 62]. SAXS relies on elastic scattering, where the energy and wavelength of the incident and scattered radiation remain unchanged.

While SAXS offers valuable insights, its lower signal-to-noise ratio (SNR) compared to XRD necessitates the use of a synchrotron source to generate high-energy, monochromatic X-ray beams [63]. Bone analysis using SAXS typically involves thin calcified sections examined in transmission mode.

Studies have shown that bone proteins, such as osteocalcin and osteopontin, play a codependent role in regulating mineral crystal size and organization [64]. Osteocalcin influences the physical properties of bone mineral, while osteopontin regulates its composition. SAXS has also been used to measure mean crystal thickness in bones from various species and trabecular bones from human iliac crest biopsies [65, 66]. Additionally, studies reveal that collagen and mineral crystals are predominantly aligned with the principal stress direction, with crystal thickness varying across skeletal sites and bone structures.

SAXS is particularly useful for detecting local variations in bone mineral properties and alignment, providing critical insights into bone nanostructure.

3.8 Solid-State Nuclear Magnetic Resonance (SSNMR)

Solid-State Nuclear Magnetic Resonance (SSNMR), also referred to as magnetic resonance spectroscopy, operates on the principle of aligning nuclear magnetic spins in a strong external magnetic field. This alignment is disrupted by an oscillating radio frequency (RF) pulse, and the electromagnetic signals emitted during relaxation are analyzed to reveal molecular structural details, including functional groups and their chemical environments. SSNMR is capable of being performed on intact bone, providing valuable compositional and structural insights [67] and offering compositional insights at picometer resolution. However, its limitations involve the requirement for a magic angle spinning system and a high-strength magnetic field to achieve well-resolved spectra.

Common isotopes used in SSNMR for bone studies include ^{1}H (proton), ^{31}P, ^{43}Ca, and ^{13}C, each providing unique insights:

- ^{1}H NMR: Differentiates between bound and mobile (pore) water, as their relaxation times produce distinct signals. Bound water concentration has been positively correlated with bone strength and toughness, highlighting its potential as a marker for fracture risk.
- ^{31}P and ^{43}Ca NMR: Probe the inorganic mineral phase, predominantly carbonated hydroxyapatite.

- ^{13}C NMR: Analyzes the organic matrix, where most carbon is located.

SSNMR spectroscopy can also investigate the interface between the bone mineral and organic matrix, making it a comprehensive tool for studying bone composition and mechanical properties [68–70].

4 Summary

X-ray-based technique (DXA) is still limited in fracture risk prediction, highlighting the need for additional bone quality parameters to more accurately identify patients at risk. This review examines various techniques for assessing bone quality, focusing on microstructural, biomechanical, and compositional properties across different length scales of bone hierarchy.

Structural characterization techniques, including QCT, pQCT, HR-pQCT, and DXA, provide in vivo assessments, while micro-CT and bone histomorphometry offer ex vivo analysis. Bone histomorphometry is unique in its ability to provide cellular indices and dynamic bone remodeling information, but for dynamic analysis, it requires prior tetracycline labelling. X-ray-based imaging techniques primarily focus on BMD and microstructure, often neglecting the organic matrix's role in bone toughness. Advances in technology are expected to address issues related to scan resolution and radiation dose.

Mechanical testing methods, such as three-point bending tests, are destructive and suited for animal studies but impractical for clinical use. Cyclic indentation (Biodent) and microindentation (Osteoprobe) offer valuable in vivo biomechanical insights. Microindentation technique is particularly useful for longitudinal studies but cannot measure deformation geometry, limiting its application. Nanoindentation provides hardness measurements but requires meticulous sample preparation to avoid artifacts.

Compositional analysis techniques, including TGA, FTIR, and Raman, are used to investigate the mineral-to-matrix ratio. Gravimetric analysis measures the total (bulk) mineral and organic content, while FTIR and Raman focus on specific bands related to phosphate and collagen. Each technique has specific sample preparation requirements, and many are not suitable for clinical use due to the need for bone biopsies. However, NMR and Raman show promise for non-invasive, in vivo assessment.

In summary, bone quality and strength are influenced by factors at various scales, including nanoscale mineral and collagen abnormalities, microscale microdamage in the form of cracks, and mesoscale structural changes. No individual method can entirely define bone fragility, necessitating a combination of methods. For comprehensive bone quality assessment in clinical settings, integrating HR-pQCT with FEA modeling, Osteoprobe, and Raman spectroscopy offers a non-invasive approach to evaluate vBMD, structural and mechanical properties, and compositional aspects. As in vivo assessment techniques like HR-pQCT, MRI, Raman imaging and Osteoprobe advance, patient-specific fracture risk diagnoses will become increasingly feasible in clinical practice.

Data Availability Not applicable.

Declaration of Competing Interest The authors have no conflicts to disclose.

References

1. Kulak, C. A. M., & Dempster, D. W. (2010). Bone histomorphometry: A concise review for endocrinologists and clinicians. *Arquivos Brasileiros de Endocrinologia e Metabologia, 54*(2), 87–98.
2. Dempster, D. W., Compston, J. E., Drezner, M. K., Glorieux, F. H., Kanis, J. A., Malluche, H., Meunier, P. J., Ott, S. M., Recker, R. R., & Parfitt, A. M. (2014). Standardized nomenclature, symbols, and units for bone histomorphometry: A 2012 update of the report of the ASBMR histomorphometry nomenclature committee. *Journal of Bone and Mineral Research, 28*(1), 2–17.
3. Wölfel, E. M., Jähn-Rickert, K., Schmidt, F. N., Wulff, B., Mushumba, H., Sroga, G. E., Püschel, K., Milovanovic, P., Amling, M., Campbell, G. M., Vashishth, D., & Busse, B. (2020). Individuals with type 2 diabetes mellitus show dimorphic and heterogeneous patterns of loss in femoral bone quality. *Bone, 140*, 115556.
4. Jobke, B., Milovanovic, P., Amling, M., & Busse, B. (2014). Bisphosphonate-osteoclasts: Changes in osteoclast morphology and function induced by antiresorptive nitrogen-containing bisphosphonate treatment in osteoporosis patients. *Bone, 59*, 37–43.
5. Parfitt, A. M., Drezner, M. K., Glorieux, F. H., Kanis, J. A., Malluche, H., Meunier, P. J., Ott, S. M., & Recker, R. R. (1987). Bone histomorphometry: Standardization of nomenclature, symbols, and units: Report of the ASBMR histomorphometry nomenclature committee. *Journal of Bone and Mineral Research, 2*(6), 595–610.
6. Link, T. M., & Kazakia, G. (2020). Update on imaging-based measurement of bone mineral density and quality. *Current Rheumatology Reports, 22*(5), 1–11. https://doi.org/10.1007/s11926-020-00892-w
7. Sihota, P., Yadav, R. N., Dhaliwal, R., Bose, J. C., Dhiman, V., Neradi, D., Karn, S., Sharma, S., Aggarwal, S., Goni, V. G., Mehandia, V., Vashishth, D., Bhadada, S. K., & Kumar, N. (2021). Investigation of mechanical, material and compositional determinants of human trabecular bone quality in type 2 diabetes. *Journal of Clinical Endocrinology and Metabolism, 106*(5), e2271–e2289. https://doi.org/10.1210/clinem/dgab027
8. Bouxsein, M. L., Boyd, S. K., Christiansen, B. A., Guldberg, R. E., Jepsen, K. J., & Müller, R. (2010). Guidelines for assessment of bone microstructure in rodents using micro-computed tomography. *Journal of Bone and Mineral Research, 25*(7), 1468–1486.
9. Sihota, P., Yadav, R. N., Poleboina, S., Mehandia, V., Bhadada, S. K., Tikoo, K., & Kumar, N. (2020). Development of HFD-fed/low dose STZ treated female Sprague Dawley rat model to investigate diabetic bone fragility at different organization levels. *JBMR Plus, 4*(10), e10379. https://doi.org/10.1002/jbm4.10379
10. Hemmatian, H., Conrad, S., Furesi, G., Mletzko, K., Krug, J., Faila, A. V., Kuhlmann, J. D., Rauner, M., Busse, B., & Jähn-Rickert, K. (2021). Reorganization of the osteocyte lacuno-canalicular network characteristics in tumor sites of an immunocompetent murine model of osteotropic cancers. *Bone, 152*, 116074. https://doi.org/10.1016/j.bone.2021.116074
11. Chiba, K., Okazaki, N., Kurogi, A., Isobe, Y., Yonekura, A., Tomita, M., & Osaki, M. (2018). Precision of second-generation high-resolution peripheral quantitative computed tomography:

intra- and intertester reproducibilities and factors involved in the reproducibility of cortical porosity. *Journal of Clinical Densitometry, 21*(2), 295–302.
12. Kroker, A., Plett, R., Nishiyama, K. K., McErlain, D. D., Sandino, C., & Boyd, S. K. (2017). Distal skeletal tibia assessed by HR-pQCT is highly correlated with femoral and lumbar vertebra failure loads. *Journal of Biomechanics, 59*(2017), 43–49.
13. Iori, G., Schneider, J., Reisinger, A., Heyer, F., Peralta, L., Wyers, C., Gräsel, M., Barkmann, R., Glüer, C. C., van den Bergh, J. P., Pahr, D., & Raum, K. (2019). Large cortical bone pores in the tibia are associated with proximal femur strength. *PLoS ONE, 14*(4), 1–18.
14. Damilakis, J., Adams, J. E., Guglielmi, G., & Link, T. M. (2010). Radiation exposure in X-ray-based imaging techniques used in osteoporosis. *European Radiology, 20*(11), 2707–2714.
15. Hunt, H. B., & Donnelly, E. (2016). Bone quality assessment techniques: Geometric, compositional, and mechanical characterization from macroscale to nanoscale. *Clinical Reviews in Bone and Mineral Metabolism, 14*, 133–149.
16. Melton, L. J., Riggs, B. L., Leibson, C. L., Achenbach, S. J., Camp, J. J., Bouxsein, M. L., Atkinson, E. J., Robb, R. A., & Khosla, S. (2008). A bone structural basis for fracture risk in diabetes. *Journal of Clinical Endocrinology and Metabolism, 93*(12), 4804–4809.
17. Vestergaard, P. (2007). Discrepancies in bone mineral density and fracture risk in patients with type 1 and type 2 diabetes—A meta-analysis. *Osteoporosis International, 18*(4), 427–444.
18. Dhaliwal, R., Cibula, D., Ghosh, C., Weinstock, R. S., & Moses, A. M. (2014). Bone quality assessment in type 2 diabetes mellitus. *Osteoporosis International, 25*(7), 1969–1973.
19. Farr, J. N., Drake, M. T., Amin, S., Melton, L. J., McCready, L. K., & Khosla, S. (2014). In vivo assessment of bone quality in postmenopausal women with type 2 diabetes. *Journal of Bone and Mineral Research, 29*(4), 787–795.
20. Furst, J. R., Bandeira, L. C., Fan, W. W., Agarwal, S., Nishiyama, K. K., Mcmahon, D. J., Dworakowski, E., Jiang, H., Silverberg, S. J., & Rubin, M. R. (2016). Advanced glycation endproducts and bone material strength in type 2 diabetes. *Journal of Clinical Endocrinology and Metabolism, 101*(6), 2502–2510.
21. An, Y. H., & Draughn, R. A. (2000). *Mechanical testing of bone and the bone implant interface* (Vol. 624).
22. Belill, K. A., Settle, T. L., Angel, C. R., Kim, S., & Rothwell, S. W. (2014). Femoral strength after induced lesions in rats. *Comparative Medicine, 64*(3), 186–192.
23. Jepsen, K. J., Silva, M. J., Vashishth, D., Guo, X. E., & Van Der Meulen, M. C. H. (2015). Establishing biomechanical mechanisms in mouse models: Practical guidelines for systematically evaluating phenotypic changes in the diaphyses of long bones. *Journal of Bone and Mineral Research, 30*(6), 951–966.
24. Silva, M. J., Brodt, M. D., Fan, Z., & Rho, J. Y. (2004). Nanoindentation and whole-bone bending estimates of material properties in bones from the senescence accelerated mouse SAMP6. *Journal of Biomechanics, 37*(11), 1639–1646.
25. Sihota, P., Kumar, S., Dhaliwal, R., Uniyal, P., Naresh, R., & Kumar, N. (2025). Multi-scale inferomedial femoral neck bone quality in type 2 diabetes patients with fragility fracture. *Bone, 192*, 117375. https://doi.org/10.1016/j.bone.2024.117375
26. Sihota, P., Yadav, R. N., Dhaliwal, R., Bose, J. C., Dhiman, V., Neradi, D., Karn, S., Sharma, S., Aggarwal, S., Goni, V. G., Mehandia, V., Vashishth, D., & Bhadada, S. K. (2021). Investigation of mechanical, material, and compositional determinants of human trabecular bone quality in type 2 diabetes. *The Journal of Clinical Endocrinology & Metabolism, 106*(5), e2271–e2289. https://doi.org/10.1210/clinem/dgab027
27. Wang, J., Zhou, B., Liu, X. S., Fields, A. J., Sanyal, A., Shi, X., Adams, M., Keaveny, T. M., & Guo, X. E. (2015). Trabecular plates and rods determine elastic modulus and yield strength of human trabecular bone. *Bone, 72*, 71–80.
28. Karim, L., & Vashishth, D. (2012). Heterogeneous glycation of cancellous bone and its association with bone quality and fragility. *PLoS ONE, 7*(4), e35047. https://doi.org/10.1371/journal.pone.0035047
29. Setters, A., & Jasiuk, I. (2014). Towards a standardized reference point indentation testing procedure. *Journal of the Mechanical Behavior of Biomedical Materials, 34*, 57–65.

30. Karim, L., Van Vliet, M., & Bouxsein, M. L. (2018). Comparison of cyclic and impact-based reference point indentation measurements in human cadaveric tibia. *Bone, 106*, 90–95.
31. Hansma, P., Turner, P., Drake, B., Yurtsev, E., Proctor, A., Mathews, P., Lelujian, J., Randall, C., Adams, J., Jungmann, R., Garza-De-Leon, F., Fantner, G., Mkrtchyan, H., Pontin, M., Weaver, A., Brown, M. B., Sahar, N., Rossello, R., & Kohn, D. (2008). The bone diagnostic instrument II: Indentation distance increase. *Review of Scientific Instruments, 79*(6), 1–8.
32. Sihota, P., Yadav, R. N., Poleboina, S., Mehandia, V., Bhadada, S. K., Tikoo, K., & Kumar, N. (2020). Development of HFD-fed/low-dose STZ-treated female Sprague-Dawley rat model to investigate diabetic bone fragility at different organization levels. *JBMR Plus, 4*(10), 10379. https://doi.org/10.1002/jbm4.10379
33. Gallant, M. A., Brown, D. M., Organ, J. M., Allen, M. R., & Burr, D. B. (2013). Reference-point indentation correlates with bone toughness assessed using whole-bone traditional mechanical testing. *Bone, 53*(1), 301–305.
34. Allen, M. R., McNerny, E. M. B., Organ, J. M., & Wallace, J. M. (2015). True gold or pyrite: A review of reference point indentation for assessing bone mechanical properties in vivo. *Journal of Bone and Mineral Research, 30*(9), 1539–1550.
35. Edward Hoffler, C., Edward Guo, X., Zysset, P. K., & Goldstein, S. A. (2005). An application of nanoindentation technique to measure bone tissue lamellae properties. *Journal of Biomechanical Engineering, 127*(7), 1046–1053.
36. Bhushan, B. (2017). Depth-sensing nanoindentation measurement techniques and applications. *Microsystem Technologies, 23*(5), 1595–1649.
37. Saini, K., Discher, D., & Kumar, N. (2018). Static and time-dependent mechanical response of organic matrix of bone. *Journal of the Mechanical Behavior of Biomedical Materials, 2019*(91), 315–325.
38. Oliver, W. C., & Pharr, G. M. (1992). An improved technique for determining hardness and elastic modulus using load and displacement sensing indentation experiments. *Journal of Materials Research, 7*(06), 1564–1583.
39. Das, R., Kumar, A., Patel, A., Vijay, S., Saurabh, S., & Kumar, N. (2016). Biomechanical characterization of spider webs. *Journal of the Mechanical Behavior of Biomedical Materials, 2017*(67), 101–109.
40. Donnelly, E., Baker, S. P., Boskey, A. L., & Van Der Meulen, M. C. H. (2004). Effects of surface roughness and maximum load on the mechanical properties of cancellous bone measured by nanoindentation. *Materials Research Society Symposia—Proceedings, 823*(2), 103–108.
41. Boskey, A., & Mendelsohn, R. (2005). Infrared analysis of bone in health and disease. *Journal of Biomedial Optics, 10*(3), 31102.
42. Spevak, L., Flach, C. R., Hunter, T., Mendelsohn, R., & Boskey, A. L. (2013). FTIRI parameters describing acid phosphate substitution in biologic hydroxyapatite. *Calcified Tissue International, 92*(5), 418–428.
43. Schmidt, F. N., Zimmermann, E. A., Campbell, G. M., Sroga, G. E., Püschel, K., Amling, M., Tang, S. Y., Vashishth, D., & Busse, B. (2017). Assessment of collagen quality associated with non-enzymatic cross-links in human bone using Fourier-transform infrared imaging. *Bone, 97*, 243–251.
44. Paschalis, E. P., Gamsjaeger, S., Tatakis, D. N., Hassler, N., Robins, S. P., & Klaushofer, K. (2014). Fourier transform infrared spectroscopic characterization of mineralizing type i collagen enzymatic trivalent cross-links. *Calcified Tissue International, 96*(1), 18–29.
45. Boskey, A. L., & Mendelsohn, R. (2005). Infrared spectroscopic characterization of mineralized tissues. *Vibrational Spectroscopy, 38*(1–2), 107–114.
46. Morris, M. D., & Mandair, G. S. (2010). *Raman assessment of bone quality*. https://doi.org/10.1007/s11999-010-1692-y
47. Taylor, E. A., & Donnelly, E. (2020). Raman and Fourier transform infrared imaging for characterization of bone material properties. *Bone, 139*, 115490.
48. Saito, M., Marumo, K., Fujii, K., & Ishioka, N. (1997). Single-column high-performance liquid chromatographic-fluorescence detection of immature, mature, and senescent cross-links of collagen. *Analytical Biochemistry, 253*(1), 26–32.

49. Saito, M., Fujii, K., Mori, Y., & Marumo, K. (2006). Role of collagen enzymatic and glycation induced cross-links as a determinant of bone quality in spontaneously diabetic WBN/Kob rats. *Osteoporosis International, 17*(10), 1514–1523.
50. Saito, M., Kida, Y., Kato, S., & Marumo, K. (2014). Diabetes, collagen, and bone quality. *Current Osteoporosis Reports, 12*(2), 181–188.
51. Saito, M., & Marumo, K. (2010). Collagen cross-links as a determinant of bone quality: A possible explanation for bone fragility in aging, osteoporosis, and diabetes mellitus. *Osteoporosis International, 21*(2), 195–214.
52. Viguet-Carrin, S., Garnero, P., & Delmas, P. D. (2006). The role of collagen in bone strength. *Osteoporosis International, 17*(3), 319–336.
53. Dragoun Kolibová, S., Wölfel, E. M., Hemmatian, H., Milovanovic, P., Mushumba, H., Wulff, B., Neidhardt, M., Püschel, K., Failla, A. V., Vlug, A., Schlaefer, A., Ondruschka, B., Amling, M., Hofbauer, L. C., Rauner, M., Busse, B., & Jähn-Rickert, K. (2023). Osteocyte apoptosis and cellular micropetrosis signify skeletal aging in type 1 diabetes. *Acta Biomaterialia, 162*, 254–265.
54. Roschger, P., Fratzl, P., Eschberger, J., & Klaushofer, K. (1998). Validation of quantitative backscattered electron imaging for the measurement of mineral density distribution in human bone biopsies. *Bone, 23*(4), 319–326.
55. Kourkoumelis, N., Balatsoukas, I., & Tzaphlidou, M. (2012). Ca/P concentration ratio at different sites of normal and osteoporotic rabbit bones evaluated by Auger and energy dispersive X-ray spectroscopy. *Journal of Biological Physics, 38*(2), 279–291.
56. Perdikouri, C., Tägil, M., & Isaksson, H. (2015). Characterizing the composition of bone formed during fracture healing using scanning electron microscopy techniques. *Calcified Tissue International, 96*(1), 11–17.
57. Rodriguez-Florez, N., Garcia-Tunon, E., Mukadam, Q., Saiz, E., Oldknow, K. J., Farquharson, C., Millán, J. L., Boyde, A., & Shefelbine, S. J. (2015). An investigation of the mineral in ductile and brittle cortical mouse bone. *Journal of Bone and Mineral Research, 30*(5), 786–795.
58. Mkukuma, L. D., Imrie, C. T., Skakle, J. M. S., Hukins, D. W. L., & Aspden, R. M. (2005). Thermal stability and structure of cancellous bone mineral from the femoral head of patients with osteoarthritis or osteoporosis. *Annals of the Rheumatic Diseases, 64*(2), 222–225.
59. Farlay, D., Panczer, G., Rey, C., Delmas, P. D., & Boivin, G. (2010). Mineral maturity and crystallinity index are distinct characteristics of bone mineral. *Journal of Bone and Mineral Metabolism, 28*(4), 433–445.
60. Turunen, M. J., Kaspersen, J. D., Olsson, U., Guizar-Sicairos, M., Bech, M., Schaff, F., Tägil, M., Jurvelin, J. S., & Isaksson, H. (2016). Bone mineral crystal size and organization vary across mature rat bone cortex. *Journal of Structural Biology, 195*(3), 337–344.
61. Vetter, U., Eanes, E. D., Kopp, J. B., Termine, J. D., & Robey, P. G. (1991). Changes in apatite crystal size in bones of patients with osteogenesis imperfecta. *Calcified Tissue International, 49*(4), 248–250.
62. Fratzl, P., Groschner, M., Vogl, G., Plenk, H., Eschberger, J., Fratzl-Zelman, N., Koller, K., & Klaushofer, K. (2009). Mineral crystals in calcified tissues: A comparative study by SAXS. *Journal of Bone and Mineral Research, 7*(3), 329–334.
63. Giannini, C., Siliqi, D., Ladisa, M., Altamura, D., Diaz, A., Beraudi, A., Sibillano, T., De Caro, L., Stea, S., Baruffaldi, F., & Bunk, O. (2014). Scanning SAXS-WAXS microscopy on osteoarthritis-affected bone—An age-related study. *Journal of Applied Crystallography, 47*(1), 110–117.
64. Poundarik, A. A., Boskey, A., Gundberg, C., & Vashishth, D. (2018). Biomolecular regulation, composition and nanoarchitecture of bone mineral. *Science and Reports, 8*(1), 1191.
65. Fratzl, P., Roschger, P., Eschberger, J., Abendroth, B., & Klaushofer, K. (1994). Abnormal bone mineralization after fluoride treatment in osteoporosis: A small-angle x-ray-scattering study. *Journal of Bone and Mineral Research, 9*(10), 1541–1549.
66. Rinnerthaler, S., Roschger, P., Jakob, H. F., Nader, A., Klaushofer, K., & Fratzl, P. (1999). Scanning small angle X-ray scattering analysis of human bone sections. *Calcified Tissue International, 64*(5), 422–429.

67. Rai, R. K., & Sinha, N. (2011). Dehydration-induced structural changes in the collagen-hydroxyapatite interface in bone by high-resolution solid-state NMR spectroscopy. *Journal of Physical Chemistry C, 115*(29), 14219–14227.
68. Granke, M., Does, M. D., & Nyman, J. S. (2015). The role of water compartments in the material properties of cortical bone. *Calcified Tissue International, 97*(3), 292–307.
69. Nyman, J. S., Ni, Q., Nicolella, D. P., & Wang, X. (2008). Measurements of mobile and bound water by nuclear magnetic resonance correlate with mechanical properties of bone. *Bone, 42*(1), 193–199.
70. Mroue, K. H., Viswan, A., Sinha, N., & Ramamoorthy, A. (2017). *Solid-state NMR spectroscopy: The magic wand to view bone at nanoscopic resolution* (1st ed.). Elsevier Ltd. https://doi.org/10.1016/bs.arnmr.2017.04.004

An Advancement of CFD in the Hemodynamics of the Cardiovascular System

Somenath Gorai

Abstract The forces and stresses generated by blood flow on the walls of the cardiovascular system play a significant role in the development of various cardiovascular diseases (CVDs). However, the underlying reasons why arterial diseases tend to occur at specific locations remain a topic of ongoing investigation. This paper offers a concise review of recent progress, particularly over the past decade, in employing computational fluid dynamics (CFD) models to simulate blood flow dynamics and its interaction with the arterial wall in the human thoracic aorta (TA). Various approaches for analyzing the human aorta and the factors contributing to disease development are outlined. Comparative details, including methodologies, limitations, results, and key findings from each study, are reviewed. The aim is to inspire computational research that advances cardiovascular health through close collaboration between engineers and clinicians.

Keywords Hemodynamics · Computational fluid dynamics · Aorta · Cardiovascular diseases

1 Introduction

Aorta, the largest artery present in the human body, serves as the primary vessel through which oxygenated blood is pumped from the left ventricle (LV) into systemic circulation. It encompasses the ascending aorta, aortic arch, and descending aorta. This segment also includes three major branches originating from the arch: the brachiocephalic artery (BA), the left common carotid artery (LCA), and left subclavian artery (LSA), which supply blood to the upper body. The thoracic aorta (TA) possesses one of the most intricate anatomies among all arteries. In addition to its

S. Gorai (✉)
Department of Mechanical Engineering, School of Engineering, Manipal University Jaipur, Jaipur, Rajasthan 303007, India
e-mail: somenath.gorai@jaipur.manipal.edu; somenathgorai22@gmail.com

K. K. Dwivedi et al. (eds.), *Mechanics and Materials Science of Biological Materials*, Materials Horizons: From Nature to Nanomaterials,
https://doi.org/10.1007/978-981-96-5366-9_2

pronounced curvature, the centerline of the arch twists, making it non-planar. Furthermore, it features branching, lumen tapering, and distensible arterial walls. Despite advancements in clinical care and public awareness, cardiovascular diseases (CVDs) continue to be the leading cause of morbidity and mortality globally. According to the World Health Organization (WHO), approximately 23.6 million people are projected to die from CVDs by 2030 [1]. The thoracic aorta (TA), with its intricate anatomy, is prone to common cardiovascular diseases (CVDs) that are often present as atherosclerosis, aneurysms, and aortic dissection (AD). Atherosclerosis, the most prevalent form of arterial disease, involves thickening of the arterial wall due to the accumulation of fatty plaques within the inner lining. Aortic aneurysms occur when progressive enlargement and dilation result from the weakening of the aortic wall. Aortic dissection (AD), by contrast, is characterized by a tear in the aortic wall, creating an intimal flap that separates a true lumen from a false one. These conditions are life-threatening because of the risk of rupture at weaker sections of the aorta, emphasizing the critical need to understand the mechanisms driving their development and progression. Over recent decades, this has been a focus of extensive research. Advances in computational fluid dynamics (CFD) have facilitated the use of three-dimensional numerical simulations to explore patient-specific hemodynamics and uncover the origins and progression of various CVDs. CFD's ability to quantify variables that cannot be measured in vivo, its reproducibility, and its potential as both a diagnostic and therapeutic tool have made it invaluable in studying these conditions. The study by Totorean et al. [2] presents a detailed CFD analysis of hemodynamic parameters in a patient-specific abdominal aorta, derived from computed tomography angiography (CTA) images, to identify and highlight the regions most susceptible to developing atherosclerosis. Tran et al. [3] carried out patient-specific computational flow simulations, uncovering notable differences in paravisceral aortic hemodynamics between fenestrated and branched endovascular aneurysm repair. An increase in wall shear stress was identified as one of the contributing factors to aneurysm repair. Bonfanti et al. [4] propose a novel method for creating personalized CFD models using non-invasive and often minimal datasets typically gathered for AD monitoring. This study demonstrates how integrating readily available clinical data with computational modeling can serve as a valuable tool to improve clinical insights into AD. Wang et al. [5] provided a detail review on biomechanical behavior of aorta categorizing, namely theoretical, in vivo, experimental and combined investigations.

Unlike in vitro studies, CFD offers greater flexibility in modifying model input parameters, such as boundary conditions. Given the complexity of blood flow and its behavior in intricate geometries, the combination of high-resolution simulation techniques with image-based measurement data has proven to be a robust and reliable approach for realistic modeling of arterial blood flow. However, the use of CFD simulations is primarily limited by the assumptions made during the modeling process, including factors such as geometry, viscosity, distensibility, and flow conditions. Despite these simplifications, they remain fundamental to understanding the physical world. Although CFD studies related to the TA have advanced significantly over the past decade, a comprehensive review of literature has yet to be undertaken. Caballero and Lain [6] reviewed the current advancements in CFD modeling of the human TA, detailing the key steps involved in simulating blood flow and its interaction with the arterial wall. They examined the methodologies used, highlighted their limitations, and identified opportunities for future research. A comprehensive review would be highly beneficial for new researchers entering the field, while experienced researchers focusing on specific sub-problems could gain valuable insights by exploring the work of others. The aim of this article is to evaluate the current advancements in CFD modeling of the human TA, focusing on the key steps involved in simulating blood flow and its interaction with the arterial wall, particularly in terms of wall shear stress and pressure variations.

2 Background of CFD in the Hemodynamics Analysis

Since the inception of CFD in cardiovascular hemodynamics began over 40 years ago, evolving from simple models to sophisticated simulations with real-world clinical applications. The application of CFD in cardiovascular hemodynamics began in the 1970s and 1980s, largely influenced by advances in computer technology and numerical methods that made simulations of fluid flow feasible. Early applications focused on simplified models of blood flow in large arteries, such as the aorta, and used relatively basic geometries and assumptions about blood as a Newtonian fluid. These early studies provided initial insights into blood flow patterns and wall shear stress, setting the foundation for understanding their roles in cardiovascular health.

Key milestones in the development of CFD for hemodynamics include:

In 1970s, the inception of CFD in cardiovascular hemodynamics started with simplified, rigid-wall models for studying blood flow, mainly in straight or slightly curved artery models. These early studies helped develop the theoretical basis for blood flow simulation in arteries.

After 1980s, computational power increased, allowing researchers to apply CFD to more complex geometries, including bifurcations and bends. Studies began to investigate the relationship between wall shear stress and atherosclerosis, leading to hypotheses about how abnormal flow patterns might influence disease development.

From 1990s, improvements in medical imaging, such as MRI and CT, enabled more accurate geometric modeling of blood vessels, allowing for realistic 3D reconstructions of the arterial system. This decade saw the development of patient-specific models and the use of CFD to study conditions like aortic aneurysms and coronary artery disease. Zhong et al. [7] presented a comprehensive review of CFD methods, highlighting their challenges and opportunities in simulating coronary and intracardiac flows. Meanwhile, Morris et al. [8] examined the challenges related to the adoption and application of CFD modeling in cardiovascular medicine.

Thereafter from 2000 onwards, it has shown advances in both hardware and software allowed for more complex simulations, including fluid–structure interaction (FSI) models that account for the elasticity of arterial walls. Karmonik et al. [9] done CFD analysis in patients with continuous-flow left ventricular assist device support, describing hemodynamic alterations in the ascending aorta. They suggest that the turbulence in blood flow increases the wall shear stress and hence the pressure. This trend of increase in wall shear stress is seen in general for pipe flows subjected to turbulence [10, 11]. This period also saw the use of non-Newtonian models to simulate the shear-thinning properties of blood in smaller vessels or low-shear conditions. Hussain et al. [12] explored the impact of thermal and hemodynamic effects on the abdominal aortic aneurysm. Their non-Newtonian blood flow simulations revealed that increases in blood temperature, pressure, and intake velocity all contribute to higher viscosity.. Since the last decade with continued advances in computational power, imaging technology, and data integration, CFD models have become more sophisticated and clinically relevant. Schroeder et al. [13] employed advanced visualization techniques to study the anatomy and hemodynamics of aortic dissection. These visualization tools enhance clinical research and decision-making by providing a thorough understanding of the aortic condition. Calo et al. [14] have recently proposed moving-boundary method (MBM) to see the impact of wall displacements on the large-scale flow coherence in ascending aorta. Today's CFD models are increasingly used in surgical planning, medical device testing, and disease prediction. Techniques like machine learning are now being integrated with CFD to improve efficiency and personalization. Recent study of Saitta et al. [15] provided a data-driven generation of four-dimensional velocity profiles in the aneurysmal ascending aorta.

3 Methodology Used in the CFD Analysis

In the field of cardiovascular hemodynamics, CFD is used to simulate blood flow within vessels, providing detailed insights into the mechanical forces that influence cardiovascular health. Key methods in CFD for hemodynamic analysis include geometric modeling, meshing, boundary condition setting, fluid–structure interaction, and post-processing. Below is a more detailed look at each of these methods and how they contribute to CFD's role in cardiovascular research.

3.1 Geometric Modeling

Creating accurate geometric models of blood vessels is foundational to CFD analysis. The goal is to replicate the anatomy of the cardiovascular structures as closely as possible, which involves several steps:

Image acquisition: High-resolution imaging techniques, such as MRI, CT angiography, and ultrasound, capture detailed anatomical information. These images provide the basis for constructing 3D models.

Segmentation: The next step is segmentation, where specific regions of interest (like the lumen or the entire vessel wall) are identified and extracted. Advanced segmentation algorithms are applied to distinguish the vessel structures from surrounding tissues and artifacts.

Reconstruction: Using segmented images, a 3D model of the vessel geometry is reconstructed. Patient-specific models are created from individual imaging data, while idealized models may be constructed based on average anatomical dimensions for generic studies.

Smoothing and refinement: The raw segmented model may contain rough edges or irregularities, so smoothing algorithms are often used to create a continuous surface suitable for simulation.

3.2 Meshing

Meshing involves dividing the geometric model into small, discrete elements, which allow the fluid flow equations to be solved within each element. This process is essential for accurate CFD analysis:

Types of meshes: Commonly used mesh types in CFD are structured (uniform, grid-like elements) and unstructured (flexible, irregular elements). Structured meshes offer computational efficiency but may struggle with complex geometries, while unstructured meshes handle complex vessel shapes better, such as bifurcations in the arterial system.

Mesh resolution: High mesh resolution (smaller elements) increases the accuracy of the simulation, especially in regions with high shear stress or complex flow patterns, such as bifurcations. However, higher resolution also increases the computational cost.

Mesh refinement: Adaptive meshing techniques are used to refine the mesh in areas where high accuracy is needed, such as near the vessel walls or where flow separation occurs [14]. Refinement allows for accurate wall shear stress calculations, essential in studying conditions like atherosclerosis.

3.3 Boundary Conditions

In cardiovascular system, setting realistic boundary conditions is crucial to simulate physiological blood flow. Boundary conditions are applied at the inlets and outlets of the model to define how blood enters and exits the system:

Inlet boundary conditions specify how blood enters the vessel, it includes:

Velocity profiles: Many models now use pulsatile flow conditions rather than using fixed inlet velocities that replicate the cyclical nature of blood flow as observed in the cardiac cycle. These conditions are often derived from Doppler ultrasound or phase-contrast MRI measurements to capture accurate inflow patterns.

Flow rate: Using a pulsatile flow rate that mimics the heart's pumping cycle. This can be prescribed as a time-dependent function.

Pressure waveforms: Pressure-based inlet conditions are used when direct velocity measurements are unavailable.

Outlet conditions: Outlet boundary conditions define how blood exits the vessel and the pressure that downstream circulation places on the vessel:

Zero pressure: Assumes minimal resistance at the outlet, zero-gauge pressure is useful in idealized models.

Resistance and compliance models: The Windkessel model is a widely used approach, incorporating elements to simulate the vascular resistance and compliance of peripheral arteries. This model better replicates the vascular impedance encountered in the systemic circulation.

Wall boundary conditions:

No-slip condition: The no-slip condition assumes that the blood velocity at the wall is zero, mimicking the interaction between blood and endothelial cells.

Moving or deformable walls: Fluid–structure interaction (FSI) methods simulate the elasticity of vessel walls, allowing them to deform under the pressure of flowing blood.

3.4 Governing Equations

CFD solves the Navier–Stokes equations, which describe fluid motion. In hemodynamic simulations, these equations are adapted to represent blood flow and consider factors like blood viscosity, pressure, and density.

Continuity equation: Ensures mass conservation in the flow, meaning that the amount of blood entering a region equals the amount exiting.

Navier–Stokes momentum equations: Represent the forces acting on blood particles, including pressure gradients and shear forces.

Blood can be modeled as:

Newtonian fluid: It assumes constant viscosity, suitable for larger vessels where shear rates are high.

Non-Newtonian fluid: Blood's viscosity varies with shear rate, particularly important in smaller vessels or regions with low shear stress, like aneurysms. Models such as the Carreau or Casson models capture this non-Newtonian behavior.

3.5 Fluid–Structure Interaction (FSI)

Blood vessels are inherently distensible, particularly large arteries like the aorta, which must stretch and contract in response to pulsatile blood flow. Modeling this dynamic interaction between blood flow (fluid) and vessel walls (structure) is essential for accurate simulations, especially in conditions involving vessel wall changes, like aneurysms. FSI models allow for the simulation of vessel wall elasticity and its response to varying blood pressures.

FSI simulations are especially useful in analyzing:

Aneurysm behavior: FSI can simulate the wall stress and deformation in aneurysms, aiding in rupture risk assessment.

Plaque buildup and progression: FSI models help understand how blood flow affects the mechanical stress on plaques, potentially leading to plaque rupture.

These models require significant computational resources but provide insights that are not possible with rigid-wall models.

3.6 Post-processing and Analysis

After the CFD simulation is completed, post-processing provides visualizations and quantitative data to interpret results:

Wall shear stress: Wall shear stress is a critical metric that influences endothelial cell function and is linked to atherosclerosis development. High or low wall shear stress regions can indicate areas susceptible to disease.

Pressure distribution: Examining pressure gradients along the vessel helps identify areas of high stress, such as in an aneurysm, which may be prone to rupture.

Velocity and flow patterns: Visualizing blood velocity and flow direction helps in understanding flow disturbances, recirculation zones, and vortices, which are important in assessing risks of thrombosis or embolism.

Flow rate and volume calculations: Quantifying flow rates in specific branches aids in assessing blood supply efficiency and potential areas of ischemia.

3.7 Hybrid Approaches

Some advanced CFD methods combine with other computational techniques to improve efficiency and accuracy:

Reduced-Order Models (ROMs): Simplified models capture key hemodynamic features at a fraction of the computational cost, useful for real-time applications.

Machine Learning integration: Machine learning (ML) algorithms can enhance boundary condition estimation, optimize mesh generation, or predict flow characteristics, allowing for faster and potentially more accurate simulations.

Multi-scale Modeling: Combines detailed CFD models of specific arteries with larger-scale cardiovascular models, providing insights into how changes in one region affect systemic circulation.

4 Advancements in CFD Modeling for Cardiovascular Hemodynamics

The sophistication of CFD models has grown considerably, with several key advancements enabling more accurate and clinically relevant simulations:

4.1 Patient-Specific Geometry Modeling

High-fidelity vessel geometries are crucial for accurate CFD simulations. Recent advancements in medical imaging, including MRI, CT angiography, and ultrasound, allow for detailed 3D reconstructions of blood vessels. These imaging techniques enable patient-specific CFD models, which capture anatomical variations and provide tailored insights into individual cardiovascular health. Advanced segmentation algorithms and image-processing software extract and reconstruct these geometries, making simulations both more precise and applicable for personalized medicine.

Additionally, models now account for small but important anatomical features such as bifurcations, bends, and curvatures that significantly impact blood flow dynamics. For example, in coronary artery disease, these geometrical features can influence plaque buildup and hemodynamic stress, which CFD can help analyze in a patient-specific context.

4.2 Non-Newtonian Blood Modeling

Blood exhibits non-Newtonian behavior, meaning its viscosity changes under different flow conditions. Modeling blood as a Newtonian fluid, which assumes constant viscosity, can lead to inaccuracies, especially in smaller vessels and areas of low shear stress. Non-Newtonian models incorporate shear-thinning behavior, where blood viscosity decreases with increased shear rates, thus better capturing the characteristics of blood flow in realistic conditions [12]. Advanced CFD models now commonly employ non-Newtonian formulations, such as the Carreau or Casson models, which are particularly important for simulating flow in capillaries, arterioles, and venules.

5 Applications of CFD in Cardiovascular Hemodynamics

CFD is applied to various aspects of cardiovascular health research, contributing to the understanding, prevention, and treatment of multiple conditions:

5.1 Atherosclerosis and Plaque Development

Regions of low wall-shear-stress are known to be prone to atherosclerosis. CFD allows researchers to identify these regions in patient-specific models, providing a non-invasive method to predict sites of plaque formation. By simulating wall-shear-stress distribution, CFD aids in studying the progression of atherosclerosis and the risk of plaque rupture, contributing valuable information for early intervention [2].

5.2 Aneurysm Growth and Rupture Risk

Aneurysms, particularly in the aorta and cerebral arteries, can pose life-threatening risks if they rupture. CFD simulations can analyze wall stress and flow patterns within aneurysms, with FSI models providing insights into how blood pressure changes impact the distensible aneurysm wall. This information helps clinicians assess rupture risk and decide on intervention strategies [12].

5.3 Surgical Planning and Medical Device Evaluation

CFD has applications in preoperative planning, where simulations can predict how different surgical interventions, such as bypass grafts or stent placements, will impact blood flow and vessel wall stress. CFD is also used in medical device testing, such as optimizing stent designs to reduce blood flow disturbances or prevent restenosis.

6 Challenges and Future Directions

While CFD has made significant strides in cardiovascular hemodynamics, several challenges remain:

Computational Load: High-fidelity CFD models, particularly those with FSI or patient-specific geometries, are computationally intensive, requiring powerful hardware and longer simulation times, which limits their clinical applicability.

Data Integration for Personalized Medicine: Integrating real-time patient data, such as continuous blood pressure measurements, with CFD models remains a challenge. Hybrid approaches that combine CFD with machine learning may improve the feasibility of real-time simulations.

Validation and Clinical Adoption: Validating CFD models against in vivo data is essential to ensure accuracy. Wider clinical adoption requires robust, validated models and user-friendly software to streamline simulations for routine use.

6.1 Future Potential in Machine Learning and Artificial Intelligence

Machine learning (ML) and artificial intelligence (AI) offer promising tools for enhancing CFD by predicting blood flow patterns and vessel responses more rapidly. ML techniques can assist in processing large datasets and improving parameter estimation for boundary conditions, enabling faster simulations [16]. Integrating ML into CFD could also enable predictive modeling, where simulations adapt to individual patient characteristics, making real-time analysis for clinical applications more achievable.

7 Conclusion

Advancements in CFD have greatly enhanced our understanding of cardiovascular hemodynamics, providing detailed insights into disease mechanisms and potential interventions. Improved modeling techniques, such as patient-specific geometries, realistic boundary conditions, FSI, and non-Newtonian blood modeling, have increased the relevance and accuracy of CFD for cardiovascular research. While challenges remain in computational demand and clinical integration, future innovations, particularly in AI and machine learning, promise to make CFD an indispensable tool for personalized cardiovascular care and improved patient outcomes.

Data Availability Not applicable.

Declaration of Competing Interest The authors have no conflicts to disclose.

References

1. W. H. Organization and others. (2013). Cardiovascular diseases (CVDs): fact sheet no. 317. *Media Centre. World Health Organization, 20*(1), 3.
2. Totorean, A. F., et al. (2022). Patient-specific image-based computational fluid dynamics analysis of abdominal aorta and branches. *Journal of Personalized Medicine, 12*(9), 1502. https://doi.org/10.3390/jpm12091502
3. Tran, K., et al. (2024). Patient-specific computational flow simulation reveals significant differences in paravisceral aortic hemodynamics between fenestrated and branched endovascular aneurysm repair. *JVS-Vascular Science, 5*, 100183. https://doi.org/10.1016/j.jvssci.2023.100183
4. Bonfanti, M., Franzetti, G., Maritati, G., Homer-Vanniasinkam, S., Balabani, S., & Díaz-Zuccarini, V. (2019). Patient-specific haemodynamic simulations of complex aortic dissections informed by commonly available clinical datasets. *Medical Engineering & Physics, 71*, 45–55. https://doi.org/10.1016/j.medengphy.2019.06.012
5. Wang, X., et al. (2023). A review on the biomechanical behaviour of the aorta. *Journal of Mechanical Behavior of Biomedical Materials, 144*, 105922. https://doi.org/10.1016/j.jmbbm.2023.105922
6. Lain, S. (2017). A review on computational fluid dynamics modelling in human thoracic aorta a review on computational fluid dynamics modelling in human thoracic aorta, June 2013. https://doi.org/10.1007/s13239-013-0146-6
7. Zhong, L., Zhang, J. M., Su, B., Tan, R. S., Allen, J. C., & Kassab, G. S. (2018). Application of patient-specific computational fluid dynamics in coronary and intra-cardiac flow simulations: Challenges and opportunities. *Frontiers in Physiology, 9*, 742. https://doi.org/10.3389/fphys.2018.00742
8. Morris, P. D., et al. (2016). Computational fluid dynamics modelling in cardiovascular medicine. *Heart, 102*(1), 18–28. https://doi.org/10.1136/heartjnl-2015-308044
9. Karmonik, C., et al. (2014). Computational fluid dynamics in patients with continuous-flow left ventricular assist device support show hemodynamic alterations in the ascending aorta. *Journal of Thoracic and Cardiovascular Surgery, 147*(4), 1326-1333.e1. https://doi.org/10.1016/j.jtcvs.2013.09.069
10. Gorai, S., Samanta, D., & Das, S. K. (2024). Experimental investigations of heat transfer in simultaneously developing transitional regime of mixed convection flows in a vertical tube. *ASME Journal of Heat and Mass Transfer, 146*(10), 102601. https://doi.org/10.1115/1.4065609
11. Gorai, S., Samanta, D., & Das, S. K. (2023). Heat transfer in simultaneously developing turbulent mixed convection flows in vertical tubes. *Heat Transfer Engineering, 45*(22), 1–20. https://doi.org/10.1080/01457632.2023.2289226
12. Hussain, A., et al. (2024). Unveiling thermal and hemodynamic effects of aneurysm on abdominal aorta using power law model and finite element analysis. *Case Studies in Thermal Engineering, 60*, 104746. https://doi.org/10.1016/j.csite.2024.104746
13. Schroeder, A., et al. (2024). Advanced visualization of aortic dissection anatomy and hemodynamics. *Computers & Graphics, 124*, 104060. https://doi.org/10.1016/j.cag.2024.104060
14. Calò, K., et al. (2023). Impact of wall displacements on the large-scale flow coherence in ascending aorta. *Journal of Biomechanics, 154*, 111620. https://doi.org/10.1016/j.jbiomech.2023.111620
15. Saitta, S., et al. (2023). Data-driven generation of 4D velocity profiles in the aneurysmal ascending aorta. *Computer Methods and Programs in Biomedicine, 233*, 107468. https://doi.org/10.1016/j.cmpb.2023.107468
16. Shelar, H., Gorai, S., & Samanta, D. (2022). Application of machine learning to predict hydrodynamic and thermal parameters in laminar mixed convection flows through vertical pipes. *ISME Journal of Thermofluids, 8*(2), 31–41. https://isme.iitd.ac.in/journals/thermofluids/index.php/announce/viewarticle/59

Experimental Technique and Constitutive Models for Skin Tissue: A Review

Krashn Kumar Dwivedi, Sachin Kumar, and Navin Kumar

Abstract This review describes the characterization techniques and constitutive models which are used to measure and capture the structural and mechanical behavior of skin. Skin is nonlinear elastic, viscoelastic, and anisotropic material, and these mechanical properties of skin are subject-specific. The measurement of skin's mechanical properties differs among the experimental techniques [1]. One main cause of variability is the lack of standardization [2]. This chapter presents the mechanical and structural characterization methods that are widely used for skin research.

1 Mechanical Characterization Techniques

The mechanical characterization techniques can vary with sample and setup size, sample attachment or nature of gripping, and the region that is tested. Further, based on the required mechanical characteristics of skin, different methods such as continuous stretching, cyclic loading, relaxation and creep, and dynamic mechanical analysis are used. The experimental testing of the skin is categorized into two distinct approaches: in vivo and ex vivo, where sample is either directly tested on the subject or excised from the subject, mounted on experimental setup and tested, respectively. Among tested species we can find mice [3], rat [4], porcine/pig [5] (often preferred for similarity of their skin with humans), and human [6]. For ethical restrictions, non-invasive in vivo characterization is usually preferred for special characterization of human skin. The ex vivo studies on human skin are also found in literature, and this approach is found capable to provide more rigorous results about the mechanical and structural characteristics of skin. The methods for skin mechanical characterization are discussed in the following sections.

K. K. Dwivedi (✉) · N. Kumar
Department of Biomedical Engineering, IIT Ropar, Rupnagar, India
e-mail: kkdwivedi159@gmail.com

S. Kumar · N. Kumar
Department of Mechanical Engineering, IIT Ropar, Rupnagar, India

K. K. Dwivedi et al. (eds.), *Mechanics and Materials Science of Biological Materials*, Materials Horizons: From Nature to Nanomaterials,
https://doi.org/10.1007/978-981-96-5366-9_3

1.1 *In Vivo Testing Methods*

The in vivo experiments are non-invasive in nature which can be performed directly on the subjects with minimal ethical restrictions. These experiments confirm the repeatability of the tests. The advantage of in vivo experiments is that they can test the skin in its natural configurations. On the other hand these experiments are limited to small deformation and damage and failure properties of skin cannot be measured. Also, these methods fail to isolate the mechanical properties of skin from other surrounding tissues. The commonly used in vivo methods are discussed in the following sections.

1.1.1 In Vivo Indentation Methods

The indentation method is used to measure the local properties of materials such as hardness, modulus, viscoelasticity, and surface tear resistance. First time Schade 1912 [7] performed the nanoindentation test on the skin using eye tonometer device. Over the time this method is improved and used to measure the effect of age on the skin elasticity [8]. This technique has the advantage of non-invasive and minimum test area preparation for the testing. Therefore, the indentation technique is extended for in vivo test to determine the mechanical properties [9]. In typical indentation test, the specimen is compressed through an intendor of the different shapes and sized to estimate the stiffness. In this method, the indenter applies the load or displacement on the specimen, and displacement or load is recorded to estimate the indentation modulus. The load displacement data along with Poisson's ratio was used to measure the stiffness of the specimens [9].

$$E^* = \frac{\sqrt{\pi/A}}{2}\frac{dF_N}{d\delta}\Big|_{F_N=F_{\max}} \tag{1}$$

where E^* is the indentation modulus, A is the projected contact area. F_N is normal force and $F_{\max}$ force at unloading. The elastic parameters can be found via the following approximation:

$$E_{\text{skin}} = E^*\left(1 - \nu_{\text{skin}}^2\right) \tag{2}$$

where E_{skin} is the uniaxial elasticity modulus of skin, and ν_{skin} is Poisson's ratio.

Apart from the elastic properties of skin, some other studies used in vivo nanoindentation test to measure the viscoelastic properties of skin [8].

Along with advantages, this technique has many limitations such as this method shows discrepancy in the calculated results for different penetration depth, indenter size, and indenter geometry. Also, the measurement of the tip contact area is a challenging task which restricts the calculation of stress–strain curve. This technique applies the compressive loading on the tissue, where the role of the collagen fibers and anisotropy cannot be estimated.

1.1.2 Torsion Method

The torsion test is usually performed by placing the high friction probe in contact with the skin and applying the torsion to the probe. This technique is more popular for in vivo compared to ex vivo testing because of the minimum preparation required for the test area [10]. Frictiometer® is a commercially available system widely used for in vivo skin elasticity measurement. This system measures the torsional resistance of the skin with angular displacements. The area near the test surface is shielded through the guard ring to reduce the contribution of the outer tissue and isolate them from the test area. Apart from elasticity, the torsion test method is also useful to measure the viscoelastic properties through the creep test [10]. This method is widely used in medical applications due to simple operation and quick measurements. On the other hand, torsion method is limited to isotropic mechanical properties of skin as it cannot be used to apply the deformation in individual direction.

1.1.3 Surface Wave Method

First time, Potts et al. [11] explored the surface wave technique to measure the mechanical properties of skin. This method overcomes the limitations of indentation and torsion techniques as this method can measure the anisotropy of skin through directional variation in wave propagation. In this method, the local harmonic vibration of a particular frequency is generated on the surface of the skin, and wave speed is measured by different contact and non-contact type systems [12]. The surface wave methods in human skin work in the frequency range of 0 to 1000 Hz, where attenuation of lower frequencies wave is associated with the stratum corneum mechanical properties, whereas attenuation of higher frequency (>500 Hz) is related to mechanical properties of dermis layer [11]. The surface wave method is capable to measure the anisotropy of skin by conducting the test in individual direction with different angular resolutions [13]. Apart from elasticity and anisotropy, this method is also found capable to measure the viscoelastic properties of skin by applying the wave of different frequencies [11, 14]. Besides, the numerous advantages, this method is time-consuming in the anisotropy and viscoelastic properties measurement as this technique utilizes the single point transducer and laser sensors. Also, most of time this technique provides surfacial mechanical properties of skin.

1.1.4 Suction Method

This method was first introduced by Grahame and Holt in 1969 [15] to investigate the effect of aging on the elastic properties of skin. In suction method, a negative pressure

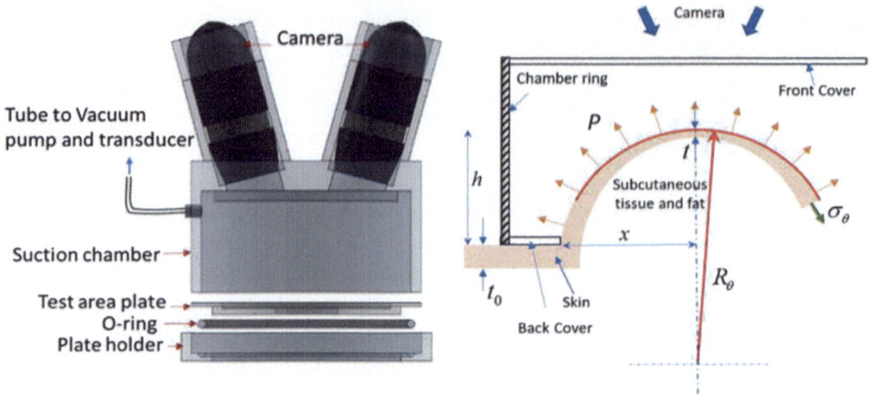

Fig. 1 Non-invasive suction device to measure the skin anisotropy and direction of Kraissl's lines: developed device (left) and method to measure the stress–strain data (right)

is applied on the skin through a cylindrical or hemispherical cup which is mounted on the tested region. The negative pressure results in an upward deformation of a dome of skin. The height of the raised dome is compared with the suction pressure, and a stress–strain curve is obtained using Tregear's equation [16] for deforming hemispherical domes:

$$\sigma = \frac{pR^2}{4xe}\left(1 + \frac{x^2}{R^2}\right) \tag{3}$$

$$\varepsilon = \frac{2x^2}{3R^2\left(1 + \frac{2x^2}{R^2}\right)} \tag{4}$$

where σ and ε are the engineering stress and strain, respectively. p applied negative pressure, R is the radius of the suction cup, e is the estimated skin thickness, and x is the height of the deformed dome. Further, modulus of the elasticity is measured as a slope of linear portion of stress–strain curve [17]. Moreover with the advancement of non-contact full-field displacement measurement devices, the suction method has been used to measure the anisotropy of skin and directions of skin tension lines [17] (Fig. 1).

1.2 Ex Vivo Testing Methods

Besides the ethical restrictions, ex vivo techniques are widely used for the mechanical characterization of skin. In ex vivo studies, either animal skin or human skin excised from cadaver or during the surgery. These methods allow the standardization of specimens which improve the consistency and reliability of the results. The ex vivo methods allow the large stretching of skin at broad range of strain rates which enables the measurement of additional mechanical characteristics such as failure properties, damage, and inelastic strain. The ex vivo methods also deliver the correlation of skin constituents with different phases of stress–strain. These methods also enable the isolation of mechanical properties of individual layers. The main complication of these methods is the slip of specimens from the grips, particularly at high strain and strain rate values, which adds error in the measurements. This issue can be sorted out by special gripping system [18].

Most of the in vivo techniques discussed in the previous sections are used for ex vivo characterization of skin. The studies on ex vivo experimental techniques are summarized in the following subsections.

1.2.1 Ex Vivo Indentation Method

The principle of ex vivo indentation is almost similar to in vivo indentation method. This method is widely used to characterize the micro- and nanomechanical properties of skin [18]. Bhushan et al. [18] explored the ex vivo indentation to characterize the tribology of skin. Other studies have also used ex vivo indentation method to characterize the elastic and viscoelastic properties of different layers of the skin [19]. Similar to in vivo indentation, ex vivo indentation is also limited to the isotropic mechanical properties of tissue. Collagen fibers are the major constituents of skin which activate in tension; therefore, a significant amount of work is based on the tensile experiments.

1.2.2 Uniaxial Tensile Testing

The uniaxial tensile test setup is commonly available tool in material characterization laboratory and uniaxial tensile testing protocol is widely implemented in skin mechanical characterization. Through this technique almost all mechanical properties of skin (elasticity, anisotropy, viscoelasticity, fracture resistance, and fatigue properties) can be measured. This method allows the standardization of specimen such as dog bone shape sample which ensures to maximize stress uniformity in the gauge region [20]. The uniaxial tensile test experiments to characterize the anisotropy of excised skin. Further with the advancement of non-contact strain measurement techniques, this ex vivo method has been widely used for broad range of experimental studies where effect of locations [21], cyclic loading [19], effect of hydration [21] and strain rate [22] on the mechanical properties of skin has been investigated. The recent studies by Yang et al. [4] and Pissarenko et al. [23] explored the uniaxial tensile test method to measure the tear resistance and fracture toughness of rabbits and pig skin, respectively. Besides the several advantages, this technique fails to elicit the multidirectional coupling mode between skin constituents, and thus, it is insufficient to inform constitutive models of the planar biaxial behavior of skin [24]. The studies where the uniaxial tensile test setup is used to measure the mechanical properties of excised skin are summarized in Table 1. Figure 2 shows a schematic presentation of uniaxial tensile test setup coupled with non-contact strain measurement techniques and a typical stress–strain diagram for pig skin.

1.2.3 Biaxial Tensile Test

Biaxial tensile test is more accurate to measure the bidirectional mechanical properties of soft tissue as it couples the two directional deformation with tissue constituents. The use of digital strain measurement technique makes this technique more accurate in stress–strain measurements. In 1974, Lanir and Fung [30] used biaxial tensile test setup to measure the three-dimensional mechanical response of rabbit skin. This setup can apply the equibiaxial stretch or off biaxial stress and sample deformations are recorded using camera. With the advancement in imaging techniques, the stress–strain results obtained from this experimental setup are used to derive the microstructural motivated constitutive models [31]. The biaxial test produces the loading closure to the in vivo condition of skin. Besides the advantages, this method fails to fully determine the constitutive equation of an anisotropic material like skin, as pointed out by Holzapfel and Ogden [31, 32].

1.2.4 Inflation Test

The inflation test is also known as bulge test and initially, this test was introduced to characterize the thin membrane of the metal. Dick 1951 [33] used this device first time to measure the deformation response of skin. The working principle of inflation test is similar to suction test except the fluid supply rather than vacuumed [5]. This device provides the accurate constitutive modeling information as it couples 360^0 deformation with tissue constituents [33]. Also, bulge test can mimic the in vivo like environmental condition by supplying the physiological equivalent fluid with body temperature [5]. The deformation of specimen during the inflation is measured using two cameras and then stretch of the specimens is measured by measuring the arc length of bulge specimen which is approximated as hemispherical cup [33]. The true stress with deformation is measured from the theory of thin spherical membrane [34, 35]. In the calculation of stress, instantaneous thickness is measured using the change in arc length of hemisphere. The major challenge of this device is that it cannot maintain constant strain rate and dynamic experiments cannot be performed.

2 Constitutive Modeling

Constitutive models are essential in understanding and predicting the mechanical behavior of biological tissues under various loading conditions. These models describe the relationship between the stress and strain in a material, incorporating

Table 1 Summary of constitutive models widely used in the field of skin biomechanics

Model name	Constitutive description/strain energy function	Parameter description
Phenomenological model		
Neo-Hookean	$W = C_1 (I_1 - 3)$	C_1 unite of stress $I_1 = \lambda_1^2 + \lambda_2^2 + \lambda_3^2$
Mooney-Rivlin	$W = C_{10}(I_1 - 3) + C_{01}(I_2 - 3) + \sum_{i=1}^{K} \frac{1}{D_i}(J - 1)^{2i}$	C_{10} and C_{01} unit of stress $I_1 = \lambda_1^2\lambda_2^2 + \lambda_2^2\lambda_3^2 + \lambda_3^2\lambda_1^2$ D_i compressibility $J = \lambda_1 \lambda_2 \lambda_3$
Ogden	$W = \sum_{i=1}^{K} \mu_i(\overline{\lambda_1}^{\alpha_i} + \overline{\lambda_2}^{\alpha_i} + \overline{\lambda_3}^{\alpha_i} - 3) + \sum_{i=1}^{K} \frac{1}{D_i}(J - 1)^{2i}$	μ_i shear modulus, α_i hardness exponent, i number of terms
Yeoh	$W = \sum_{i=1}^{3} C_{i0}(I_i - 3)^i + \sum_{i=1}^{K} \frac{1}{D_i}(J - 1)^{2i}$	C_{i0} unit of stress, i terms
Polynomial	$W = \sum_{i,j=0}^{n} C_{ij}(I_1 - 3)^i (I_2 - 3)^j + \sum_{i=1}^{K} \frac{1}{D_i}(J - 1)^{2i}$	C_{ij} unit of stress
Semi-structural constitutive models		
Arruda-Boyce Eight-chain model	$W(\lambda_{chain}) = n k_B \theta\left[\lambda_{lock} \lambda_{chain}\beta - \lambda_{lock}^2 \ln\left(\frac{\sinh\beta}{\beta}\right)\right]\beta =$ $L^{-1}(\lambda_{chain}/\lambda_{lock}),\ L(\beta) = \coth\beta - 1/\beta$	$\lambda_{lock} = \sqrt{N}$ N—chain density $\lambda_{lchain} = \sqrt{I_1/3}$
Gent model	$W = -\frac{\mu}{2}J_m \ln(1 - \frac{I_1-3}{J_m}), I_1 < J_m + 3$ $\mu = nkT, J_m = 3(N\text{-}3)$	n molecular chain density k Boltzmann's constant T Temperature, J_m limiting chain extensibility, N chain density

(continued)

Table 1 (continued)

Model name	Constitutive description/strain energy function	Parameter description
Limbert model	$W_{Lim} = W_{vol}(J) + \sum_{i=1}^{2} W^i_{dev}(\bar{\lambda}_i) + W^i_1(\alpha_{i1})$ $+W^i_2(\alpha_{i2}, \bar{\lambda}_i)$ First term, volumetric, second term deviatoric associated with fiber stretch, third term corresponding isotropic shear energy, fourth term associated with fiber–fiber and fiber matrix interaction	*The detail about each term expression is disrobed by lambert* [25]
Lanir model Angular integration approach	$W_{lanir} = (1 - \phi_c - \phi_e)\frac{\mu}{2}(I_1 - 3) + \phi_c W_c(\lambda) + \phi_e W_e(\lambda)$ Each fiber is oriented at an particular angle which follows a statistical distribution $R(\theta)$ $W_c(\lambda) = \int_0^{\pi} R_c(\theta) \int_1^{\lambda(\theta)} D_c(x)\frac{k_c}{2}\left(\frac{\lambda(\theta)}{x} - 1\right)^2 dx\, d\theta$ $W_e(\lambda) = \int_0^{\pi} R_e(\theta)\frac{k_c}{2}(\lambda(\theta) - 1)^2 d\theta$	Description of material parameters are given elsewhere [26] Complicated in computational implementation [26]
Holzapfel-Gasser-Ogden model (HGO)	$\Psi = \Psi_g(\boldsymbol{C}) + \sum_{i=1}^{w} \Psi_{fiber}(\boldsymbol{C}, \boldsymbol{H}_i^{fiber})$ Ψ_g and Ψ_f are the strain energy function corresponding to the contribution of ground matrix and collagen fibers, respectively Generalized structure tensor $\boldsymbol{H}_i = A^a \boldsymbol{I} + B\, m_i \otimes m_i + (1 - 3A - B)\, m_z \otimes m_z; \quad (i = 1, 2),$ m_i mean direction of collagen fibers This is three-dimensional model which incorporate in-plane and out-plane dispersion of collagen fibers	*A and B* are constants and expressed as $A = 2\,\kappa_{in}\,\kappa_{out}$ $B = 2\kappa_{out}(1 - \kappa_{out})$ where κ_{in} *and* κ_{out} are in-plane and out-of-plane dispersion parameters measured from imaging [27, 28]

(continued)

Table 1 (continued)

Model name	Constitutive description/strain energy function	Parameter description
Viscoelastic models		
Generalized Maxwell model	$\boldsymbol{G}(t) = G_\infty + \sum_{i=1}^{n} G_i \exp\left(-\frac{t}{\tau_i}\right)$ where G_∞ is the long-term relaxation modulus	G_∞ and G_i are the dimensionless constants $G_\infty + \sum_{i=1}^{n} G_i = 1$
Discrete Fung model	$G(t) = \frac{1+\int_{\tau_1}^{\tau_2} (C/\tau)e^{-t/\tau} d\tau}{1+\int_{\tau_1}^{\tau_2} (C/\tau) d\tau}$	G(t) is relaxation function
Schapery nonlinear viscoelastic model	$\sigma(\varepsilon, t) = h_e(\varepsilon)E_e\varepsilon + \ldots$ $h_1(\varepsilon) \int_0^t \Delta E(\rho(t) - \rho\prime(\tau)) \frac{dh_2(\varepsilon)\varepsilon}{d\tau} d\tau$ $\rho = \int_0^t \frac{dt\prime}{a_e[\varepsilon(t')]}$	E is modulus h strain-dependent material properties

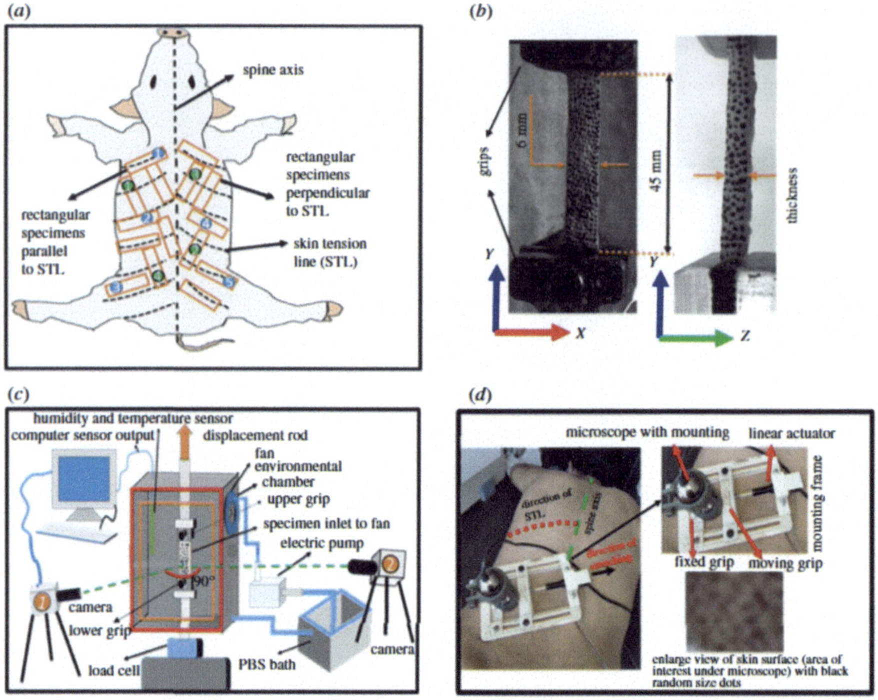

Fig. 2 **a** Schematic shows the rectangle specimens extracted from the dorsal portion of the pig skin with marked orientations to skin tension line (STL). Circle numbers 1–5 in blue circles represent the random locations of five specimens that were excised parallel to STLs, whereas circle numbers 1–4 in green circles represent the random locations of four specimens that were excised perpendicular to STLs. These specimens were used for microstructural assessment. **b** A typical specimen with dimensions (gauge length and width) and random-sized black dots. **c** In vivo stretching experiment on human back skin. Note: here, the direction of stretching and direction perpendicular to stretching were nomenclatures as yy and xx, respectively [29]

the material's properties, structure, and the mechanical environment. Soft tissues, including skin, cartilage, muscle, and tendons, present unique challenges for constitutive modeling due to their complex, anisotropic, nonlinear, and time-dependent behavior. In the context of soft tissue mechanics, constitutive models help to bridge the gap between the biological complexity of tissues and the need for accurate predictions in both clinical and engineering applications. Table 1 demonstrates some important constitutive models for skin tissue.

3 General Framework for Constitutive Modeling

The Cauchy stress from the invariant-based strain energy function can be derived by the differentiation of W with respect to its invariants of left Cauchy–Green deformation tensor (B) [35, 36],

$$\sigma = 2J^{-1}\boldsymbol{B}\left(\frac{\partial \mathrm{W}}{\partial I_1}\frac{\partial I_1}{\partial \boldsymbol{B}} + \frac{\partial \mathrm{W}}{\partial I_2}\frac{\partial I_2}{\partial \boldsymbol{B}} + \frac{\partial \mathrm{W}}{\partial I_3}\frac{\partial I_3}{\partial \boldsymbol{B}}\right) \tag{5}$$

The derivatives of invariants are defined as,

$$\begin{aligned} \frac{\partial I_1}{\partial \boldsymbol{B}} &= \boldsymbol{I} \\ \frac{\partial I_2}{\partial \boldsymbol{B}} &= I_1\boldsymbol{I} - \boldsymbol{B} \\ \frac{\partial I_3}{\partial \boldsymbol{B}} &= I_2\boldsymbol{I} - I_1\boldsymbol{B} + \boldsymbol{B}^2. \end{aligned} \tag{6}$$

After solving Eqs. (5) and (6) with the assumption of incompressibility $J = 1$ (as soft tissues contain a large volume of water), we obtain

$$\sigma = -p\boldsymbol{I} + 2\left(\frac{\partial \mathrm{W}}{\partial I_1} + I_1\frac{\partial \mathrm{W}}{\partial I_2}\right)\boldsymbol{B} - 2\frac{\partial \mathrm{W}}{\partial I_2}\boldsymbol{B}^2 \tag{7}$$

where I is identity tensor and p is the Lagrange multiplier represents the hydrostatic pressure which can be determined from the boundary conditions.

3.1 Constitutive Relation for Uniaxial, Shear, and Equibiaxial Model of Deformation

For the calibration of the material parameters and prediction of the mechanical behavior of soft tissues under different modes of deformation, the constitutive framework should be in terms of the principal stretch ratio. Therefore, the left Cauchy–Green deformation tensor is given in terms of the principal stretch ratio.

3.2 Simple Uniaxial Extension

For uniaxial extension, the stretch ratio is applied in direction-1 ($\lambda_1 = \lambda$), and the stretch ratios in the remaining two directions can be calculated using the assumption of incompressibility. The stretch ratio in direction-1&2 areas. $\lambda_2 = \lambda_3 = \lambda^{-1/2}$. For these values of stretch ratios, the left Cauchy–Green tensor is given as,

$$\boldsymbol{B} = \begin{bmatrix} \lambda^2 & 0 & 0 \\ 0 & \lambda^{-\frac{1}{2}} & 0 \\ 0 & 0 & \lambda^{-\frac{1}{2}} \end{bmatrix} \tag{8}$$

using Eq. (8), Eq. (7) is deduced as,

$$\begin{aligned} \boldsymbol{\sigma}_{11} &= -p^e + 2A\lambda^2\left[\exp(-a(I_1 - 3)) + b\ln(I_1 - 2)\right] \\ \boldsymbol{\sigma}_{22} = \sigma_{33} &= -p^e + 2A\lambda^{-1}\left[\exp(-a(I_1 - 3)) + b\ln(I_1 - 2)\right] \end{aligned} \tag{9}$$

After putting the value of $I_1 = \lambda^2 + 2\lambda^{-1}$ and boundary conditions ($\boldsymbol{\sigma}_{11} = \boldsymbol{\sigma}$ and $\boldsymbol{\sigma}_{22} = \boldsymbol{\sigma}_{33} = 0$), the value of p^e can be obtained, and the final expression of Cauchy stress (true stress) for a simple uniaxial extension is given in Eq. (10).

$$\boldsymbol{\sigma} = 2A\left[\exp\left(-a(\lambda^2 + 2\lambda^{-1} - 3)\right) + b\ln(I_1 - 2)\right]\left(\lambda^2 - \frac{1}{\lambda}\right) \tag{10}$$

3.3 *Pure Shear Test*

In the pure shear test, the stretch ratios are $\lambda_1 = \lambda$, $\lambda_2 = 1$, and $\lambda_3 = \lambda^{-1}$. The stresses in the directions of 1 and 2 are positive ($\boldsymbol{\sigma}_{11} > 0$, $\boldsymbol{\sigma}_{22} > 0$) value, whereas the third direction is stress-free ($\boldsymbol{\sigma}_{33} = 0$). Using the constraints, the left Cauchy–Green deformation tensor and Cauchy stress tensor for pure shear deformation are given in Eqs. (11) and (12), respectively.

$$\boldsymbol{B} = \begin{bmatrix} \lambda^2 & 0 & 0 \\ 0 & 1 & 0 \\ 0 & 0 & \lambda^{-2} \end{bmatrix} \tag{11}$$

$$I_1 = \lambda^2 + 1 + \lambda^{-2} \tag{12}$$

$$\boldsymbol{\sigma} = 2A\left[\exp\left(-a(\lambda^2 + \lambda^{-2} - 2)\right) + b\ln(\lambda^2 + \lambda^{-2} - 1)\right]\left(\lambda^2 - \frac{1}{\lambda^2}\right) \tag{13}$$

3.4 Equibiaxial Extension

In the equibiaxial test, the specimen is equally stretched in two directions with stretch ratios of $\lambda_{11} = \lambda_{22} = \lambda$, and the value of stresses is $\boldsymbol{\sigma}_{11} = \boldsymbol{\sigma}_{22} = \boldsymbol{\sigma}$, and $\boldsymbol{\sigma}_{33} = 0$. With the assumption of incompressibility, the left Cauchy–Green deformation tensor and Cauchy stress tensor for equibiaxial extension mode can be expressed as given in Eqs. (14 and 15).

$$\boldsymbol{B} = \begin{bmatrix} \lambda^2 & 0 & 0 \\ 0 & \lambda^2 & 0 \\ 0 & 0 & \lambda^{-4} \end{bmatrix} \tag{14}$$

$$I_1 = 2\lambda^2 + \lambda^{-4} \tag{15}$$

$$\boldsymbol{\sigma}_{11} = \boldsymbol{\sigma}_{22} = 2A\left[\exp\left(-a(2\lambda^2 + \lambda^{-4} - 3\right) + b\ln(2\lambda^2 + \lambda^{-4} - 2)\right]\left(\lambda^2 - \frac{1}{\lambda^4}\right) \tag{16}$$

4 Summary and Discussion

The in vivo mechanical tests are non-invasive and capture mechanical properties of skin under its physiological conditions. Therefore, these methods are implemented in clinical practices for diagnosis the alteration in skin with different pathological conditions. Besides these advantages, these techniques fail in full mechanical characterization of skin as they are limited to particular type of mechanical stimulation with low amplitude. Also, these techniques do not provide the structural insight of the tissue, therefore fail to couple the structural response with the mechanical behavior of the tissue. On the other hand, the ex vivo methods capture the complex mechanical response of skin such as finite stress–strain response, dynamic response, failure properties, tear resistance, damage, long-term creep, and stress relaxation. These mechanical properties are imported to simulate the skin mechanical response against over physiological loading condition such as trauma, tissue expansion, and large wound closure. Moreover, these techniques are capable to investigate the correlation between structural response and mechanical behavior of skin. These techniques also provide the sufficient data for the constitutive modeling of skin, where imaging techniques provide structural information and mechanical test provide finite stress–strain data. Therefore, these techniques are popular for skin biomechanical characterization. With the advancement of imaging techniques, the available techniques are sufficient to quantify and qualify the structural characteristics of skin tissue. Among the microstructural characterization techniques, non-destructive techniques such as Micro-CT and MPM are the interest of researchers as they can capture three-dimensional arrangement of collagen fibers in a single shot.

Further, to model the complex mechanical response of skin, microstructurally motivated constitutive models are widely used as they can capture the skin anisotropy by incorporating the fiber dispersion and orientation. Basically, two constitutive models are used first: angular integration (AI) model, proposed by Lanir [26] and second generalized structural tensor (GST) model proposed by Holzapfel-Gasser-Ogden [28]. Among these two models, the AI model is computationally expensive due to complex integration, whereas GST model is simple and its computational implementation is state forward [37]. Therefore, GST model is found more popular for skin tissues [37]. For more experimental and computation approaches and data, please see work by Kumar et al. [5, 34, 36, 38–42]

Overall, this chapter summarizes the in vivo and ex vivo mechanical characterization techniques, structural characterization techniques, and constitutive models which are commonly used in skin biomechanics research. Some of these discussed techniques are used in this research work.

References

1. Rodrigues, L. (2001). EEMCO guidance to the in vivo assessment of tensile functional properties of the skin. Part 2: Instrumentation and test modes. *Skin Pharmacology and Applied Skin Physiology, 14*(1), 52–67. https://doi.org/10.1159/000056334. PMID: 11174091.
2. Pissarenko, A., & Meyers, M. A. (2020). The materials science of skin: Analysis, characterization, and modeling. *Progress in Materials Science, 1*(110), 100634.
3. Groves, R. B., Coulman, S. A., Birchall, J. C., & Evans, S. L. (2013). An anisotropic, hyperelastic model for skin: Experimental measurements, finite element modelling and identification of parameters for human and murine skin. *Journal of the Mechanical Behavior of Biomedical Materials, 18*, 167–180. https://doi.org/10.1016/j.jmbbm.2012.10.021. Epub 2012 Nov 19 PMID: 23274398.
4. Yang, W., Sherman, V. R., Gludovatz, B., Schaible, E., Stewart, P., Ritchie, R. O., & Meyers, M. A. (2015). On the tear resistance of skin. *Nature Communications, 27*(6), 6649. https://doi.org/10.1038/ncomms7649. PMID: 25812485; PMCID: PMC4389263.
5. Lakhani, P., Dwivedi, K. K., & Kumar, N. (2020). Directional dependent variation in mechanical properties of planar anisotropic porcine skin tissue. *Journal of the Mechanical Behavior of Biomedical Materials, 104*, 103693. https://doi.org/10.1016/j.jmbbm.2020.103693. Epub 2020 Feb 11 PMID: 32174437.
6. Lakhani, P., Dwivedi, K. K., Parashar, A., & Kumar, N. (2021). Non-invasive in vivo quantification of directional dependent variation in mechanical properties for human skin. *Frontiers in Bioengineering and Biotechnology, 22*(9), 749492.
7. Schade, H. (1912). Untersuchungen zur Organfunction des Bindegewebes. *Zeitschrift für experimentelle Pathologie und Therapie, 11*(3), 369–399.
8. Pailler-mattei, C., Bec, S., & Zahouani, H. (2008). In vivo measurements of the elastic mechanical properties of human skin by indentation tests. *Medical Engineering & Physics, 30*, 599–606.
9. Escoffier, C., de Rigal, J., Rochefort, A., Vasselet, R., Lévêque, J.-L., & Agache, P. G. (1989). Age-related mechanical properties of human skin: An in vivo study. *The Journal of Investigative Dermatology, 93*(3), 353–357.

10. Potts, R. O., Chrisman, D. A., Jr., & Buras, E. M., Jr. (1983). The dynamic mechanical properties of human skin in vivo. *Journal of Biomechanics, 16*(6), 365–372.
11. Li, C., Guan, G., Reif, R., Huang, Z., & Wang, R. K. (2012). Determining elastic properties of skin by measuring surface waves from an impulse mechanical stimulus using phase-sensitive optical coherence tomography. *Journal of The Royal Society Interface, 9*(70), 831–841.
12. Analyzer, V. S., Vexler, A., Polyansky, I., & Gorodetsky, R. (1999). Evaluation of skin viscoelasticity and anisotropy by measurement of speed of shear wave propagation with. *Journal of Investigative Dermatology [Internet], 113*(5), 732–739. https://doi.org/10.1046/j.1523-1747.1999.00751.x
13. Zhang, X., & Greenleaf, J. F. (2007). Estimation of tissue's elasticity with surface wave speed. *Journal of the Acoustical Society of America, 122*(5), 2522.
14. Grahame, R., & Holt, P. J. L. (1969). The influence of ageing on the in vivo elasticity of human skin. *Gerontology, 15*(2–3), 121–139.
15. Tregear, R. T. (1966). *Physical functions of skin* (Vol. 5). Academic.
16. Laiacona, D., Cohen, J. M., Coulon, K., Lipsky, Z. W., Maiorana, C., Boltyanskiy, R., et al. (2019). Non-invasive in vivo quantification of human skin tension lines. *Acta Biomaterialia [Internet], 88*, 141–148. https://doi.org/10.1016/j.actbio.2019.02.003
17. Remache, D., Caliez, M., Gratton, M., & Dos Santos, S. (2018). The effects of cyclic tensile and stress-relaxation tests on porcine skin. *Journal of the Mechanical Behavior of Biomedical Materials, 77*, 242–249.
18. Bhushan, B. (2012). Nanotribological and nanomechanical properties of skin with and without cream treatment using atomic force microscopy and nanoindentation. *Journal of Colloid and Interface Science [Internet], 367*(1), 1–33. https://doi.org/10.1016/j.jcis.2011.10.019
19. Crichton, M. L., Chen, X., Huang, H., & Kendall, M. A. F. (2013). Biomaterials elastic modulus and viscoelastic properties of full thickness skin characterised at micro scales. *Biomaterials, 34*, 2087–2097.
20. Pissarenko, A., Yang, W., Quan, H., Brown, K. A., Williams, A., Proud, W. G., et al. (2019). Tensile behavior and structural characterization of pig dermis. *Acta Biomaterialia, 86*, 77–95.
21. Liu, X., Cleary, J., & German, G. K. (2016). The global mechanical properties and multi-scale failure mechanics of heterogeneous human stratum corneum. *Acta Biomaterialia, 43*, 78–87.
22. Ottenio, M., Tran, D., Ní Annaidh, A., Gilchrist, M. D., & Bruyère, K. (2015). Strain rate and anisotropy effects on the tensile failure characteristics of human skin. *Journal of the Mechanical Behavior of Biomedical Materials, 41*, 241–250.
23. Pissarenko, A., Yang, W., Quan, H., Poyer, B., Williams, A., Brown, K. A., et al. (2020). The toughness of porcine skin: Quantitative measurements and microstructural characterization. *Journal of the Mechanical Behavior of Biomedical Materials, 109*, 103848.
24. Sacks, M. S. (2000). Biaxial mechanical evaluation of planar biological materials. *Journal of Elasticity and the Physical Science of Solids, 61*(1), 199–246.
25. Limbert, G. (2011). A mesostructurally-based anisotropic continuum model for biological soft tissues—Decoupled invariant formulation. *Journal of the Mechanical Behavior of Biomedical Materials [Internet], 4*(8), 1637–1657. https://doi.org/10.1016/j.jmbbm.2011.07.016
26. Lanir, Y. (1983). Constitutive equations for fibrous connective tissues. *Journal of Biomechanics, 16*(1), 1–12.
27. Tonge, T. K., Voo, L. M., & Nguyen, T. D. (2013). Full-field bulge test for planar anisotropic tissues: Part II–A thin shell method for determining material parameters and comparison of two distributed fiber modeling approaches. *Acta Biomaterialia, 9*(4), 5926–5942.
28. Gasser, T. C., Ogden, R. W., & Holzapfel, G. A. (2006). Hyperelastic modelling of arterial layers with distributed collagen fibre orientations. *Journal of the Royal Society Interface, 3*(6), 15–35.
29. Dwivedi, K. K., Lakhani, P., Kumar, S., & Kumar, N. (2022). Effect of collagen fibre orientation on the Poisson's ratio and stress relaxation of skin: An *ex vivo* and *in vivo* study. *Royal Society of Open Science, 9*(3), 211301. https://doi.org/10.1098/rsos.211301. PMID: 35345435; PMCID: PMC8941416.

30. Lanir, Y., & Fung, Y. C. (1974). Two-dimensional mechanical properties of rabbit skin—II. Experimental results. *Journal of Biomechanics, 7*(2), 171–182.
31. Holzapfel, G. A., Gasser, T. C., & Ogden, R. W. (2000). A new constitutive framework for arterial wall mechanics and a comparative study of material models. *Journal of Elasticity and the Physical Science of Solids, 61*(1), 1–48.
32. Engineering, B., Mechanics, S., Sciences, E., & Backe, O. (2007). On planar biaxial tests for anisotropic nonlinearly elastic solids. *Mathematics and Mechanics of Solids, 14*(5), 474–489.
33. Dick, J. C. (1951). The tension and resistance to stretching of human skin and other membranes, with results from a series of normal and oedematous cases. *Journal of Physiology, 112*(1–2), 102–113.
34. Dwivedi, K. K., Lakhani, P., Kumar, S., & Kumar, N. (2020). Frequency dependent inelastic response of collagen architecture of pig dermis under cyclic tensile loading: An experimental study. *Journal of the Mechanical Behavior of Biomedical Materials, 112*, 104030. https://doi.org/10.1016/j.jmbbm.2020.104030. Epub 2020 Aug 15 PMID: 32858398.
35. Dwivedi, K. K., Lakhani, P., Sihota, P., Tikoo, K., Kumar, S., & Kumar, N. (2023). The multiscale characterization and constitutive modeling of healthy and type 2 diabetes mellitus Sprague Dawley rat skin. *Acta Biomaterialia, 1*(158), 324–346. https://doi.org/10.1016/j.actbio.2022.12.037. Epub 2022 Dec 21 PMID: 36565785.
36. Dwivedi, K. K., Lakhani, P., Kumar, S., & Kumar, N. (2022). A hyperelastic model to capture the mechanical behaviour and histological aspects of the soft tissues. *Journal of the Mechanical Behavior of Biomedical Materials, 126*, 105013. https://doi.org/10.1016/j.jmbbm.2021.105013. Epub 2021 Dec 3 PMID: 34920323.
37. Gah, Ã. (2006). Determination of material models for arterial walls from uniaxial extension tests and histological structure. *Journal of Theoretical Biology, 238*, 290–302.
38. Dwivedi, K. K., Lakhani, P., Sihota, P., Kumar, S., & Kumar, N. (2023). The multiscale characterization and constitutive modeling of healthy and type 2 diabetes mellitus (t2dm) Sprague Dawley. *Acta Biomaterilia, 158*, 324–346. https://doi.org/10.1016/j.actbio.2022.12.037)
39. Dwivedi, K. K., Lakhani, P., Yadav, A., Mehta, D., Kumar, S., & Kumar, N. (2023). Location specific multi-scale characterization and constitutive modeling of pig aorta. *Journal of the Mechanical Behavior of Biomedical Materials, 142*, 105809. https://doi.org/10.1016/j.jmbbm.2023.105809
40. Dwivedi, K. K., Lakhani, P., Parashar, A., & Kumar, N. (2022). Effect of collagen fiber orientation on the poisson's ratio and stress relaxation of skin: An ex-vivo and in-vivo study. *Journal of Royal Society Open Science, 9*, 211301. https://doi.org/10.1098/rsos.211301
41. Lakhani, P., Dwivedi, K. K., Parashar, A., & Kumar, N. (2021). Non-invasive in vivo quantification of directional dependent variation in mechanical properties for human skin. *Frontiers in bioengineering and biotechnology, 979*, * Contributed equally. https://doi.org/10.3389/fbioe.2021.749492
42. Dwivedi, K. K., Lakhani, P., Kumar, S., & Kumar, N. (2020). Effect of strain rate on the stress relaxation of the pig dermis: A hyper-viscoelastic approach. *Journal of Biomechanical Engineering, 142*(9), 091006. https://doi.org/10.1115/1.4046205

Experimental and Mechanical Analysis of Aorta and Diaphragm Tissues Under High Strain Rate Loading

Piyush Gaur

Abstract The increasing number of road traffic accidents (RTAs) worldwide has underscored the need to better understand the mechanisms of trauma, particularly in vulnerable thoracic tissues like the aorta and diaphragm. This chapter investigates the mechanical behaviour of these tissues under high strain rate loading conditions, which are typical during automotive collisions. The experimental characterization of the aorta and diaphragm provides insights into their viscoelastic properties and rate-dependent failure mechanisms, which are critical for enhancing the accuracy of human body finite element (FE) models used in crash safety simulations. Advanced constitutive models were developed using data from dynamic testing methods, including uniaxial and biaxial tensile tests, as well as Split Hopkinson Pressure Bar (SHPB) experiments. The findings reveal significant variations in tissue stiffness, elasticity, and failure thresholds at elevated strain rates, which contribute to the understanding of soft tissue injury mechanisms during high-speed impacts. This work aims to improve biofidelity in FE models of the human body, thereby offering more accurate predictions of injury risks and aiding in the development of better vehicle safety systems.

Keywords High strain rate loading · Aorta mechanical properties · Diaphragm viscoelasticity · Soft tissue injury mechanisms · Finite element modelling · Automotive crash simulation

1 Introduction

Increased mobility has resulted in an increase in the number of fatalities caused by road traffic accidents (RTAs). Every year, RTAs result in the injuries of over 50 million individuals and the deaths of over 1 million individuals at global level [1–4]. Approximately 30–50% of all injury related fatalities are caused by motor

P. Gaur (✉)
Mechanical Cluster, School of Advanced Engineering, UPES, Dehradun, India
e-mail: piyush.gaur@ddn.upes.ac.in

K. K. Dwivedi et al. (eds.), *Mechanics and Materials Science of Biological Materials*, Materials Horizons: From Nature to Nanomaterials,
https://doi.org/10.1007/978-981-96-5366-9_4

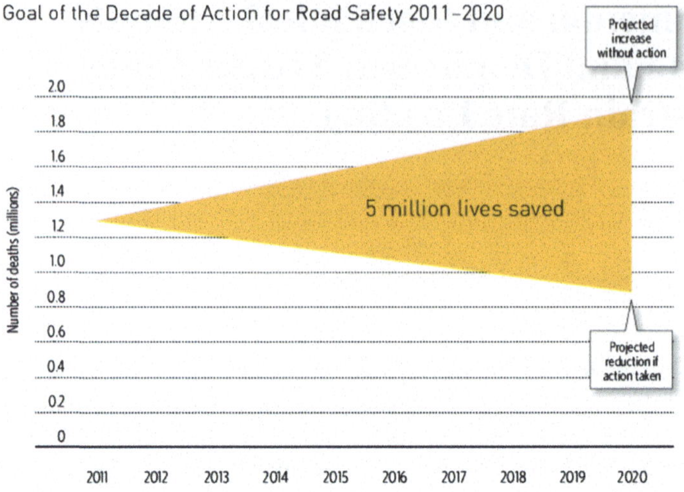

Fig. 1 The United Nation decade for action for road safety 2011–2020

vehicle crashes (MVCs) [5]. In 2013–2014, the global number of road fatalities was approximately 1.24 million [6]. Vehicle car occupants account for about 31% of all the fatalities worldwide. Furthermore, between 20 and 50 million individuals experience non-fatal injuries. In 2005, the number of individuals injured because of RTAs worldwide exceeded 78 million, with 9.2 million necessitating medical treatment. The World Health Organization (WHO) [7] estimated that the RTAs are the ninth most common cause of death worldwide by 2015, unless no substantial measures are taken, RTAs will increase from the ninth to fifth common cause of fatality by 2030. Such alarming statistics have prompted the United Nations to designate 2011–2020 as the "***Decade of Action for Road Safety***" due to significant global public health concern that RTAs pose (Fig. 1).

The escalating surge in RTAs in developing nations is a burgeoning apprehension. Although the fatality rates have decreased in developed countries, road traffic injuries continue to be a significant contributor to high mortality, physical harm, and impairment [8]. According to Dinesh Mohan research [9], RTAs in India showed an annual increase of approximately 4% from 1997 to 2005. In 2007–2010, India recorded 1,14,950 deaths [10]. Without taking preventive measures, the number of death by 2018–2020 could be estimated to 1,50,000–1,80,000 [3, 4]. Thus, it is crucial to decrease both quantity and intensity of injuries in road accidents. Automotive trauma is a notable health issue that is frequently disregarded [8, 11]. An imperative collaboration among manufacturers, researchers, and the government is crucial in order to mitigate the injuries resulting from RTAs [12]. Improved road design and thew implementation of safer vehicles can effectively decrease RTAs, with particular emphasis on enhancing safety for pedestrians.

Automotive collisions cause primary, secondary, and tertiary impacts [8]. Primary collisions occur between vehicle and external object, while secondary collision occurs between occupant and vehicle interiors. Internal organs and body structure collide in tertiary collision. When biological tissues are deformed beyond their physiological capacity, anatomical damage occurs [13]. RTAs often cause blunt trauma from contact with the vehicle or other passengers. Secondary and tertiary impacts on the rib cage and soft tissues like the heart, lungs, diaphragm and aorta can cause fatal thoracic trauma. Thoracic injuries cause the second most deaths and serious injuries after the traumatic brain injuries [14]. These injuries account for 19–20% of the Injury Priority Rating (IPR) [14, 15]. Frontal impacts often injure the aorta and diaphragm, with aortic injuries being fatal. Thoracic injuries mechanisms must be understood to improve the vehicle occupant safety. This necessity gives rise to studying the experimental properties of aorta and diaphragm tissue and develop the appropriate mathematical models to describe their behaviour particularly under dynamic loads, i.e. high strain rate loading along with their implementation in human body models to improve their overall biofidelity.

Human body finite element (FE) models help evaluate alternative safety systems to reduce injury risks by understanding vehicle occupant and pedestrian loading mechanisms during crashes [16]. FE models, with their ability to compute materials states like strains, stresses and failure [17], have become an essential part of vehicle design process due to increase in computational power with decrease in hardware costs. Overtime, human body FE models that mimic external and internal anatomical structures were developed to predict internal organ injuries, which dummy models cannot. For accurate human body models, realistic geometry and strain rate-dependent materials' properties must reflect dynamic responses under different loading conditions [18, 19]. Due to the lack of appropriate constitutive models that accurately represent soft tissues dynamics, which are crucial for occupant safety, these models are limited. Thoracic soft tissues like the aorta, diaphragm, heart, and lungs, respond differently at high strain rates, like in RTAs and traditional experimental data obtained at quasi-static loading don't capture this. Thus, to develop more accurate and biofidelic human body models, soft tissues must be experimentally studied across a wide range of strain rates and develop constitutive models to capture their dynamic behaviour at high strain rates. Validated constitutive models help human body FE models predict injury mechanisms, improving vehicle safety and reducing crash injuries. This book chapter focuses on the experimental and mechanical analysis of aortic and diaphragm tissues subjected to high strain rate loading conditions. It aims to provide a comprehensive overview of the behaviour of these biological tissues under dynamic conditions, highlighting their mechanical properties and response to impact. The chapter integrates experimental data with advanced constitutive models to offer valuable insights for applications in biomedical and impact biomechanics research.

2 Anatomy and Physiology of Aorta and Diaphragm

2.1 Human Aorta

The aorta is the main trunk of vessels that carry oxygenated blood from the left ventricle of the heart (Figs. 2 and 3). It starts at the base of the left ventricle, called the aortic root, with a diameter of about 30 mm. The ascending aorta rises obliquely for around 50 mm, curving anteriorly and to the right, continuing as the aortic arch, which arches postero-superiorly and leftward over the left pulmonary hilum. Three branches—the innominate artery (supplying the right arm and head), the left common carotid artery (supplying the head), and the left subclavian artery (supplying the left arm)—originate from the superior convexity of the arch, collectively known as the superior vasculature. The ligamentum arteriosum extends from the pulmonary trunk and attaches to the inferior convexity of the aortic arch. The descending aorta begins at the lower border of the twelfth thoracic vertebrae, passing through the diaphragmatic aortic hiatus. It is securely anchored to the thoracic spine by the pleura, intercostal arteries, and paravertebral fascia, while the ascending and arch regions remain relatively mobile. The area between the left subclavian artery's anastomosis and the junction of the aortic arch and descending aorta is referred to as the peri-isthmus.

The aorta is a tubular structure composed of three tissue layers: the tunica intima (inner), tunica media (middle), and tunica adventitia (outer), as illustrated in Fig. 2c. The tunica intima contains a lining of endothelial cells exposed to blood, supported by a thin layer of fibrous tissue with collagen and elastic fibres. The tunica media, the thickest layer, consists primarily of smooth muscle fibres and some collagen. The tunica adventitia is a thin outer layer composed mainly of collagen with a few elastic fibres. The aorta can be viewed in two anatomical directions: the longitudinal (axial) axis parallel to the vessel's centre and the circumferential (transverse) axis, which is radial. As the largest blood vessel, the aorta is subject to pulsating pressure, with systolic pressure being the highest during the cycle and diastolic pressure the lowest. In young adults, typical systolic and diastolic pressures are 120 mm Hg (16 kPa) and 80 mm Hg (11 kPa), respectively, increasing to around 150 mm Hg (20 kPa) and 90 mm Hg (12 kPa) with age. Cardiac output, or the volume of blood the heart pumps, is about 5 L per minute at rest and can rise to 25–35 L per minute during intense exercise.

2.2 Human Diaphragm

The diaphragm is a muscular and fibrous structure with a bi-domed shape. It functions to separate the thoracic cavity from the abdominal cavity. The diaphragm muscle consists of two components: a central fibrous tendinous aponeurosis and peripheral

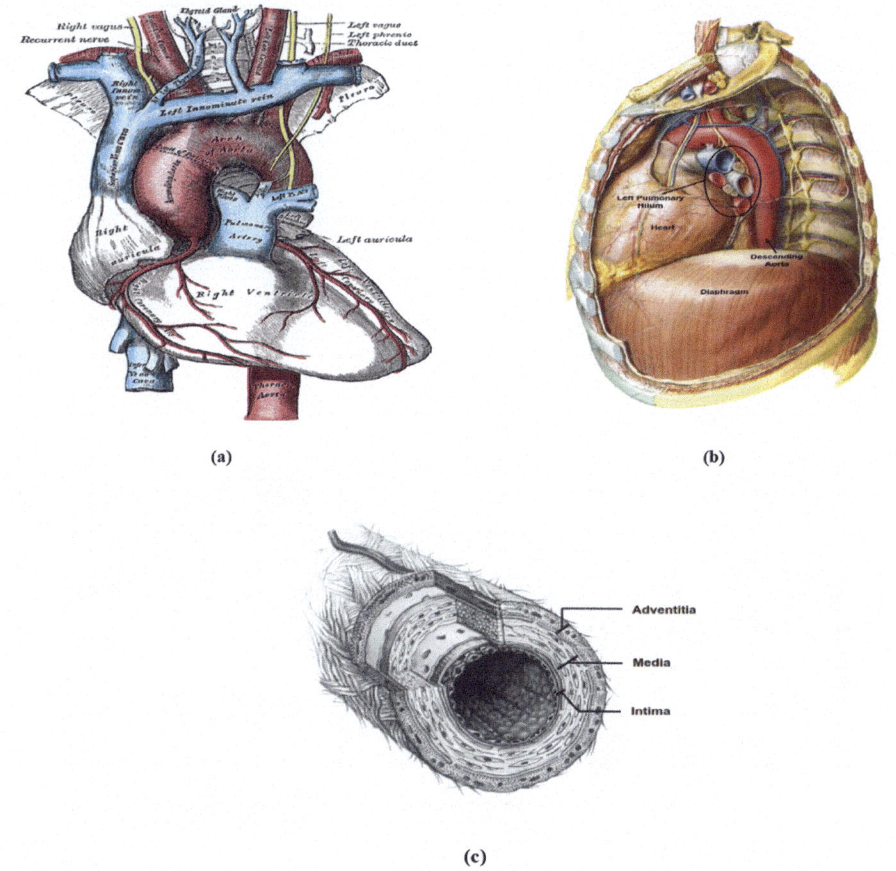

Fig. 2 **a** Anatomy of human aorta (reproduced from Gray, 2000), **b** layers of aorta [20]

muscle bundles. The diaphragm is anatomically comprised of a thin central aponeurosis connected to the outer striated muscle. It is further divided into two sections: the lumbar diaphragm and the costal diaphragm. The origin of the structure is located at the front of the lumbar vertebrae, specifically through the left and right crura. The right crus is connected to the lumbar vertebrae L1, L2, and L3, while the left crus is connected to L1 and L2. The diaphragm attaches to the central tendon, which is fused with the pericardium. The inferior vena cava and the right phrenic nerves traverse an aperture at the T8 level, while the oesophagus and the vagus nerve traverse an aperture at the T10 level. Additionally, the abdominal aorta, thoracic duct, and azygous vein traverse an aperture at the T12 level. The diaphragm receives its nerve supply from the phrenic nerves (C3, C4, and C5), which provide both sensory and motor innervation. The actions it performs during respiration account for 70% of the effort required for inspiration. The diaphragm is enveloped by the parietal pleura and peritoneum, with the exception of the liver's bare area [21–23].

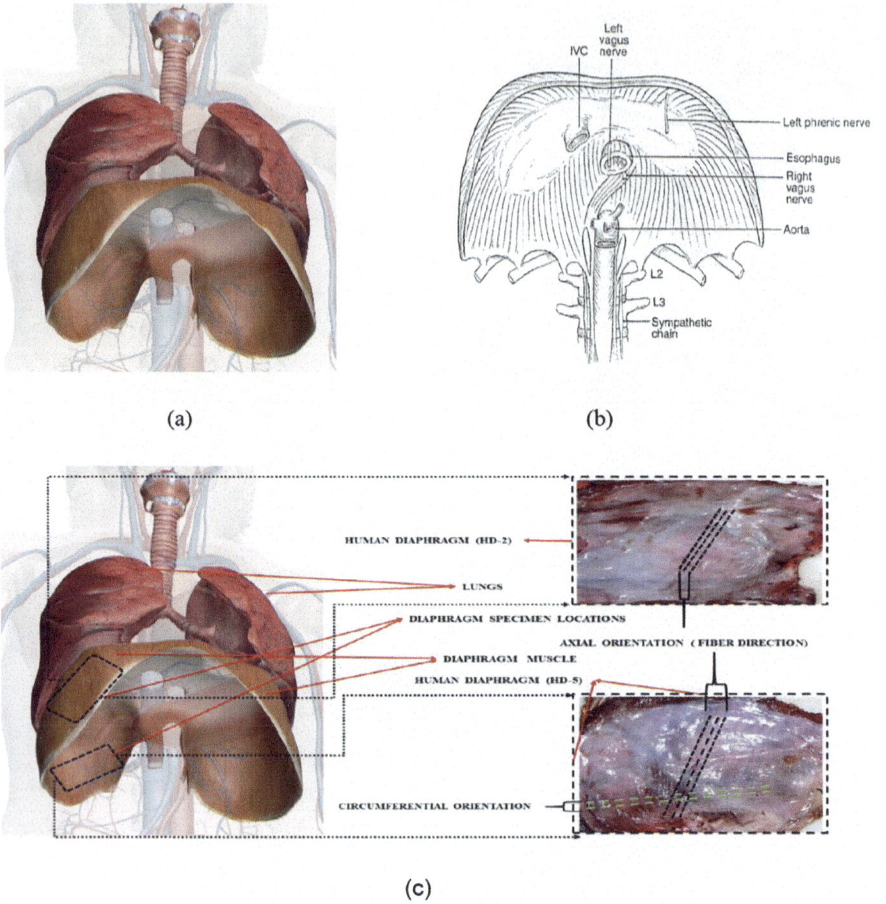

Fig. 3 Schematic of human diaphragm muscle, **a** enlarged view of different openings and parts (Adapted from https://www.innerbody.com/image_chest1/musc82.html), **b** fibre orientations of diaphragm muscle [24]

2.3 Physiological Relevance in High Strain Rate Conditions

Under high strain rate conditions, commonly observed in automotive crashes, soft tissues including muscles, ligaments, tendons, and internal organs exhibit distinctive mechanical behaviour. Soft tissues exhibit viscoelasticity, nonlinearity in their stress–strain relationship, and significant strain rate sensitivity at rates as high as 100–500/s or more [11, 20, 25]. This implies that tissues become more susceptible to damage and become stiffer when subjected to higher loading rates, thereby increasing the risk of injury in collisions. Soft tissues including the heart, lungs, aorta, and liver are particularly susceptible to failure in high-impact situations, including internal bleeding, tears, and ruptures. These injuries can lead to complications such as internal bleeding or organ failure, which can have serious physiological repercussions. Similarly, soft tissue injuries of the musculoskeletal system, including muscle injuries and ligament tears, can significantly impair mobility and function. The brain is susceptible to traumatic brain injury (TBI), including diffuse axonal injury (DAI) and blunt force injuries [26–28], which are exacerbated by high stress levels, due to the protective structure of the skull, which places it under different loads.

High-speed incidents such as vehicular collisions generate failure strains within milliseconds. Human body organs exhibit varying responses to alterations in loading rates. It is essential to input material data acquired at strain rates near vehicular collisions to achieve an adequate response from finite element models. In recent years, researchers have concentrated on testing isolated body organs at loading rates characteristic of crash events, primarily employing viscosity-based models to elucidate the effects [11, 29, 30]. The comprehensive human body model can be utilized in vehicle crash simulations to replicate injuries resulting from the motor vehicle crashes. Currently, a major constraint in predicting human body behaviour during crash events with finite element human models is the inadequacy of constitutive models to accurately depict the dynamic response of soft tissues [31]. The mechanics of soft tissues under impact loading conditions remain inadequately comprehended. Therefore, it is essential to characterize soft tissues of abdomen and thorax region to enhance the development of comprehensive finite element models of the human body. The physiological significance of soft tissue responses under high strain rate conditions is critical for the advancement of high-fidelity human body models (HBMs) [8, 31–33]. These models enable more accurate simulation of crash scenarios and are essential for studying injury mechanisms and improving automobile crash safety. Traditional crash dummies primarily assess bone fractures and occupant kinematics, while human body models (HBMs) provide a more sophisticated representation of soft tissue dynamics, enabling predictions of internal organ deformation and crash injury severity. Advanced HBMs not only simulate the mechanical properties of soft tissues, but also consider interactions between tissues, organs, and protective safety systems such as airbags and seatbelts. These models provide important insights into injury types, including aortic ruptures, pulmonary contusions, and liver lacerations, by analysing soft tissue deformation and failure during collisions, particularly severe accidents.

3 Experimental Characterization of Soft Tissues at High Strain Rates

Experimental characterization of the soft tissues is a crucial step in understanding their mechanical behaviour during high dynamic loading conditions, particularly in motor vehicle crashes (MVCs) [34, 35]. At these rates which can reach up to 100–500/s or higher, soft tissues exhibit viscoelastic, non-Hookean behaviour in their stress–strain relationship, and possess significant strain rate sensitivity. Unlike bone, which has been extensively studied under various quasi-static, and dynamic loading conditions, soft tissues like tendons, ligaments, and other internal organs shows complex mechanical behaviour under various mechanical loadings [36–39], and also most of the soft tissues were studied either from the animal tissues or from the cadaveric tissues [11, 40–48]. At high strain rates, soft tissues like the aorta, heart, muscles, tendons, ligaments, and other internal organs tend to stiffen and their capability to absorb energy without failing becomes significantly reduced. In order to develop biofidelic human body models for accurate injury predictions and analysis, it is essential to experimentally characterize the mechanical properties of soft and hard tissues at high strain rates [30, 31]. This experimental data serves as a foundation for development of more detailed constitutive models that can enhance vehicle safety design and improve the ability to mitigate various traumatic injuries.

One of the most commonly used techniques for characterizing the soft tissues behaviour at high strain rates is the use of Split Pressure Hopkinson Bar (SHPB) [49]. This is the ideal method for capturing the mechanical response of various tissues under high-rate dynamic loading. In an SHPB tests, a soft tissue specimen is positioned between the two bars, and high-speed impact from the gas gun generates a compressive or tensile wave that passes through the tissue and deformed it. By measuring the force and displacement on both sides of the tissues, a stress–strain curve is obtained that reflect the behaviour of the tissues at strain rate far exceeding the physiological range of the tissue. This testing modality is very crucial for identifying the strain rate sensitivity of different tissues and helpful in developing strain rate-dependent constitutive models. For instance, tendons and ligaments, which typically experience low strains in normal physiological activities, stiffens dramatically at high strain rates. Similarly, organs like the liver or spleen, which are relatively soft and highly deformable under normal conditions, exhibit complex mechanical responses which exposed to high-rate conditions, leading to a much higher likelihood of rupture or tearing in crash scenarios. High-speed tensile testing is also one of the other essential methods for characterizing soft tissues under high-rate loading conditions. This technique entails the elongation of tissue specimens, including skin, muscle, or internal organ tissues, while consistently quantifying the forces applied to the tissue. High-speed tensile testing yields useful critical data regarding tissue strength, failure stress and strain, and the velocity of tissue damage onset. The aorta, frequently exposed to intense forces during collision, can be analysed through high-speed tensile tests to determine the conditions and timing of the rupture. Likewise, the integumentary system and the subjacent muscular tissues can be evaluated to estimate certain tissue thresholds, commonly useful in the predictions of injuries like abrasions, lacerations, tears, or ruptures. This testing is essential in the design of restraint systems. The outcome from the tensile tests directly affects the calibration of safety systems such as airbags and set belts to mitigate rapid high energy impacts. Incorporating high strain rate data of soft tissues allows for the optimization of restrain systems to minimize the tissue deformation, decrease the occurrence of internal injuries, and improve overall occupant protection during a collision.

The experimental data of tissues generated from high strain rate tests is essential for enhancing the precision of human body models [8, 34, 50]. The mechanical properties, such as elastic modulus, yield strength, and failure strain, derived from these tests are utilized to parameterize constitutive models that characterize soft tissue behaviour in simulations [31]. These models must consider the rate-dependent characteristics of tissues, as their response varies with the loading velocity. This is especially critical in crash simulations, where elevated strain rates lead to substantial tissue stiffening. Furthermore, these experiments assess the energy necessary for soft tissue tears or ruptures, contributing to the comprehension of severe crash injuries such as liver lacerations, brain contusions, and lung ruptures. Integrating this data into HBMs enables researchers to forecast injury patterns and devise more efficient crash mitigation strategies [34]. Also, the incorporation of high strain rate data from soft tissues has resulted in significant improvements in seatbelt systems, which now distribute forces more evenly across the body to avert concentrated pressure on vulnerable tissues such as the abdomen or thorax [51]. Airbags can be designed to deploy at speeds that more effectively reduce occupant impact, thereby lowering the risk of injuries such as rib fractures or lung contusions. The experimental characterization of soft tissues at high strain rates is essential for developing more precise injury criteria [52]. Understanding tissue failure at different strain rates allows engineers to determine new injury risk thresholds, thereby improving crash test safety standards.

3.1 Quasi-Static Versus Dynamic Testing

The rate at which soft tissues are loaded has a significant impact on their mechanical behaviour, which normally varies significantly. For example, walking and bending are examples of everyday physiological activities are normally mimicked and characterized by quasi-static testing [26, 53–55]. When tissues are subjected to such a lower strain rate, they tend to behave in a more elastic manner, manifesting gradual deformation and limited stiffness. The baseline mechanical properties like elastic modulus, tensile strength, and failure strain are determined through this type of testing under conditions that involve no time dependent effects. The soft tissues response in quasi-static conditions [56–58] is characterized by a high deformation (higher failure strain) and without premature failure. Because of these features, it is suitable for understanding how soft tissues behave under normal physiological conditions. Dynamic testing, on the other hand, involves subjecting tissues to significantly higher strain rates, which simulates the rapid loading that occurs during traumatic events such as traumatic rupture of aorta or TBI. Under higher strain rates, tissues display a more rigid and brittle behaviour, and become less capable of withstanding deformation when damaged [11]. Due to their viscoelastic behaviour, tissues tend to resist loading more aggressively, and this is the main cause of rate dependent in tissues. Consequently, tissues that typically endure stretching under slow loading may sustain tearing or rupture during high-velocity impacts [11, 59, 60]. For instance, organs like the aorta, liver, or brain exhibit markedly different responses to dynamic loading, revealing an increased susceptibility to injuries such as ruptures, lacerations, or contusions. Due to the discrepancy between quasi-static and dynamic responses, it is crucial to comprehend injury mechanisms during high-velocity incidents. This disparity underscores the necessity for specialized testing methodologies to accurately assess these behaviours.

The experimental setups employed to evaluate soft tissues at quasi-static and dynamic strain rates are intended to elucidate the specific mechanical properties pertinent to each loading conditions [61]. Quasi-static testing normally employs mechanical testing equipment's, such as universal testing machines (UTMs). These machines applied gradual, regulated loads on tissue samples, facilitating accurate assessment of the stress–strain response across a board spectrum of deformations [62]. The procedure generally entails the incremental stretching or compressing of tissue samples, while force and displacement sensors monitor the mechanical response. Quasi-static tests are uncomplicated to perform and yield significant data regarding the essential mechanical properties of soft tissues, such as their elastic limits, ultimate tensile strength, and failure mechanisms under low-loading conditions [63, 64]. Uniaxial tensile testing is particularly useful for examining the long-term resilience of soft tissues and their deformation under continuous stresses. However, the uniaxial tensile testing of soft tissues suffers from few limitations. One of the limitations is that uniaxial tensile testing assesses the tissue behaviour in a single directions, failing to take into account the multidirectional forces that tissues encounters within the body [24, 31, 65, 66]. This may result in partial characterization to intricate the multidimensional loading conditions. Also, for tissues like heart, aorta, skin, diaphragm, etc., uniaxial tests fail to account for the anisotropic behaviour. Hence, solely relying on the uniaxial data may adequately represent the behaviour of soft tissues in real-world scenarios, thereby diminishing its efficacy in formulating precise models for crashworthiness and injury prediction. Biaxial testing, thus, becomes essential for evaluating soft tissues behaviour under multiloading conditions, accurately mimicking the real-life loading conditions, particularly in organs such as the heart or lungs that endure complex forces. Biaxial tests involve applying simultaneous loads to tissue samples in two orthogonal directions, yielding enhanced understanding of their mechanical properties under more realistic conditions. Biaxial testing is an essential experimental method used to evaluate the mechanical behaviour of soft tissues when subjected to forces in two perpendicular directions at the same time [67, 68, p. 202]. This technique is particularly important for tissues such as the skin, heart, lungs, and blood vessels, which naturally encounter multidirectional stresses within the body. In contrast to uniaxial testing, biaxial tests allow researchers to more precisely replicate the intricate physiological loading conditions that tissues undergo in reality, including those associated with respiration, blood circulation, or organ motion. Biaxial testing offers a comprehensive insight into the mechanical properties of tissues, including stiffness, strength, strain energy density, and failure characterizes under multidirectional stresses [69]. In a standard biaxial test, a square or rectangular tissue sample is placed on a testing apparatus and stretched simultaneously in both the x and y directions using precise force or displacement controls (Fig. 4). Sensors measure the tissue's stress and strain responses, capturing its deformation under multiaxial loading. This method is crucial for understanding

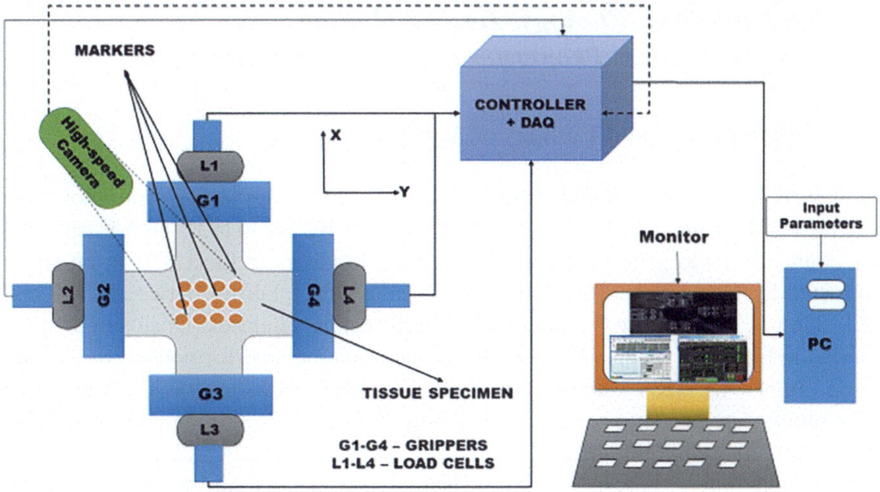

Fig. 4 Typical biaxial device used in soft tissue testing [70]

the multiaxial mechanical behaviour of tissues. Pulsatile blood flow induces circumferential and axial stresses in arterial tissues, while multidirectional tension arises in the skin during bodily movement. Biaxial testing is particularly crucial for tissues that exhibit non-uniform behaviour in various directions, as their material properties can differ markedly between the two axes.

Conversely, dynamic testing necessitates specialized apparatus to simulate the high-velocity impacts encountered during RTAs. The Split Hopkinson Pressure Bar is a frequently employed device for dynamic testing, enabling researchers to evaluate tissue responses at strain rates from several hundred to over a thousand [71]. In a Split Hopkinson Pressure Bar test, a tissue specimen is positioned between two bars and subjected to a rapid impact or compressive force, producing strain waves within the tissue. The stress–strain response is assessed through wave propagation, rendering the SHPB optimal for evaluating tissue properties at elevated strain rates and supplying essential information on tissue failure during rapid loading. Another approach involves high-speed tensile machines that elongate tissue samples at elevated velocities to replicate rapid deformations [59]. These configurations facilitate quantitative evaluation of soft tissues properties during rapid loading rates, including dynamic stiffness, fracture toughness, and energy absorption. Quasi-static and dynamic testing yield critical information regarding the behaviour of soft tissues under varying strain rates. By integrating data from quasi-static, dynamic, and biaxial testing methodologies, researchers can develop more precise human body models that mimics the complete spectrum of tissue responses, from everyday activities to extreme trauma such as automotive collisions/accidents.

3.2 *Epidemiology, Etiology, Testing Methods, and Mechanical Properties of Aorta and Diaphragm*

3.2.1 Epidemiology and Etiology of Aortic Injuries

Parmley et al. [72] reported that almost 57% of all injuries involving rupture of aorta were the result of motor vehicle impacts. In a review by Viano [73], TRA is reported to account for 10 to 20% of automotive related fatalities in the USA. In an study conducted by Newman and Rastogi [74] for all traffic fatalities between 1979 and 1981, TRA is responsible for about 17% of all traffic fatalities. Dosios et al. [75] reported in his study using autopsies from 11,446 patients performed over a 3 years period (1994–96). TRA was reported for 13% cases consisting of 251 victims of either blunt or penetrating aortic injury. About 72% of the total fatalities were from RTAs, while the remaining injuries were accounted due to falls, and/or compression by heavy machinery, blocks, or stones. Katyal et al. [33] found evidence of TRA in 21% of crash related autopsies in a study in Toronto, Canada, between 1991 and 1995. Out of these cases, 50% of the cases involved lateral impact suggesting that side impact is just as important to TRA as frontal impact. Drivers and passengers were found to be affected equally, but 80% of the side impact cases of TRA involved near side impact, whereas 26% of TRA in frontal crashes were experienced by drivers and 24% of the passengers.

Mcgwin et al. [76] examined the 1995 to 2000 NASS database to calculate the relation between risk of TRA and occupant, vehicle, and collisions characteristics. Of the 20,191 cases of TRA were identified approximately 31 million occupants in crashes. Only 11.4% of TRA cases survived. Augenstein et al. [77] found that TRA can occur both with and without seat belt use. According to Viano [73], aortic ruptures normally occur at three sites in motor vehicle crashes: the peri-isthmus, the descending aorta near the arcuate ligament at the diaphragmatic hiatus, and the aortic root. He also reported that among these three sites, peri-isthmus site is the one which often ruptures the most. The abdominal and the thoracic aorta accounted for 4.7 and 12.7% of TRA. As per notified by Dosios et al. [75], 59% of the victims have died due to isthmus laceration. Hunt et al. [78] report that 44% of all TRA victims who arrive at the emergency department will not survive, and 70–90% will die at the accident scene. Smith and Chang [79] estimate the mortality rate for motor vehicle accident victims with aortic injury to be 84% at the site of injury. One-third of patients with lacerations will die within the first six hours, and if untreated, 90% are anticipated to die from their injuries within three months.

The research indicates various mechanisms causing aortic ruptures. The majority of the proposed systems focused on specific loading modes that induce mechanical failure in aortic tissue [80]. Letterer [81] posited that the rupture at the aortic root may result from falls and the subsequent downward tension on the heart. Shennan [82] focused on the hinged aortic fixation at the isthmus; Lim et al. [83] suggested that TRA arose from unequal acceleration of the mediastinal components. Cammack et al. [84] asserted that shear and pressure are the primary contributors to aortic rupture, while twisting induced by deceleration (inertia) is a secondary one. Hossack [85] posited that the aortic spindle and indentation ridge were significant, suggesting that lacerations from bone resulted in non-circumferential tears. Three primary hypotheses elucidate the mechanism of aortic rupture due to laceration: (a) Traction or shear forces arising from the relative motion between the aorta and its anchor points; (b) direct compression between the sternum and spinal column, referred to as the shovelling effect; and (c) the accumulation of blood pressure within the aorta [86]. The compressive force from a chest impact with the steering wheel elevates the sternum and abdomen in a car. This influence is commonly termed the "shovelling effect." When the chest is compressed, the heart is constricted between the sternum and the spine, thereby propelling blood from the heart into the aorta and swiftly elevating blood pressure in both the aorta and pulmonary trunk. Furthermore, it is anticipated that at the location where the aorta traverses the diaphragm, it will bend and obstruct blood flow through the aorta, hence elevating blood pressure. Arenberg [87] proposed that the hemodynamic consequences of acceleration may also lead to an increase in aortic pressure. Elevated blood pressure may consequently result in the rupture of the vessel. While the descending aorta, connected to the spine via the lungs, remains immobile, the shovelling motion propels the heart and aortic arch upward. The isthmus of the descending aorta, situated at the intersection of the fixed descending and upward displaced aortic arch, endures both torsional and tensional stress. The inherent vulnerability of the isthmus results in the aorta tearing at that location. The first mechanism typically leads to a laceration of the aortic isthmus, the aortic root, or the thoracic aorta at the diaphragm's entrance, characterized by vertical, lateral, or oblique heart displacement. Regardless of the involvement of the adventitia, a significant elevation in aortic blood pressure typically leads to the rupture of the intima and media [88].

3.2.2 Aorta Testing Methods and Mechanical Properties

To understand the mechanisms of aortic ruptures which seem to be complex, knowledge of mechanical properties of the aorta becomes crucial. Most of the experimental investigation on aorta has been conducted within the normal physiological operation range of stresses and strain [18, 89–95]. Some of the experimental efforts include pressure–volume, inflation, and uniaxial and biaxial tensile testing. Studies investigating the mechanical and failure behaviour of human aorta under high-impact conditions [20, 47, 96] are fewer.

Pressure–volume experiments were among the initial techniques employed to investigate the mechanical properties of cadaveric aortas. The applied pressure and volume variation were employed to evaluate the mechanical properties of the aorta. Figure 5 illustrates a pressure–volume setup established by Biewener [97]. Hallock and Benson [98] utilized cylindrical aorta specimens, sequentially pressured with saline solution, to ascertain the pressure–volume relationship of cadaver aortas. The tissue's rigidity was demonstrated to diminish with the advancement of specimen age. Bergel [99] conducted quasi-static testing on canine aortas to assess pressure–volume

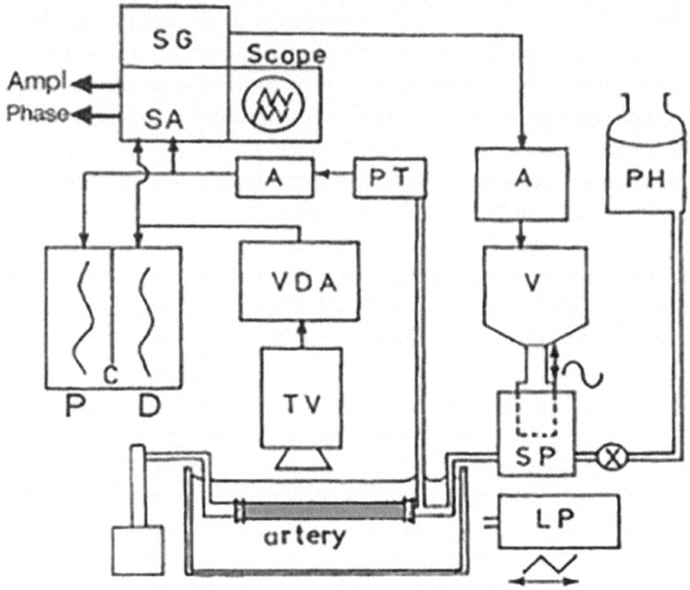

Fig. 5 Pressure–volume apparatus for arterial inflation [97]

relationships and introduced an incremental modulus for the identified nonlinear stress–strain relationship. The incremental modulus varied from 0.12 to 1.81 MPa within a pressure range of 5.2–29.3 kPa.

Mohan and Melvin (1983) conducted aortic biaxial bubble inflation experiments with nominal strain rates of 0.01–1 and 20 /s.

Figure 6 illustrates a circular grid along with the longitudinal orientation of circular aortic specimens measuring 50 mm in diameter. Quasi-static (0.01/s) and dynamic (20/s) failure tests were conducted. The distortions of the circular grids were quantified by utilizing camera frames and a pressure transducer. They consistently documented longitudinal rupture, which corresponded with data from actual vehicular collisions. Pressure–volume experiments simulated a generalized loading condition that closely resembles the in vivo behaviour of the aorta. Clamping and irregular bulging or barreling of the aorta were among the challenges observed during inflation experiments. The regional properties along each axis of the aorta were indeterminate, necessitating the execution of uniaxial and biaxial tests. Lundevall [100] found that the isthmus and descending aorta exhibited 63 and 80% of the baseline strength, respectively. The mechanical behaviour of aortic tissue was found to be non-Hookean, anisotropic, and sensitive to strain rate. Collins and Hu [101] investigated the rate dependence of the modulus of porcine aortic segments by applying internal pressure to the aortic specimens at two different rates, corresponding to nominal strain rates of 0.005 and 1 to 3.5 s^{-1}. A monotonic increase in longitudinal failure stress was observed as the strain rate was increased from 300 to 450 kPa.

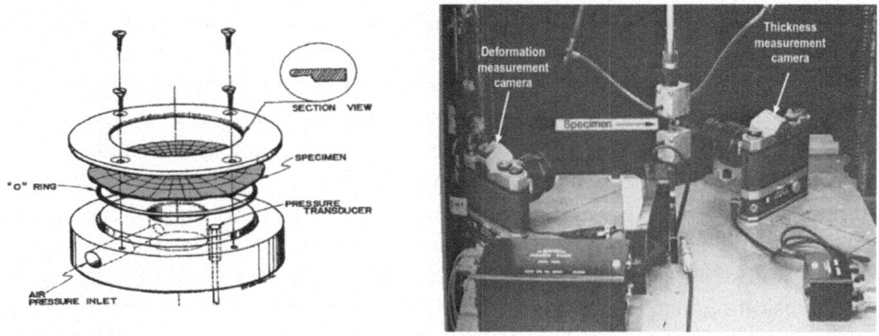

(a) (b)

Fig. 6 Bubble inflation experiment on circular aorta specimen (top) [96], and uniaxial test setup used by [47] (bottom)

With increasing strain rates, a reduction in failure strain and an increase in stiffness were reported; however, the specific values of the observed failure strains were not reported. Mohan and Melvin [47] performed circumferential and longitudinal uniaxial tensile tests (Fig. 6b) at nominal strain rates of 0.01 and 100 s^{-1} on descending aortic specimens to evaluate the failure characteristics of human aortic tissue. They observed increases in longitudinal and circumferential fracture stresses, ranging from 1.47 and 1.72 MPa to 3.59 and 5.07 MPa. Stemper et al. [102] performed uniaxial tensile tests on porcine aorta specimens at four nominal strain rates: 0.06, 0.6, 6, and 30 s^{-1}. At the maximum strain rate, the results indicated significantly reduced failure strain. The elastic modulus and ultimate stress at failure have also been found to increase with strain rate. Other uniaxial experiments conducted at physiological loading rates include [59, 60, 90, 91, 103–106]. Despite their simplicity, uniaxial tests cannot provide the three-dimensional material properties of the tissue [92]. Biaxial tests represent a significant first step in the comprehensive characterization of the mechanical properties of anisotropic tissues [95, 107, 108].

Bass et al. (2001) performed biaxial loading tests on porcine aorta. Aortic specimens measuring 25 mm × 25 mm were subjected to harmonic vibrations at nominal frequencies of 20 Hz and 65 Hz. The failure properties of aortic tissue were investigated both in situ and ex vivo in intact human aorta following a rapid increase in intraluminal pressure. However, they failed to identify a preferred direction of aortic rupture, in contrast to autopsy results from car accident victims [88, 96]. Their results showed that the circumferential strain ratios at failure for the distal and proximal aorta were 1.16 and 1.19, respectively, while the longitudinal strain ratio was 1.13. The circumferential and longitudinal failure stress values (0.794 and 0.828 MPa and 0.398 and 0.414 MPa, respectively) at the distal and proximal aorta are consistent with the values documented by Collins and Hu [101]. Shah et al. [20] performed planar biaxial experiments to determine the failure properties of human aortic specimens from the ascending aorta, peri-isthmus region, and descending aorta at nominal loading rates of 1 and 5 m/s. The two groups showed remarkable overlap in strain rates, with Lagrangian strain rates ranging from 23 to 206/s at low rate and 52 to 230/s at high strain rates. The mean strain rates showed a significant difference (77.86 ± 43.27 s^{-1} vs. 136 ± 55 s^{-1}). The persistent failure in all aortic samples corresponds to the actual TRA results. The longitudinal stress and strain to failure were 1.96 MPa and 0.25, respectively, with an average longitudinal modulus of elasticity of 9 and 10.57 MPa, respectively. Vande Geest et al. [109] investigated age-related variations in the behaviour of the human abdominal aorta. Samples with an area of 4 cm^2 were tested biaxially. The fabric was stretched over a series of nylon loops using a tension-regulated protocol. No significant differences were observed between the top stems in the circumferential and longitudinal direction. The samples of the abdominal aorta in advanced age showed increased stiffness. Numerous significant biaxial experiments have been performed on aortic tissue [95, 109–116]. Recently, Gaur et al. [70] performed constrained biaxial testing on porcine aortic tissue, a new method that elucidates the behaviour of soft tissue under directional constraints. In this method, one direction (either circumferential or longitudinal) was constrained while the other was subjected to a load. This was to study the effect of Poisson's ratio on load transfer, particularly tissue deformation in the unloaded direction as a result of perpendicular loading. This testing is important to understand the relationship between the anisotropic behaviour of tissue and its mechanical properties, especially at high strain rates. The stress–strain curves are shown in Fig. 7a–d. The results showed that failure stress, strain, and elastic modulus showed significant variations depending on the loading direction and applied strain rate (Fig. 8a–c). The Poisson ratio decreased with the increased strain rates, as the constrained direction showed significant deformation despite the absence of direct loading. The elastic modulus and failure stress showed an increased trend as the strain rates escalated, highlighting the strain rate-dependent behaviour. The strain to failure decreased with increasing strain rates, indicating a shift towards more viscoelastic behaviour. These constrained biaxial tests provide a new perspective on soft tissue behaviour, particularly with regard to stress transfer between fixed and loaded directions. The unpublished results are critical for improving human body models (HBMs) used in automotive crashworthiness research and provide important insights to improve the understanding of soft tissue mechanics in actual trauma situations. Summary of the past literature on impact experiments conducted on aortic tissue is tabulated in Table 1 [70].

3.2.3 Constitutive Models for Aorta

The aorta, like many other soft tissues, exhibits a non-Hookean viscoelastic stress–strain response that is influenced by the strain rate [20, 47, 59, 96, 101, 102]. Shah et al. [20] suggested that aortic tissue exhibits anisotropy and possesses different mechanical properties in both the longitudinal and circumferential directions, as demonstrated by previous experiments [47, 96]. Research on arterial incompressibility has shown that aortic tissue is incompressible [18, 19]. These properties must

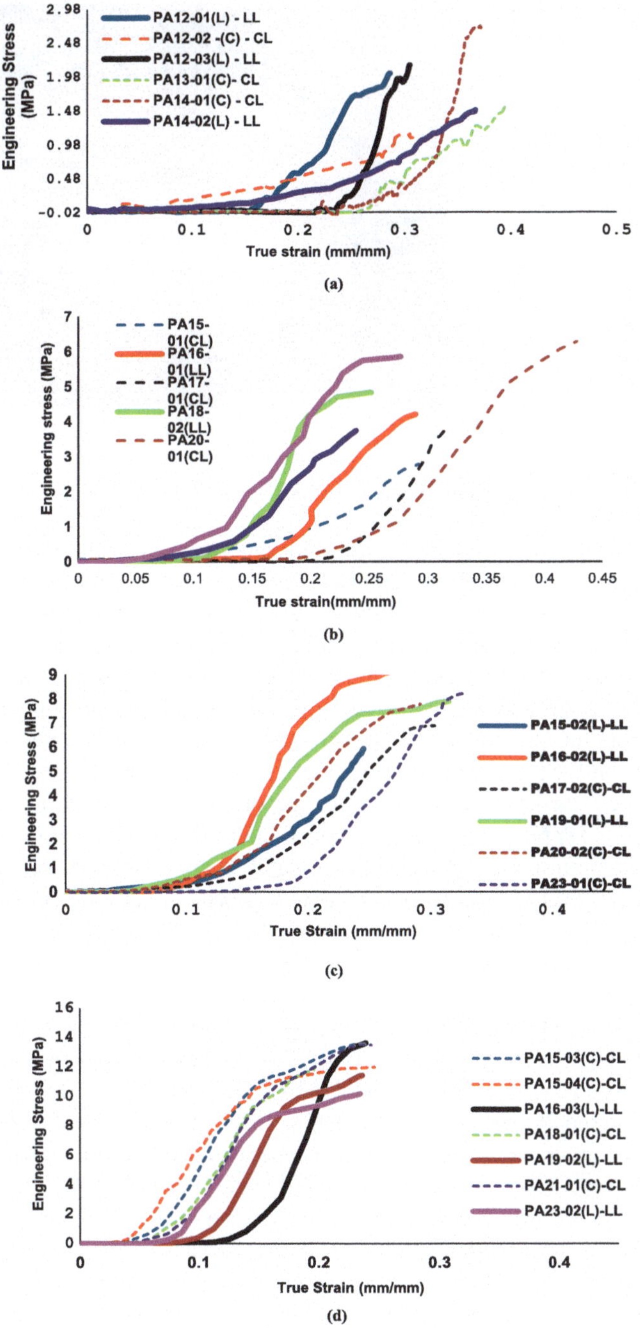

Fig. 7 Constrained biaxial stress–strain curves at different strain rates at **a** 0.01/s, **b** 70/s, **c** 150/s, and **d** 250/s strain rates (note—*L—longitudinal direction, C—circumferential direction, LL—loaded in longitudinal direction [70]

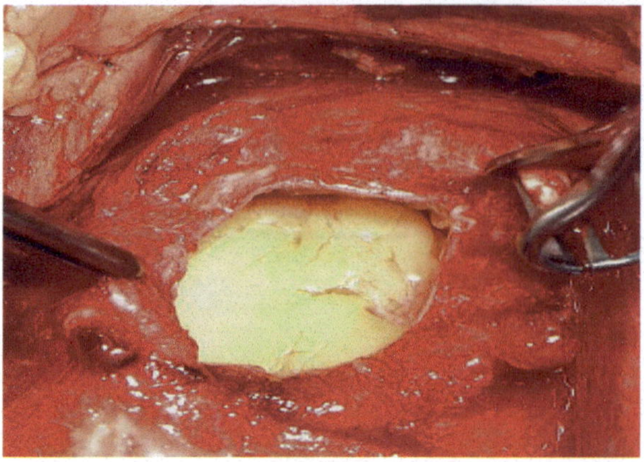

Fig. 8 Radial tear within the left hemi-diaphragm [128]

be considered in any constitutive model of aortic tissue for development of human body models.

Deng [119] characterized the aortic and superior vena cava tissues as elastic materials surrounding blood, which was represented as an elastic fluid. Material properties were determined using experiments performed by Abé et al. [120]. Richens et al. [121] used the Blatz-Ko rubber model (MAT_007 in LS-Dyna) to simulate aortic tissue in their thoracic model, maintaining a constant Poisson's ratio of 0.463. Richens et al. [121] found that using isotropic linear elastic, orthotropic linear elastic, and Ogden hyperelastic material models to simulate the aorta resulted in numerical instability in the LS-DYNA environment at strains greater than 10%. Darvish et al. [122] simulated aortic tissue using a second-order Mooney-Rivlin rubber model (MAT_77 in LS DYNA). The material constants of this model were determined by a least-squares optimization performed with a 20 Hz biaxial vibration displacement and a single data point of the aortic tissue in the aortic arch region. Quasi-static and dynamic test data presented by Mohan and Melvin [47] were compared with the response of the finite element model.

Deng (1999) modelled the aorta and the superior vena cava tissues as an elastic material enclosing with blood modelled as elastic fluid. The material properties were derived from experiments by Abé et al. (1996). A thorax model by David Richens et al. (2004) used Blatz-Ko rubber model (MAT_007 in LS-Dyna) to represent aortic tissue with a fixed Poisson's ratio of 0.463. David Richens et al. (2004) attempt using isotropic linear elastic, orthotropic linear elastic, and Ogden hyperelastic material models to model aorta resulted numerical instability in LS-DYNA environment beyond 10% strain. Darvish et al. (2004) used a second-order Mooney-Rivlin (MAT_77 in LS DYNA) rubber model for an aortic tissue. Least square optimization obtained the material constants for this model from only one aortic tissue data point from the arch region of the aorta and subjected to the biaxial oscillatory displacement of 20 Hz. Their FE model response was compared against quasi-static and dynamic tests data reported by Mohan and Melvin (1982).

Table 1 Comparison to published aorta properties [70]

References	Species	Region	Technique (sample type)	Direction L = Long. C = Circ	Strain rate (s^{-1})	Peak stress (MPa)	Modulus (MPa)
Collins and Hu [101]	Porcine	Descending	Uniaxial (cylinder)	L	3.5	4.0–4.5	0.4–1.0
Bass et al. [117]	Porcine	Isthmus	Biaxial (planar)	L	3–7		0.25–0.40
				C	2–4		0.5–0.7
Mohan and Melvin [47, 96]	Human	Descending	Uniaxial (dumbbell)	L	100	3.53	1.1–10.6
				C	100	5.07	2.2–23.0
		Descending	Biaxial inflation (planar)		20	1.96	6.7–31.0
Vande Geest et al. [109]	Human	Abdominal	Biaxial (planar)	L		0.08	0.1–4.0
				C		0.07–0.09	0.08–4.00
Vorp et al. [118]	Human	Ascending	Biaxial (strip)	L		1.75	2.5
				C		2	3.25
Shah et al. [20]	Human	Descending	Biaxial (cruciate)	L	37	1.64	7.54
				C	37	1.92	9.26
Shah et al. [20]	Human	Ascending	Biaxial (cruciate)	L	101	1.69	6.92
				C	101		14.91
	Human	Descending	Biaxial (cruciate)	L	73	1.87	7.62
				C	73		10.05
	Human	Isthmus	Biaxial (cruciate)	L	90	2.32	8.82
				C	90		10.25
	Human	Overall	Biaxial (cruciate)	L	85	1.96	7.79
				C	85		11.37
Gaur et al. [59]	Human	Overall	Uniaxial (rectangular)	L	0.01	0.86	6.4
					65	1.23	9.74
					130	1.48	13.51
					200	1.86	15.93
				C	0.01	1.1	4.54
					65	2.18	10.29
					130	2.9	21.8
					200	2.67	20.93
Herein**	Porcine	Overall	Uniaxial (rectangular)	L	0.01	1.64	9.42
					70	2.98	15.91
					150	3.62	23.16
					250	4.84	31.62

(continued)

Table 1 (continued)

References	Species	Region	Technique (sample type)	Direction L = Long. C = Circ	Strain rate (s^{-1})	Peak stress (MPa)	Modulus (MPa)
				C	0.01	1.98	12.33
					70	3.56	20.75
					150	6.37	37.71
					250	7.05	51.51
			Constrained biaxial (cruciate)	L	0.01	1.85	11.71
					70	4.82	22.40
					150	6.94	31.07
					250	12.61	50.78
				C	0.01	1.81	14.95
					70	4.45	24.38
					150	7.01	35.91
					250	13.03	48.92

In addition, various aortic finite element models have been created and documented for application in clinical and medical settings to study the behaviour of aortic tissue in conditions such as intraluminal thrombus and abdominal aortic aneurysms. In these scenarios, strain energy density functions (SEDFs) are often used to model the aorta, typically as isotropic, incompressible hyperelastic material. These models include explicit representations of the stress–strain curve, including the Holzapfel model [92, 123], the Fung exponential orthotropic model [106], anisotropic models [95, 109], and the Ogden models [124, 125]. Many researchers have proposed various continuum-based constitutive models for aortic tissue as well. Elad et al. [126] suggest that nonlinear elasticity can be characterized as "incremental elasticity" or "finite elasticity." Incremental elasticity considers small deformations superimposed on a finite equilibrium state. In contrast, the finite elasticity method uses a strain energy density function (SEDF) to model soft tissues using a single set of material parameters, resulting in a thorough nonlinear elastic response. The traditional SEDF formulation was introduced by Fung [127] to model arteries. Fung's SEDF type is rarely used in finite element analysis because it cannot handle high strain rates and large deformations.

The table below summarizes the aortic constitutive models identified in the literature. Most hyperelastic models for aortic tissue effectively captured physiological behaviour and demonstrated considerable accuracy, as previously discussed. Challenges associated with significant deformations and elevated strain rates. Shah et al. (2006) introduced a linear orthotropic model that failed to consider non-Hookean behaviour or strain rate-dependent effects observed in experimental studies. Prior research has overlooked these effects.

3.2.4 Traumatic Diaphragmatic Rupture: Etiology and Injury Mechanisms

The traumatic rupture of diaphragm (TRD) is broadly classified into blunt and penetrating traumatic diaphragm rupture. Along with diaphragm rupture, iatrogenic injury can is always considered due to respiratory complications [128, 129]. RTAs are one of the most common cause of blunt traumatic rupture of diaphragm muscle, while knife attacks and gunshot wounds are the common mechanisms of penetrating trauma [130, 131]. Patients with significant trauma have reported the diaphragm injuries ranging from 1 to 7% in all of the blunt incidences while 10 to 15% for penetrating tears/ruptures [132]. TRD is relatively an uncommon scenario, but few autopsy studies have reported the incidents of these injuries between 5.2 and 17% [133, 134]. Asensio et al. [130] have conducted a review of 53,031 patients admitted to various trauma centres and reported 592 people have TRD (**3% overall**). In another review conducted by Feliciano et al. [135] on 952,242 patients in the USA, 6038 patients have diaphragmatic injuries. Both of the authors in their surveys have reported the increasing trend in the occurrence of diaphragm injuries due to the rise in the numbers of the RTAs [128, 130, 136].

Two mechanisms have been proposed in literature for the occurrence of diaphragmatic rupture. For blunt trauma, abrupt changes in intra-abdominal pressure is the common mechanism for inducing the diaphragm injury accompanied by simultaneous shearing and avulsion [137]. Normal intra-abdominal pressure varies from +20 to +100 mm H_2O during inspiration. A pressure gradient of about 1000 mm of H_2O can also be induced at the moment of diaphragm injury [138]. This pressure gradient is also responsible for the herniation of abdominal contents through a diaphragm injury. Motor vehicle crashes with lateral intrusions into the vehicle are the other causes of this rupture [23, 134, 139, 140]. Direct impact on the chests compresses the side of the ribcage and cause a tear in the diaphragm rib attachments, eventually leads to transverse rupture of diaphragm.

Various research has shown a greater occurrence of left side diaphragmatic injuries following blunt trauma. Moreover, studies on bursting pressure in cadavers have shown consistent weak left side [138]. This is due the protective effect of the liver on the right side and the location of lumbocostal trigone on the left. It has also been proposed that the oesophageal hiatus also contributes in the weak left hand side [136, 138]. In a review of 13 cases conducted by Bisgaard et al. [141], all of the 13 people have suffered left sided ruptures/tears, further highlighting the natural weakness of the left hemi-diaphragm. A diaphragm rupture is actually a sign/marker of severe trauma. Tears of the diaphragm caused be penetrating trauma usually measured 5–15 cm in length and are typically radial as shown in Fig. 8. The diaphragm muscle is rarely injured alone, with its comitative injuries approaching 100% [142]. This injury is commonly linked to intra-abdominal injuries, thoracic injuries, rib fractures, pelvis and long bones, head injuries, and rarely aortic injuries.

3.2.5 Properties and Failure Thresholds of Diaphragm Tissue

During the last two to three decades, there has been a research focus on the relevant biomechanical studies of human body orangs. These studies play a significant role in the development of human body models which provides a better understanding of injury quantifications and analysis, not only mechanically, but also medically. This section discusses the diaphragm as an active muscle tissue with its physiological and mechanical properties obtained from the literature. However, to the author's knowledge, data on the failure and mechanical properties of diaphragm is very little.

Biaxial mechanical response of active diaphragm tissue has been studied in a limited number of studies, due to significant challenges associated in keep the diaphragm muscle viable and stable during the tests. The first study on the biaxial response of diaphragm tissue was proposed by Strumpf et al. [143]. Ease of making the tissue tetanized, the architecture, vascular supply makes the diaphragm muscle an ideal structure subjected to multidimensional loading in active and passive state. The specimens with dimensions of 3×3 cm in size were tested in a single dominant fibre direction. Nonlinear behaviour was seen in the passive state in both fibre directions. The specimens showed anisotropic behaviour, with the cross-fibre direction being stiffer than the axial fibre direction. The diaphragm being a tri-layered structure, consisting of an inner thick muscular layer and two thin collagenous membrane. Strumpf et al. [133] disrupted the membranes covering each surface, to determine the contribution of each outer membrane layers. The reported higher extensibility and decreased anisotropy, thereby suggesting that diaphragm membrane bear most of the mechanical load and as a result has high stiffness and anisotropy in cross-fibre direction. Boriek et al. [129] have conducted biaxial tests on canine square shaped samples (5×5 cm^2). Four markers were placed at the centre of the specimen creating a 1cm^2 area for measuring tissue deformation. Stepper motors and force transducers were attached to both ends of each axis. Ten threaded fishhooks placed at 5 mm intervals connected each specimen edge. The authors reported that tensile force varies with region and the direction, with the larger tensile forces acting parallel to the fibres in the muscular region having high material anisotropy. Diaphragm tension force was noted to be 3.5–6 Pa along the muscle fibre direction and 1.0–2.5 Pa across the fibres in the plane of the canine diaphragm.

Other authors [129, 144–146] have reported the mechanics of active and tetanized canine diaphragm. The authors, in their studies have concluded that muscle fibres of the diaphragm form a curved sheets which extends from the wall of the chest to central tendon and balances the trans-diaphragmatic pressure. In the last decade, only one study by Gaur et al. [11] has reported the high strain rate (until 200/s)-dependent properties of the passive diaphragm tissue. This work provides valuable insights into the biomechanical properties of human diaphragm tissue under uniaxial tensile loading across a range of strain rates. While it offers a comprehensive analysis of diaphragm's failure properties and its response uptill 200/s strain rates, the study has its implications for injury mechanisms in the real-world scenarios such as MVCs. The use of only uniaxial tensile testing, although informative, limits the understanding of how the diaphragm might behave under multiaxial loads. Additionally, the study reliance on engineering stress and strain-based calculations, while practically, may have introduced some simplifications that could obscure more nuanced viscoelastic behaviour. Despite this, the work establishes a strong foundation for developing strain rate-dependent constitutive models, crucial for improving the bio fidelity of HBMs used in injury prediction, particularly in context of human safety.

3.2.6 Computational Studies on Human Diaphragm Muscle

Numerous studies exist in the open literature on the computational modelling of human diaphragm tissue. Most of the proposed studies were focused on the geometrical modelling and trauma rupture analysis utilizing radiology, MRI, and X-ray techniques. Limited research has attempted to characterize the constitutive models for diaphragm tissue. The diaphragm is a curved anisotropic sheet, that experiences stress in both directions within its plane. Consequently, diaphragm mechanics are pertinent to stress–strain relationship.

Behr et al. [147] examined the diaphragm tissue and integrate its geometrical model into the comprehensive computational model of the human body (RadiossTM Humos project). The authors aim to reproduce the diaphragmatic ruptures that occurs in vehicle crashes. A 3D geometry of the diaphragm was reconstructed using an MRI scan of postmortem human subject (PMHS), and mechanical properties were assigned to all its components (Table 2), including all muscular, tendinous, and pillar sections, along with interactions among these components and adjacent organs. The PMHS was a European male, preserved at −23 °C, and was positioned in a driving seat within a car cockpit. This facilitated the precise segmentation of each anatomical region. The diaphragm was solely analysed in the thoracic and abdominal areas because of its anatomical position. The resultant slices were digitized into a standard image format following a slicing process at intervals of 7.5 mm using a saw with a thickness of 2.5 mm. The authors also encountered complications in identifying and contouring the diaphragm in each slice due to the use of non-automatic procedure. The final phase of the reconstruction entailed the creation of a surface derived from contours and subsequent refinements, including smoothing. However, the generation of a surface was a very challenging task for them. Marching cube algorithm was used to generate these surfaces. The tendinous structure of the inferior vena cava required further modifications concerning its hiatus. Distinct components needed to be established in the reconstructed surface because of the diaphragm's unique mechanical properties. As a result, the muscular, tendinous, and pillar elements fused autonomously. The muscular component was characterized by elastic orthotropic properties, whereas the tendinous component was represented as an elastic isotropic material. Figure 9 depicts the diaphragm's final FE model.

Condurache et al. [148] used X-rays to observe and determine the movement of the diaphragm during inspiration. A temporal two-dimensional contour of the diaphragm is derived from radiographic images of the diaphragm captured from multiple angles, although these images are not used for reconstruction purposes. However, researchers have encountered considerable difficulties in distinguishing the

Table 2 Aorta and diaphragm properties used in human body simulations [70]

Case	Case description	Simulations done
A	Original GHBMC properties	1. Frontal impact
		2. Side impact
B	Aorta properties replaced by longitudinal properties obtained in [174]	3. Frontal impact
		4. Side impact
C	Aorta properties replaced by aorta circumferential properties obtained in [174]	5. Frontal impact
		6. Side impact
D	Diaphragm properties replaced by Diaphragm properties obtained in [24]	7. Frontal impact
		8. Side impact

diaphragm from adjacent soft tissues in X-ray images. Segars et al. [149] modelled the upper body torso and respiratory motion using Mathematical Cardiac Torso (MCAT). The authors used the diaphragm to represent the upper torso. They then integrated respiratory motion into a dynamic 4D MCAT phantom, which was modelled using NURBS. NURBS-based surfaces can be represented at any resolution due to their continuous characteristics. The mathematical complexity of this work constitutes one of its main shortcomings. Timc-imag and Gallard [150] have developed a three-dimensional model of the diaphragm muscle which facilitates the measurement of motion from functional representations. They created six models for various applications: five of these simplified geometric models illustrate the mechanical principles of contraction, evaluate the modelling methodology, and emulate the motion of the diaphragm piston. A sixth model provides data on the location and motion of the piston. However, no constitutive models of human diaphragmatic tissue have been developed using mechanical analysis.

The finite element method was used by Martins et al. [151] and D'Aulignac et al. [152, 153] to study the pelvic floor muscles. The one-dimensional Hill-type muscle model is the basis for a three-dimensional muscle model that was proposed, refined, and evaluated. This new model improved the cardiac muscle model proposed by Humphrey and Yin [154]. The diaphragm consists of skeletal muscles and tendon tissue. Tendon tissue exhibits linear elasticity and isotropy. The diaphragm muscle exhibits transverse isotropy [151]. Skeletal muscles exhibit complex mechanical behaviour due to their incompressibility and the unidirectional orientation of muscle fibres. Skeletal muscle fibres extend the entire length of the muscle. Tendons attached to bones serve as attachment points for the fibres. There are subunits within each

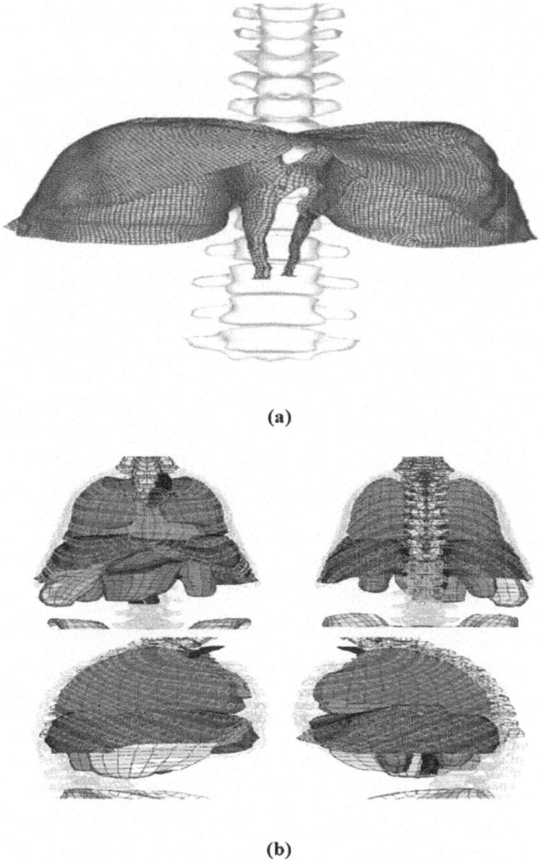

Fig. 9 **a** Diaphragm final FE model, **b** diaphragm FE model integration in RadiossTM human body model [147]

muscle fibre. Actin and myosin are used by the smallest unit, the sarcomere, to facilitate muscle contraction. The Hill muscle model consists of an active contractile element (CE) arranged in series with a passive elastic element (SE), which is in parallel with an additional passive element (PE) [155] (Blümel et al., 2012). The parallel elastic element refers to the connective tissue surrounding the contractile unit, while the SE elastic element refers to the connective tissue arranged in series with the sarcomeres, including the tendon. The active force of the contractile element depends on the number of active bridges formed between the filaments. The Hill model evaluates longitudinal muscle tension by synthesizing parallel and series tensions [151].

$$T = T_{PE} + T_{CE}$$

while the stress in the contractile element is equal to the stress in the series element

$$T_{CE} - T_{SE}$$

and

$$\Delta L_{SE} = \Delta L_{PE} - \Delta L_{CE}$$

Diaphragmatic rupture data was collected by applying a shock tube load to the entire lung, including the diaphragm. Modelling efforts was focused on evaluating the elastic properties of the human diaphragm only [156]. Roberts and Chen [157] created a finite element model of the rib cage that includes the sternum, ribs, costal cartilage, and spine. The diaphragm and aorta were represented as shell elements with associated elastic properties. Plank and Eppinger [158] formulated a model that includes the sternum, ribs, spine, diaphragm, muscles, and cartilage and ligament elements, all of which have linear elastic properties. Lee and Yang [159] developed a model that represents a unified mesh segmented into three regions to account for different stiffnesses. Most current finite element models of the human body, such as THUMS (I and II), HUMOS, GHBMC, and Wayne State HBM, consider the diaphragm as a separate entity. The membrane was modelled using shell elements and given the appropriate elastic properties. Shah et al. [20] used the following elastic properties to describe membrane failure: elastic modulus (E): 65.5 MPa; density: 1×10^{-6} kg/m^3; Poisson's ratio: 0.46.

4 Applications in Biomechanical Modelling

4.1 Finite Element Modelling of Human Body

The human body is an extraordinarily complex system. Many researchers have attempted to model specific regions or the entire organism. Models have ranged from extremely simple to relatively complex. This section quickly describes in brief the development of various human body models along with Wayne State University full-body finite element models, as well as finite element models developed by other researchers.

Roberts and Chen (1970) created a finite element model of the human thoracic skeleton to examine biomechanical responses of the chest. In their model, the sternum, ribs, costal cartilage, and vertebral column were represented by beam elements, i.e. Fig. 10. Sudaram and Feng [160] extended the beam element method to a three-dimensional framework. The rib cage, sternum, and musculature were represented as beam, plate, and membrane elements, respectively. Internal organs were modelled using solid elements. Different material properties were assigned to the heart and lungs. The assumption was that internal organs had contiguous boundaries, rather than a complex representation of their geometry.

Plank and Eppinger [158] developed a more accurate model of the human thorax more than a decade later (Fig. 11). It was an improved iteration of the model created by Plank and Eppinger (1989). It included the sternum, ribs, spine, and musculature, as

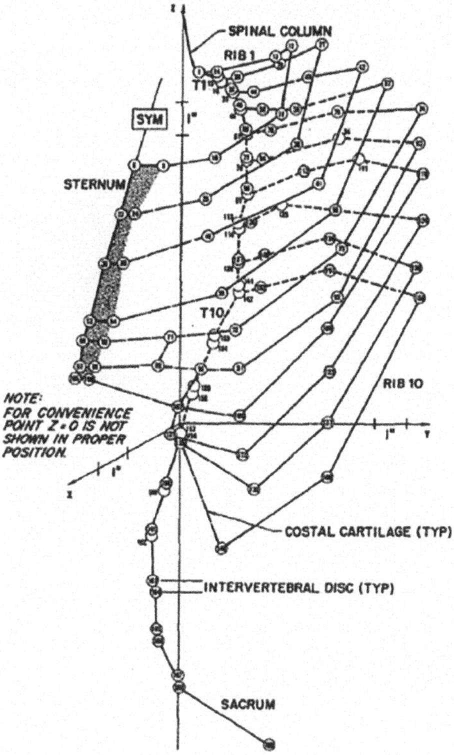

Fig. 10 FE human thorax model by Roberts and Chen [157]

well as certain cartilage and ligament elements. All components were modelled using eight-node brick elements. The thoracic viscera were represented as a unified whole, without detailed descriptions of individual organs. All elements were assumed to have linear elastic material properties, with the exception of the viscoelastic viscera. The experimental data of Neathery [161] on frontal pendulum impacts were used to validate the model. It was then incorporated into a Hybrid III dummy model to evaluate thorax responses to out-of-position airbag deployment [162].

Huang et al. [163] developed a simplified human body model optimized for predicting injury parameters during side impacts, including Thoracic Trauma Index (TTI), Spinal Average Acceleration (ASA), Compression (C), and Viscosity Criteria (V *C). The ribcage and sternum were modelled with shell elements. The internals have been recreated with flexible rigids and damper assemblies. The head and neck were depicted as rigid and without a defined cervical spine. The skin was constructed from rigid solid elements, while the shoulder complex was represented by isotropic solid elements. Figure 12 illustrates several features of the model. The model was developed by Cavanaugh et al. [164] using sled test data and Viano [13] using pendulum impact test data.

Happee et al. [165] conducted a study analysing the responses of small and medium-sized women and men to frontal, lateral, and rear impacts using whole-body

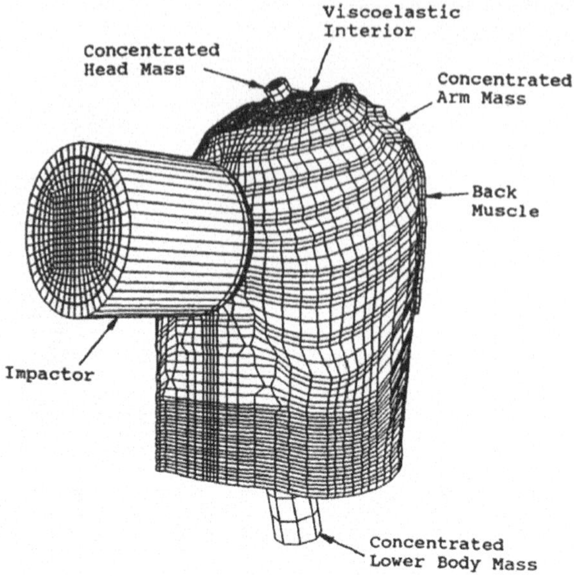

Fig. 11 A 3D thorax model by Plank and Eppinger [158]

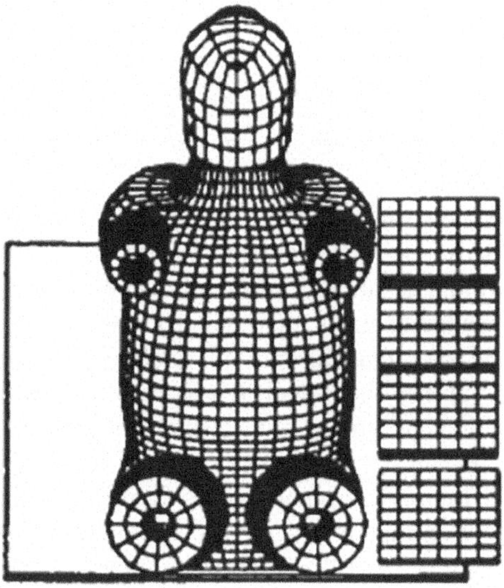

Fig. 12 Whole-body FE model by Huang et al. [163]

finite element models. The RAMSIS model was used as a source of anthropometry data. The finite element model presented the human body as an assembly of rigid bodies connected by kinematic joints. The skin is represented as triangular surfaces. Jost and Nurick [166] developed a finite element model of the human body to reproduce lateral impacts. The model illustrated the bony structures of the skeletal shell elements. Solid elements were used to represent the internal organs, muscles, adipose

Fig. 13 Facet human body model by Happee et al. [165]

tissue, and adjacent structures. The internal organs of the chest and abdominal cavity were not adequately represented in their models. Jost and Nurick [167] refined the model to broadly include the head and neck complex, spine, limbs, and synovial joints. The heart, spleen, liver, and other organs were represented as solid elements, while the lungs were represented as a low-density region (Fig. 13).

While these simplified models are computationally efficient, none of them were capable of predicting injuries at individual organ level (for instance aortic injury, or injury to the liver, spleen, lung or heart, etc.) in automotive crash environments. Hence, there calls for the development of more refined human body model with improved biofidelity and incorporate detailed constitutive models in them.

The Total Human Model for Safety (THUMS) was developed by Toyota Central R&D Lab, Inc. and is based on the 50th percentile of adult males [168]. The THUMS model has detailed skeletal structures, although the internal organs of the torso have been simplified to improve computational efficiency in crash simulations. Ruan et al. [169] created a detailed whole-body model for the Ford Motor Company representing an average adult male, comprising 103,000 nodes and 119,000 elements. The Ford model has elaborate details on the torso, head, neck, and extremities. The thorax geometry of the Ford and THUMS models was based on the original thorax model by Wang [119]. In the 1990s, Wayne State University (WSU) began creating finite element models for anthropomorphic test equipment [170, 171]. This was followed by the creation of finite element models of human anatomical regions. In 1995, Wang created a complete finite element model of the human rib cage, which included the heart, lungs, and aorta. To the author's knowledge, this is the first attempt to accurately recreate a human rib cage. Subsequently, Shah et al. [172] improved this model to study aortic injury.

Lee and Yang [159] also developed a comprehensive finite element model of the human abdomen. The detailed finite element shoulder model was developed by Iwamoto et al. (2000) at Washington State University. The goal was to integrate these models. **Subsequently, with reduced difficulty,** organ injuries can be predicted by component models in some impact scenarios, such as swing impacts. A model of an occupant coming into contact with vehicle interior structures and/or restraint systems is essential for realistic crash simulations. Shah et al. (2004; 2005) [35, 173] effectively merged the previously mentioned component models. This section provides a complete integration mechanism and a concise summary of the previously mentioned component models. This study used the Wayne State Human Model (WSHM) to examine aortic injuries. Car accidents can cause damage to organs, including the heart, lungs, liver and spleen. Summarily, finite element (FE) models of the human body have evolved from simple beam element representations of the rib cage to complex multicomponent systems designed to improve biofidelity in crash simulations. Initial models, such as those by Roberts and Chen [157] and Sudaram and Feng [160], focused on skeletal structures and adjacent internal organs with minimal material complexity. Conversely, later developments by Plank and Eppinger [158] and Huang et al. [163] integrated more accurate components for injury prediction while maintaining generalization of internal structures. Toyota's THUMS models, global human body modelling consortium (GHBMC), and Ford's whole-body EF models are examples of modern designs that feature improved skeletal and organ detail while streamlining internal components to maintain computational efficiency. However, these models are limited by inadequate individual biofidelity, computational constraints, simplified organ geometry, and rudimentary material properties, reducing the accuracy of organ-level injury prediction. Therefore, the need for sophisticated models with complex representations of the components is emphasized to accurately describe the collision dynamics in the real world.

4.2 Integration and Implementation of Tissue Data in GHBMC Human Body Models for Impact Loads

This section demonstrate the implementation strain rate-dependent constitutive model developed using experimental data of human aorta and diaphragm reported in Gaur et al. [11], Gaur et al. [174], and Gaur et al. [24]. Simulations of the frontal impact and the side impact on GHBMC human body model have been done with the original material model and the proposed models, and the response has been compared. Variation in the force–time, force–deflection, and the strain–time responses has been studied in these cases.

Simulations of frontal and lateral thoracic pendulum impacts were performed using the GHBMC model [1, 175, 176]. Impact cases were selected based on the frequency of aortic and diaphragm injuries. The frontal impact was simulated at 6.5 m/s and the side impact at 4.5 m/s, corresponding to the test conditions of Jansova et al. [175] and Shaw et al. [176]. In both cases, the aorta and diaphragm were modelled with different material configurations, as shown in Table 2. The GHBMC model uses the elastic material model for both the aorta and diaphragm (case A). In contrast, the bilinear strain rate-dependent material model (cases B, C, and D) specified in Gaur et al. [24, 174] is as follows: The eight simulations performed were representative of frontal and side impact scenarios, as detailed in Table 3.

Table 3 Properties used in frontal and side impact simulations [70]

Case A: Original GHBMC elastic model							**Cases simulated**
Human aorta	**Elastic modulus (MPa)**	**Density (kg/ mm^{-3})**	**Poisson's ratio**				**1—frontal impact, 1—side impact**
	8.9	1.00E-06	0.4				
Diaphragm	3.3	1.00E-06	0.4				
Case B: Aorta longitudinal properties used in UMAT material model (obtained from [174])							
Strain rate (/s)	**E1 (MPa)**	**E2 (MPa)**	**E3 (MPa)**	**Toe strain**	**Poisson's ratio**	**Failure strain**	**1—Frontal Impact, 1—Side Impact**
0.0015	1.085	10.85	5.43	0.1	0.408	0.6	
200	2.04	22.04	11.02	0.21			
400	4.683	46.83	23.45	0.32			
600	7.065	70.65	35.32	0.43			
Case C: Aorta circumferential properties used in UMAT material model (obtained from [174])							
Strain rate (/s)	**E1 (MPa)**	**E2 (MPa)**	**E3 (MPa)**	**Toe strain**	**Poisson's ratio**	**Failure strain**	**1—frontal impact, 1—side impact**
0.0015	0.96	9.66	4.83	0.08	**0.429**	0.6	
200	4.05	40.49	20.25	0.22			
400	6.78	67.88	33.94	0.3			
600	10.51	105.04	52.51	0.47			
Case D: Diaphragm _properties used in UMAT material model (obtained from [24])							
Strain rate (/s)	**E1 (MPa)**	**E2 (MPa)**	**E3 (MPa)**	**Toe strain**	**Poisson's ratio**	**Failure strain**	**1frontal impact, 1—side impact**
0.0015	2.337	23.37	11.52	0.052	**0.443**	0.7	
150	6.998	69.98	34.5	0.18			
300	17.66	176.58	88.29	0.31			
600	26.81	268	134	0.41			

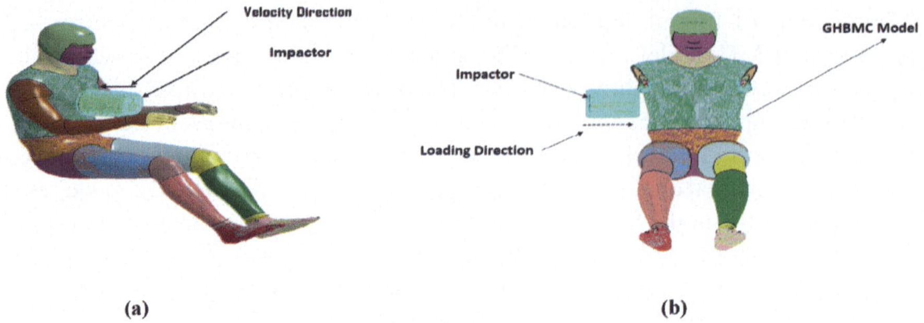

Fig. 14 GHBMC setup for frontal and side impact test setup [70]

4.2.1 Chest Frontal and Side Impact Loading

Kroell et al. [1] performed frontal impact tests on human cadavers by applying pendulum strikes to the mid-sagittal plane near the fourth intercostal space (Fig. 14a). It was applied from a 150 mm (6 in.) unfilled flat wooden mould with a half-inch radius and a mass of 23.1 kg at the mid-sagittal plane. Impact occurred approximately in the fourth intercostal space, starting with an initial velocity of 6.5 m/s, whereas in the side impact models, the impactor was positioned in the sagittal plane midway between the sternum and the back at two heights rotated 90 degrees from its original frontal orientation. The impactor was placed on the sixth rib and followed the lateral impact configuration described by Shaw et al. (2006). The impactor had a mass of 23.4 kg and an outer diameter of 150 mm. The impact velocity was 4.5 m per second. The simulation setup for this scenario is shown in Fig. 14b. For comparison purposes, force–time and strain–time variations were measured in the diaphragm and aorta.

4.2.2 Results and Discussions

Frontal Impact

Figure 15a and b compares the force–time and force–deflection results for the four cases. The maximum load in case A (original GHBMC) is about 7 kN, suggesting a relatively stiffer response. The maximum load varies between 6.85 kN, while cases B, C, and D exhibit UMAT characteristics, resulting in reduced stiffness. However, the peak load was constant in all four cases, exactly 11.92 ms. The loading time shows minimal variation of only 2%. The force–deflection responses are closely matched. The GHBMC model (case A) exhibits a deflection of 53 mm, while cases B and C exhibit deflections of 55 and 54 mm in case D, respectively. The differences between case A and cases B, C, and D are 3.6, 1.8, and 1.8%, respectively. UMAT features have a negligible influence in this context.

Green's strains were also measured for the aorta and diaphragm in all four cases. Figure 16a–b shows the Green's strain response over time for the aorta and diaphragm.

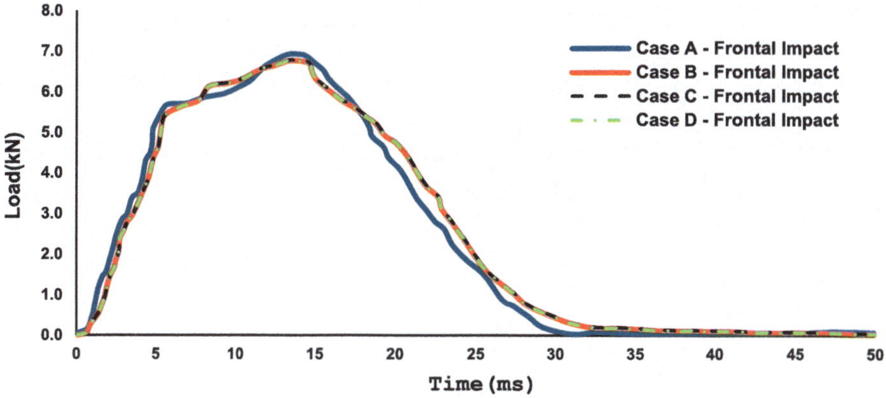

(a)

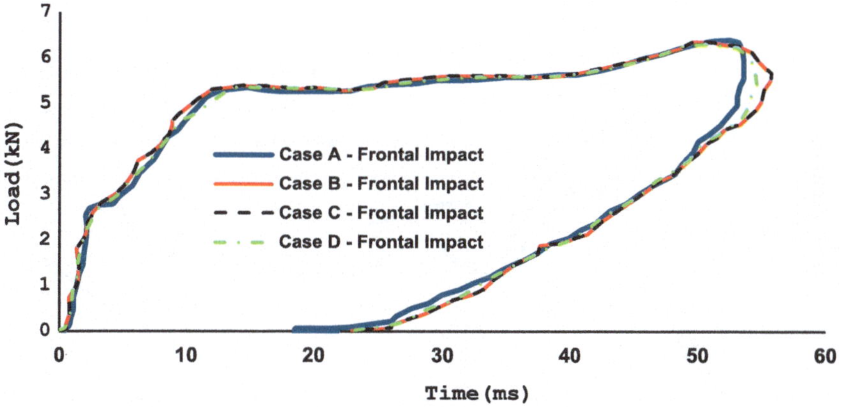

(b)

Fig. 15 Frontal impact (configuration of Kroell et al. 1971, 1974) [1, 177] response comparison. **a** Force–time and **b** force–deflection comparison

In case D, the maximum deformation of the membrane is 13%, while in cases A, B, and C, it is 22% (Fig. 16a). Due to the identical properties of the membrane, the deformation of the membrane is uniform in cases A, B, and C. However, the implementation of the UMAT functions in case D leads to a deviation of 40%. The maximum Green's strain for the aorta is 18%, 16%, 11%, and 18% in cases A, B, C, and D, respectively. This means that cases B and C deviate from the reference case A by 11 and 38%, respectively. The aortic stress remains unchanged because the aortic properties in case D are identical to those in case A. Consequently, the UMAT constitutive model developed for the aorta and diaphragm was found to significantly affect the strain in the respective organ and may influence the prediction of damage.

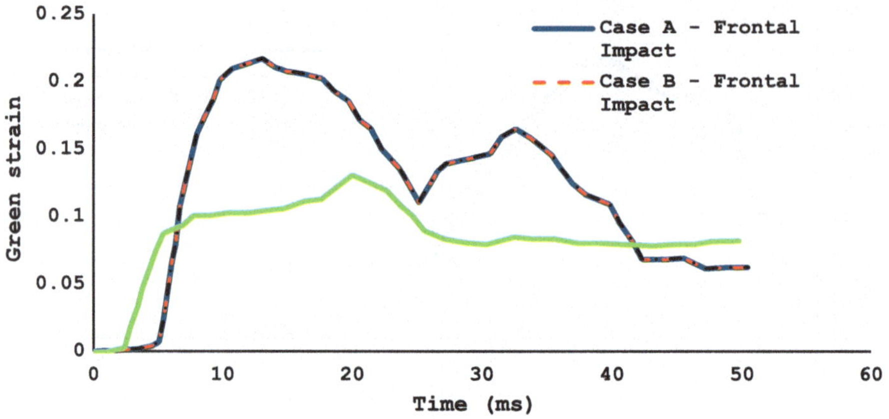

(a)

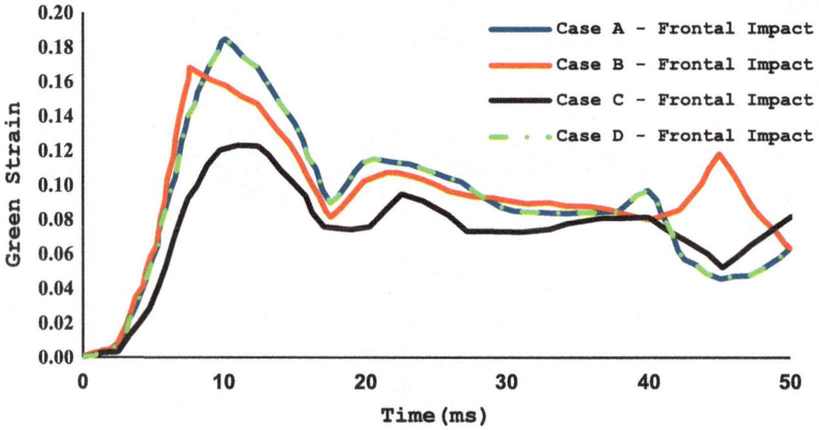

(b)

Fig. 16 Comparison of maximum green strain variation with time in frontal impact for **a** diaphragm tissue, **b** human aorta defined in Cases A–D [70]

Side Impact

Figure 17 shows the force–time response for all cases in the side impact scenario. There is no perceptible difference between the four scenarios. However, the maximum green stresses for the aorta and diaphragm are different in each case. Figure 18a, b shows the green strain–time curves for the diaphragm and aorta. The maximum stress for the diaphragm tissue in cases A-C is 27%, while in case D it is 30%, a discrepancy of 10% (Fig. 18a). The properties of the diaphragm are consistent in cases A and C. The variation in stress in cases A and D is the same for the aorta. This is to be expected, since in these cases the properties of the aorta remain unchanged.

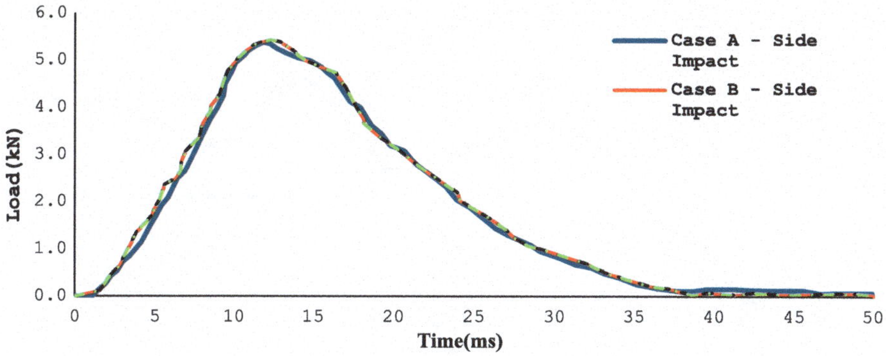

Fig. 17 Load versus time response for side impact response

In scenarios A and D, the maximum stress is 14%. This indicates a deviation of 21 and 35% from the original model (Case A) for cases B and C, i.e. 11 and 9%, respectively. Thus, for side impact also, it is observed that the UMAT constitutive model developed for the aorta and the diaphragm significantly affects the strain in the respective organ and can affect the prediction of injury.

4.3 Summary

The response of the GHBMC model of the human body to impact loading is illustrated by applying the strain rate-dependent properties obtained in [24, 174]. Two impact scenarios from the literature are simulated: frontal and lateral impacts. The response of the finite element simulations with the UMAT model properties and the properties defined by GHBMC is the subject of this investigation. The force–time and force–deflection responses of the two models are relatively similar. However, the green strains showed significant differences in the aorta and diaphragm. The differences in strain thresholds between elastic and UMAT models can be very useful when studying tissue damage. Therefore, the developed simulations show that the rate-dependent model has a significant influence on the strain response, thus influencing the expected injuries in various organs.

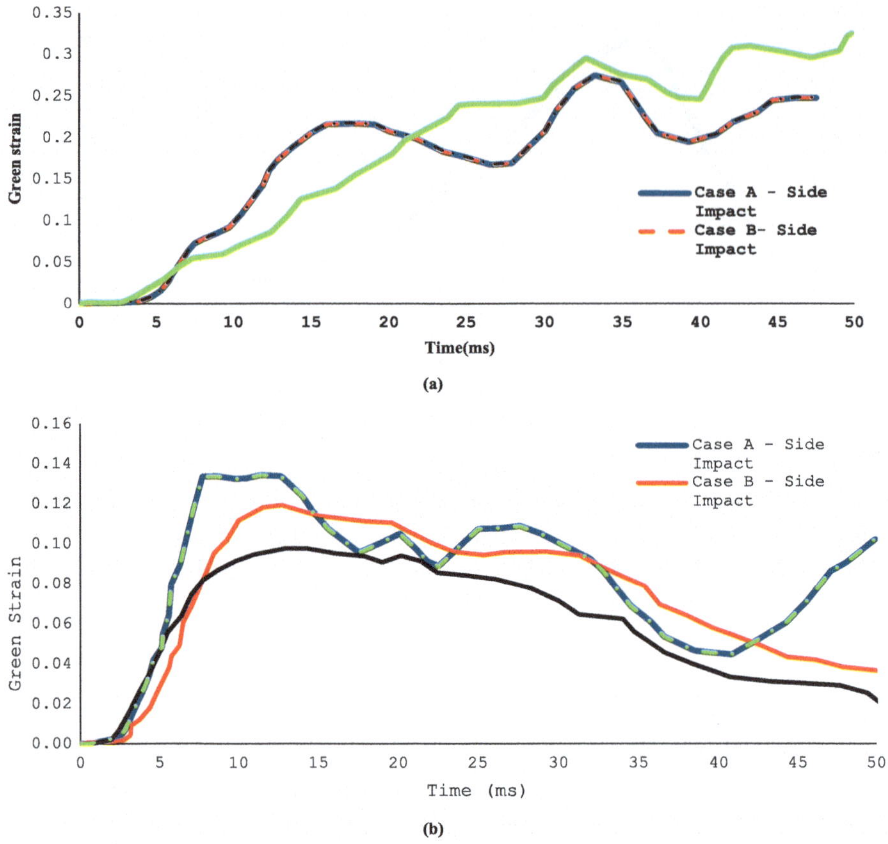

Fig. 18 Comparison of maximum green strain variation with time in side impact for **a** diaphragm tissue, **b** human aorta in Cases A–D [70]

5 Conclusions

5.1 Summary of Key Findings

- The experimental study revealed that the aorta and diaphragm tissues are strongly influenced by strain rate, as indicated by an increase in stiffness and a reduction in failure strain under high strain rates typical of car crash situations. This behaviour highlights the importance of integrating strain rate-sensitive features into human body models to improve their biofidelity.
- Mechanical testing revealed that the aorta exhibited distinct responses in circumferential and longitudinal orientations. Anisotropic characteristics were assessed by biaxial testing, essential for predicting tissue behaviour under multidirectional loading.

5.2 *Implications for Automotive Safety*

- By integrating load-dependent features of the aorta and diaphragm into high-fidelity human body models (HBMs), vehicle safety systems such as airbags and seatbelts can be optimized to offload forces to vital organs, reducing the risk of severe injury.
- The results provide a foundation for improving occupant restraint systems to more effectively distribute forces throughout the body, reducing injuries to sensitive soft tissues, including the aorta and diaphragm, particularly during high-impact, high-speed collisions.
- Improved human body models (HBMs) with refined tissue features enable more accurate predictions of injury patterns and severity and enable design changes that improve overall occupant protection.

5.3 *Recommendations for Future Research*

- The overall accuracy of HBMs used in vehicle safety can be improved by extending the research to other thorax and abdominal tissues under similar high strain rate conditions.
- In order to more accurately reproduce real-world crash scenarios, further research on the membrane behaviour under multiaxial loading at different strain rates is essential, as existing data is mainly based on uniaxial tests.
- Improvement of HBMs is achieved by creating sophisticated constitutive models that accurately represent the nonlinear, anisotropic, and viscoelastic properties of these tissues over a wide range of strain rates. This would improve their effectiveness in injury prediction and promote safer vehicle design.

References

1. C. K. Kroell, D. C. Schneider, and A. M. Nahum. (1974). Impact tolerance and response of the human thorax II. *SAE International*. https://doi.org/10.4271/741187
2. Naci, H., Chisholm, D., & Baker, T. D. (2009). Distribution of road traffic deaths by road user group: A global comparison. *Injury Prevention: Journal of the International Society for Child and Adolescent Injury Prevention, 15*(1), 55–59. https://doi.org/10.1136/ip.2008.018721
3. Ruikar, M. (2013). National statistics of road traffic accidents in India. *Journal of Orthopaedics, Traumatology and Rehabilitation, 6*(1), 1. https://doi.org/10.4103/0975-7341.118718
4. Singh, S. K. (2017). ScienceDirect-review under responsibility of world conference on transport research society. ScienceDirect road traffic accidents in India: Issues and challenges. *Transportation Research Procedia, 25*, 10–15. https://doi.org/10.1016/j.trpro.2017.05.484
5. Blincoe, L., Miller, T. R., Zaloshnja, E., & Lawrence, B. A. (2014, May). The economic and societal impact of motor vehicle crashes, 2010. In *Motor vehicle crashes: Economic and societal impact, In Depth* (Vol. 2015, pp. 1–285).

6. Manuel, V. (2014). Thoracic injuries in frontal car crashes: Risk assessment using a finite element human body model.
7. United Nations Road Safety Collaboration. (2011). Global plan for the decade of action for road safety 2011–2020 (p. 25). Geneva WHO. https://doi.org/10.1016/j.injury.2011.11.002
8. Forbes, P. (2005). Development of a human body model for the analysis of side impact automotive thoracic trauma (Vol. Maestría, p. 224). Departamento de Ingeniería Mecánica.
9. Mohan, D. (2002, October). Social cost of road traffic crashes in India. In *Proceedings of first safe community conference on cost injective* (pp. 33–38).
10. Mohan, M., Tsimhoni, D., Sivak, O., & Flannagan, M. (2009, January) Road safety in India: Challenges and opportunities.
11. Gaur, P., et al. (2016). Characterization of human diaphragm at high strain rate loading. *Journal of the Mechanical Behavior of Biomedical Materials, 60*, 603–616. https://doi.org/10.1016/j.jmbbm.2016.02.031
12. Murray, C. J. L., & Lopez, A. D. (1996) The global burden of disease: a comprehensive assessment of mortality and disability from deceases, injuries and risk factors in 1990 and projected to 2010 (Vol. 1, pp. 1–35). Harvaed University Press. https://doi.org/10.1186/1471-2458-13-863
13. Viano, D. C. (1989). Biomechanical responses and injuries in blunt lateral impact. *SAE International*. https://doi.org/10.4271/892432
14. Nahum, A., & Melvin, J. (2002). *Accidental injury*. Springer Science & Business Media. https://doi.org/10.1007/978-1-4757-2264-2
15. Melvin, J. W. (1988) Review of biomechanical impact response and injury in the automotive environment (pp. 15–18). Task B final report, Rep. No DOTHS 807 042.
16. Gindre, J., Takaza, M., Moerman, K. M., & Simms, C. K. (2013). A structural model of passive skeletal muscle shows two reinforcement processes in resisting deformation. *Journal of the Mechanical Behavior of Biomedical Materials, 22*, 84–94. https://doi.org/10.1016/j.jmbbm.2013.02.007
17. Bathe, K., Ramm, E., & Wilson, E. L. (1975). Finite element formulations for large deformation dynamic analysis. *International Journal for Numerical Methods in Engineering, 9*(2), 353–386.
18. Carew, T. E., Vaishnav, R. N., & Patel, D. J. (1968). Compressibility of the arterial wall. *Circulation Research, 23*(1), 61–68.
19. Patel, D. J., Janicki, J. S., & Carew, T. E. (1969). Static anisotropic elastic properties of the aorta in living dogs. *Circulation Research, 25*(6), 765–779.
20. Shah, C. S., et al. (2006). Dynamic biaxial tissue properties of the human cadaver aorta. *Stapp Car Crash Journal, 50*, 217–246.
21. Shackleton, K. L., Stewart, E. T., & Taylor, A. J. (1998). Traumatic diaphragmatic injuries: Spectrum of radiographic findings. *Radiographics: A Review Publication of the Radiological Society of North America Inc, 18*(1), 49–59. https://doi.org/10.1148/radiographics.18.1.9460108
22. Adegboye, V. O., Ladipo, J. K., Adebo, O. A., & Brimmo, A. I. (2002). Diaphragmatic injuries. *African Journal of Medicine and Medical Sciences, 31*(2), 149–153.
23. Ahmad, R., et al. (2015, March). Isolated blunt traumatic diaphragmatic rupture in a case of situs inversus (pp. 133–137).
24. Gaur, P., et al. (2020). A bilinear structural constitutive model for strain rate-dependent behaviour of human diaphragm tissue. *International Journal of Crashworthiness, 25*(3), 284–298. https://doi.org/10.1080/13588265.2019.1583423
25. Pioletti, D. P., Rakotomanana, L. R., & Leyvraz, P.-F. (1999). Strain rate effect on the mechanical behavior of the anterior cruciate ligament–bone complex. *Medical Engineering & Physics, 21*(2), 95–100. https://doi.org/10.1016/S1350-4533(99)00028-4
26. Brooks, C. A., Gabella, B., Hoffman, R., Sosin, D., & Whiteneck, G. (1997). Traumatic brain injury: Designing and implementing a population-based follow-up system. *Archives of Physical Medicine and Rehabilitation, 78*(8 SUPPL. 4), S26–S30. https://doi.org/10.1016/S0003-9993(97)90152-0

27. Cernak, I., & Noble-Haeusslein, L. J. (2010). Traumatic brain injury: An overview of pathobiology with emphasis on military populations. *Journal of Cerebral Blood Flow & Metabolism Official Journal of International Society of Cerebral Blood Flow Metabolism, 30*(2), 255–266. https://doi.org/10.1038/jcbfm.2009.203.
28. Cernak, I., et al. (2011). The pathobiology of blast injuries and blast-induced neurotrauma as identified using a new experimental model of injury in mice. *Neurobiology of Diseases, 41*(2), 538–551. https://doi.org/10.1016/j.nbd.2010.10.025
29. Salzar, R. S., et al. (2011). Fracture tolerance of the patellofemoral joint in frontal knee impacts of 75 and 35 year-old males. *International Journal of Crashworthiness, 16*(4), 397–409. https://doi.org/10.1080/13588265.2011.596676
30. Sudipto, M., et al. (2011). Dynamic properties of the shoulder complex bones. Soc. Automot. Eng.
31. Mukherjee, S., Chawla, A., Karthikeyan, B., & Soni, A. (2007). Finite element crash simulations of the human body: Passive and active muscle modelling. *Sadhana—Academy Proceedings in Engineering Sciences, 32*(4), 409–426. https://doi.org/10.1007/s12046-007-0032-8
32. Grimal, Q., Gama, B. A., Naili, S., Watzky, A., & Gillespie, J. W., Jr. (2004). Finite element study of high-speed blunt impact on thorax: Linear elastic considerations. *International Journal of Impact Engineering, 30*(6), 665–683. https://doi.org/10.1016/j.ijimpeng.2003.08.002
33. Katyal, D., McLellan, B. A., Brenneman, F. D., Boulanger, B. R., Sharkey, P. W., & Waddell, J. P. (1997). Lateral impact motor vehicle collisions: significant cause of blunt traumatic rupture of the thoracic aorta. *Journal of Trauma, 42*(5), 769–772.
34. Watanabe, I., Furusu, K., Kato, C., Miki, K., & Hasegawa, J. (2001). Development of practical and simplified human whole body FEM model. *JSAE Review, 22*(2), 189–194. https://doi.org/10.1016/S0389-4304(01)00092-3
35. Shah, C. S., Lee, J. B., Hardy, W. N., & Yang, K. H. (2004). A partially validated finite element whole-body human model for organ level injury prediction (Vol. 47225, pp. 71–79).
36. Korac, Z., Kelenc, D., Baskot, A., Mikulic, D., & Hancevic, J. (2001). Substitute ellipse of the permanent cavity in gelatin blocks and debridement of gunshot wounds. *Military Medicine, 166*(8), 689–694.
37. Ragsdale, B. D., & Josselson, A. (1988). Predicting temporary cavity size from radial fissure measurements in ordnance gelatin. *Journal of Trauma, 28*(1 Suppl), S5–S9.
38. Bartlett, C. S. (2003). Clinical update: Gunshot wound ballistics. *Clinical Orthopaedics, 408*, 28–57.
39. Snedeker, J. G., et al. (2005). Strain-rate dependent material properties of the porcine and human kidney capsule. *Journal of Biomechanics, 38*(5), 1011–1021. https://doi.org/10.1016/j.jbiomech.2004.05.036
40. Bir, C., Viano, D., & King, A. (2004). Development of biomechanical response corridors of the thorax to blunt ballistic impacts. *Journal of Biomechanics, 37*(1), 73–79.
41. Brunon, A., Bruyere-Garnier, K., & Coret, M. (2010). Mechanical characterization of liver capsule through uniaxial quasi-static tensile tests until failure. *Journal of Biomechanics, 43*(11), 2221–2227. https://doi.org/10.1016/j.jbiomech.2010.03.038
42. Butler, D. L., Kay, M. D., & Stouffer, D. C. (1986). Comparison of material properties in fascicle-bone units from human patellar tendon and knee ligaments. *Journal of Biomechanics, 19*(6), 425–432. https://doi.org/10.1016/0021-9290(86)90019-9
43. Kemper, A. R., Santago, A. C., Stitzel, J. D., Sparks, J. L., & Duma, S. M. (2010). Biomechanical response of human liver in tensile loading. In *Annals of advances in automotive medicine association for the advancement of automotive medicine. Annual scientific conference. Automotive medicine science conference* (Vol. 54, pp. 15–26).
44. Agah, M. R., Laksari, K., Assari, S., & Darvish, K. (2017). Mechanical behavior of porcine thoracic aorta in physiological and supra-physiological intraluminal pressures. *Proceedings of the Institution of Mechanical Engineers. Part H, Journal of Engineering in Medicine, 231*(4), 326–336. https://doi.org/10.1177/0954411917695577

45. Carter, F. J., Frank, T. G., Davies, P. J., McLean, D., & Cuschieri, A. (2001). Measurements and modelling of the compliance of human and porcine organs. *Medical Image Analysis, 5*(4), 231–236.
46. Mattucci, S. F. E., Moulton, J. A., Chandrashekar, N., & Cronin, D. S. (2012). Strain rate dependent properties of younger human cervical spine ligaments. *Journal of the Mechanical Behavior of Biomedical Materials, 10*, 216–226. https://doi.org/10.1016/j.jmbbm.2012.02.004
47. Mohan, D., & Melvin, J. W. (1982). Failure properties of passive human aortic tissue. I–uniaxial tension tests. *Journal of Biomechanics, 15*(11), 887–902.
48. Ottenio, M., Tran, D., Ní Annaidh, A., Gilchrist, M. D., & Bruyère, K. (2015). Strain rate and anisotropy effects on the tensile failure characteristics of human skin. *Journal of the Mechanical Behavior of Biomedical Materials, 41*, 241–250. https://doi.org/10.1016/j.jmbbm.2014.10.006
49. Lee, O. S., & Kim, M. S. (2003). Dynamic material property characterization by using split Hopkinson pressure bar (SHPB) technique. *Nuclear Engineering and Design, 226*(2), 119–125. https://doi.org/10.1016/S0029-5493(03)00189-4
50. Kerrigan, J. R., et al. (2003). Rate-sensitive constitutive and failure properties of human collateral knee ligaments. In *Proceedings of the IRCOBI international research counce on the biomechanics of IMPACT conference HELD LISBON port. September 2003—Session 3—Biomechanics low. LIMBS*. Retrieved April 11, 2016, from https://trid.trb.org/view.aspx?id=843398
51. Untaroiu, C. D., & Lu, Y.-C. (2013). Material characterization of liver parenchyma using specimen-specific finite element models. *Journal of the Mechanical Behavior of Biomedical Materials, 26*, 11–22. https://doi.org/10.1016/j.jmbbm.2013.05.013
52. Rhule, H. H., et al. (2002, November). *Development of a new biofidelity ranking system for anthropomorphic test devices 46th stapp car crash conference* (Vol. 46).
53. Navindaran, K., Kang, J. S., & Moon, K. (2023). Techniques for characterizing mechanical properties of soft tissues. *Journal of the Mechanical Behavior of Biomedical Materials, 138*, 105575. https://doi.org/10.1016/j.jmbbm.2022.105575
54. Omari, E. A., Varghese, T., Kliewer, M. A., Harter, J., & Hartenbach, E. M. (2015). Dynamic and quasi-static mechanical testing for characterization of the viscoelastic properties of human uterine tissue. *Journal of Biomechanics, 48*(10), 1730–1736. https://doi.org/10.1016/j.jbiomech.2015.05.013
55. Remus, R., Sure, C., Selkmann, S., Uttich, E., & Bender, B. (2024). Soft tissue material properties based on human abdominal in vivo macro-indenter measurements. *Frontiers in Bioengineering and Biotechnology, 12*, 1384062. https://doi.org/10.3389/fbioe.2024.1384062
56. Kang, W., Zhang, Y., Bu, W., Zhao, Y., Wang, L., & Liu, S. (2023). Statistical analysis of mechanical properties of biological soft tissue under quasi-static mechanical loading. *Medicine in Novel Technology and Devices, 17*, 100202. https://doi.org/10.1016/j.medntd.2022.100202
57. Shen, G., Fang, F., & Kang, C. (2018). Tribological performance of bioimplants: A comprehensive review. *Nanotechnology and Precision Engineering, 1*(2), 107–122. https://doi.org/10.13494/j.npe.20180003
58. Nitta, N., Shiina, T., & Ueno, E. (2003, October). Hysteresis parameter imaging of soft tissue under quasi-static deformation. In *IEEE symposium on ultrasonics* (Vol. 2, pp. 1606–1609). https://doi.org/10.1109/ULTSYM.2003.1293216.
59. Gaur, P., et al. (2018). High rate failure properties of human aortic tissue under longitudinal extension. *International Journal of Experimental and Computational Biomechanics, 4*(2–3), 125–151. https://doi.org/10.1504/IJECB.2018.092273
60. Gaur, P., et al. (2019). High rate failure properties of porcine aortic tissue under uniaxial tension. *International Journal of Crashworthiness, 26*, 1–11. https://doi.org/10.1080/13588265.2019.1688454
61. Belkoff, S. M., & Haut, R. C. (2008). Experimental methods in biological tissue testing. In W. N. Sharpe (Ed.), *Springer handbook of experimental solid mechanics* (pp. 871–890). Boston, MA: Springer US. https://doi.org/10.1007/978-0-387-30877-7_31.

62. Griffin, M., Premakumar, Y., Seifalian, A., Butler, P. E., & Szarko, M. (2016). Biomechanical characterization of human soft tissues using indentation and tensile testing. *Journal of Visualized Experiments JoVE, 118*, 54872. https://doi.org/10.3791/54872
63. Scholze, M., et al. (2020). Standardized tensile testing of soft tissue using a 3D printed clamping system. *HardwareX, 8*, e00159. https://doi.org/10.1016/j.ohx.2020.e00159
64. Sanusi, O. M., Benelfellah, A., & Aït Hocine, N. (2020). Clays and carbon nanotubes as hybrid nanofillers in thermoplastic-based nanocomposites—A review. *Applied Clay Science, 185*, 105408. https://doi.org/10.1016/j.clay.2019.105408
65. Delalleau, A., Josse, G., Lagarde, J.-M., Zahouani, H., & Bergheau, J.-M. (2006). Characterization of the mechanical properties of skin by inverse analysis combined with the indentation test. *Journal of Biomechanics, 39*(9), 1603–1610. https://doi.org/10.1016/j.jbiomech.2005.05.001
66. Morin, C., & Avril, S. (2015). Inverse problems in the mechanical characterization of elastic arteries. *MRS Bulletin, 40*(04), 317–323. https://doi.org/10.1557/mrs.2015.63
67. Pasta, S., Catalano, C., Crascì, F., & Scuoppo, R. (2023). A custom-built planar biaxial system for soft tissue material testing. *HardwareX, 16*, e00475. https://doi.org/10.1016/j.ohx.2023.e00475
68. Jiang, M., Sridhar, R. L., Robbins, A. B., Freed, A. D., & Moreno, M. R. (2021). A versatile biaxial testing platform for soft tissues. *Journal of the Mechanical Behavior of Biomedical Materials, 114*, 104144. https://doi.org/10.1016/j.jmbbm.2020.104144
69. Takada, J., Hamada, K., Zhu, X., Tsuboko, Y., & Iwasaki, K. (2023). Biaxial tensile testing system for measuring mechanical properties of both sides of biological tissues. *Journal of the Mechanical Behavior of Biomedical Materials, 146*, 106028. https://doi.org/10.1016/j.jmbbm.2023.106028
70. Gaur, P. (2019). Characterization of aorta and diaphragm at high strain rate loading.
71. Chawla, A., Mukherjee, S., Marathe, R., & Karthikeyan, B. (2006, September). Determining strain rate dependence of human body soft tissues using a split Hopkinson pressure bar, pp. 173–182.
72. Parmley, L. F., Manion, W. C., & Mattingly, T. W. (1958). Nonpenetrating traumatic injury of the heart. *Circulation, 18*(3), 371–396. https://doi.org/10.1161/01.CIR.18.3.371
73. Viano, D. C. (1983). Biomechanics of nonpenetrating aortic trauma: A review. In *SAE technical paper*. SAE International. https://doi.org/10.4271/831608.
74. Newman, R. J., & Rastogi, S. (1984). Rupture of the thoracic aorta and its relationship to road traffic accident characteristics. *Injury, 15*(5), 296–299. https://doi.org/10.1016/0020-1383(84)90049-4
75. Dosios, T. J., Salemis, N., Angouras, D., & Nonas, E. (2000). Blunt and penetrating trauma of the thoracic aorta and aortic arch branches: An autopsy study. *Journal of Trauma Acute Care Surgery, 49*(4). http://journals.lww.com/jtrauma/Fulltext/2000/10000/Blunt_and_Penetrating_Trauma_of_the_Thoracic_Aorta.18.aspx
76. McGwin, G. J., Metzger, J., Moran, S. G., & 3rd Rue, L. W. (2003). Occupant- and collision-related risk factors for blunt thoracic aorta injury. *Journal of Trauma, 54*(4), 652–655. https://doi.org/10.1097/01.TA.0000055548.84109.C2
77. Augenstein, J., et al. (1999). Injuries in near-side collisions. *Annual Proceedings. Association for the Advancement of Automotive Medicine, 43*, 139–158.
78. Hunt, J. P., et al. (1996). Thoracic aorta injuries: management and outcome of 144 patients. *Journal of Trauma, 40*(4), 547–555. Discussion 555–6.
79. Smith, R. S., & Chang, F. C. (1986). Traumatic rupture of the aorta: still a lethal injury. *American Journal of Surgery, 152*(6), 660–663.
80. Sevitt, S. (1977). The mechanisms of traumatic rupture of the thoracic aorta. *British Journal of Surgery, 64*(3), 166–173.
81. Letterer, E. (1924). Beiträge zur Entstehung der Aortenrupturen an typischer Stelle. *Virchows Archiv für Pathologische Anatomie und Physiologie und für Klinische Medizin, 253*, 534–544. https://doi.org/10.1007/BF01994390

82. Shennan, T. (1929). Traumatic (false) aneurysm of the aorta. *The Journal of Pathology and Bacteriology, 32*(4), 795–798. https://doi.org/10.1002/path.1700320409
83. Lim, R. C., Sanderson, R. G., Hall, A. D., & Thomas, A. N. (1968). Multiple traumatic thoracic aneuryms after nonpenetrating chest injury. *Annals of Thoracic Surgery, 6*(4), 377–383. https://doi.org/10.1016/S0003-4975(10)66040-7
84. Cammack, K., Rapport, R. L., Paul, J., & Baird, W. C. (1959). Deceleration injuries of the thoracic aorta. *AMA Archives of Surgery, 79*(2), 244–251. https://doi.org/10.1001/archsurg.1959.04320080080010
85. Hossack, D. W. (1980). Rupture of the aorta in road crash victims. *Australian and New Zealand Journal of Surgery, 50*(2), 136–137. https://doi.org/10.1111/j.1445-2197.1980.tb06650.x
86. Creasy, J. D., Chiles, C., Routh, W. D., & Dyer, R. B. (1997). Overview of traumatic injury of the thoracic aorta. *Radiographics, 17*(1), 27–45. https://doi.org/10.1148/radiographics.17.1.9017797
87. Arenberg, H. (2017). Spontaneous rupture of the aorta. *American Heart Journal, 8*(2), 217–225. https://doi.org/10.1016/S0002-8703(32)90276-2
88. Bertrand, S., et al. (2008). Traumatic rupture of thoracic aorta in real-world motor vehicle crashes. *Traffic Injury Prevention, 9*(2), 153–161. https://doi.org/10.1080/15389580701775777
89. Choudhury, N., et al. (2009). Local mechanical and structural properties of healthy and diseased human ascending aorta tissue. *Cardiovascular Pathology, 18*(2), 83–91. https://doi.org/10.1016/j.carpath.2008.01.001
90. Duprey, A. (2008). Mechanical properties of the aorta. *European Journal of Vascular and Endovascular Surgery* 1–54.
91. Ferrara, A., Morganti, S., Totaro, P., Mazzola, A., & Auricchio, F. (2016). Human dilated ascending aorta: Mechanical characterization via uniaxial tensile tests. *Journal of the Mechanical Behavior of Biomedical Materials, 53*, 257–271. https://doi.org/10.1016/j.jmbbm.2015.08.021
92. Holzapfel, G. A. (2006). Determination of material models for arterial walls from uniaxial extension tests and histological structure. *Journal of Theoretical Biology, 238*(2), 290–302. https://doi.org/10.1016/j.jtbi.2005.05.006
93. Sokolis, D. P. (2007). Passive mechanical properties and structure of the aorta: Segmental analysis. *Acta Physiologica Oxford England, 190*(4), 277–289. https://doi.org/10.1111/j.1748-1716.2006.01661.x
94. Thubrikar, M. J., Labrosse, M., Robicsek, F., Al-Soudi, J., & Fowler, B. (2001). Mechanical properties of abdominal aortic aneurysm wall. *Journal of Medical Engineering & Technology, 25*(4), 133–142.
95. Vande Geest, J. P., Sacks, M. S., & Vorp, D. A. (2006). The effects of aneurysm on the biaxial mechanical behavior of human abdominal aorta. *Journal of Biomechancis, 39*(7), 1324–1334. https://doi.org/10.1016/j.jbiomech.2005.03.003
96. Mohan, D., & Melvin, J. W. (1983). Failure properties of passive human aortic tissue. II–Biaxial tension tests. *Journal of Biomechanics, 16*(1), 31–44.
97. Biewener, A. A. (1992). *Biomechanics–structures and systems: a practical approach* (Vol. 94). USA: Oxford University Press.
98. Hallock, P., & Benson, I. C. (1937). Studies on the elastic properties of human isolated aorta. *The Journal of Clinical Investigation, 16*(4), 595–602.
99. Bergel, D. H. (1961). The static elastic properties of the arterial wall. *Journal of Physiology, 156*(3), 445–457.
100. Lundevall, J. (1964). The mechanism of traumatic rupture of the aorta. *Acta Pathologica et Microbiologica Scandinavica, 62*(1), 34–46. https://doi.org/10.1111/apm.1964.62.1.34
101. Collins, R., & Hu, W. C. L. (1972). Dynamic deformation experiments on aortic tissue. *Journal of Biomechanics, 5*(4), 333–337. https://doi.org/10.1016/0021-9290(72)90062-0
102. Stemper, B. D., Yoganandan, N., & Pintar, F. A. (2007). Mechanics of arterial subfailure with increasing loading rate. *Journal of Biomechanics, 40*(8), 1806–1812. https://doi.org/10.1016/j.jbiomech.2006.07.005

103. Yamada, H., & Evans, F. G. (1970). Strength of biological materials.
104. Iliopoulos, D. C., Kritharis, E. P., Giagini, A. T., Papadodima, S. A., & Sokolis, D. P. (2009). Ascending thoracic aortic aneurysms are associated with compositional remodeling and vessel stiffening but not weakening in age-matched subjects. *Journal of Thoracic and Cardiovascular Surgery, 137*(1), 101–109. https://doi.org/10.1016/j.jtcvs.2008.07.023
105. Sato, M., Niimi, H., Okumura, A., Handa, H., Hayashi, K., & Moritake, K. (1979). Axial mechanical properties of arterial walls and their anisotropy. *Medical and Biological Engineering and Computing, 17*, 170–176. https://doi.org/10.1007/BF02440925
106. Zhou, J., & Fung, Y. C. (1997). The degree of nonlinearity and anisotropy of blood vessel elasticity. *Proceedings of the National Academy of Sciences, 94*(26), 14255–14260. https://doi.org/10.1073/pnas.94.26.14255
107. Humphrey, J. D., & Na, S. (2002). Elastodynamics and arterial wall stress. *Annals of Biomedical Engineering, 30*(4), 509–523. https://doi.org/10.1114/1.1467676
108. Lanir, Y., Lichtenstein, O., & Imanuel, O. (1996). Optimal design of biaxial tests for structural material characterization of flat tissues. *Journal of Biomechanical Engineering, 118*(1), 41–47.
109. Vande Geest, J. P., Sacks, M. S., & Vorp, D. A. (2004). Age dependency of the biaxial biomechanical behavior of human abdominal aorta. *Journal of Biomechanical Engineering, 126*(6), 815–822.
110. Cooney, G. M., Lake, S. P., Thompson, D. M., Castile, R. M., Winter, D. C., & Simms, C. K. (2016). Uniaxial and biaxial tensile stress–stretch response of human linea alba. *Journal of the Mechanical Behavior of Biomedical Materials, 63*, 134–140. https://doi.org/10.1016/j.jmbbm.2016.06.015
111. Delgadillo, J. O. V., Delorme, S., Mora, V., DiRaddo, R., & Hatzikiriakos, S. G. (2010). Effect of deformation rate on the mechanical properties of arteries. *Journal of Biomedical Science and Engineering, 03*(02), 124–137. https://doi.org/10.4236/jbise.2010.32018
112. Liman, S., Kuzucu, A., Tastepe, A. I., Ulasan, G. N., & Topcu, S. (2003). Chest injury due to blunt trauma. *European Journal of Cardio-Thoracic Surgery, 23*(3), 374–378. https://doi.org/10.1016/s1010-7940(02)00813-8
113. Gasser, T. C., Ogden, R. W., & Holzapfel, G. A. (2006). Hyperelastic modelling of arterial layers with distributed collagen fibre orientations. *Journal of the Royal Society Interface, 3*, 15–35. https://doi.org/10.1098/rsif.2005.0073
114. Matsumoto, T., et al. (2009). Biaxial tensile properties of thoracic aortic aneurysm tissues. *Journal of Biomechanical Science and Engineering, 4*(4), 518–529. https://doi.org/10.1299/jbse.4.518
115. Zeinali-Davarani, S., Chow, M. J., Turcotte, R., & Zhang, Y. (2013). Characterization of biaxial mechanical behavior of porcine aorta under gradual elastin degradation. *Annals of Biomedical Engineering, 41*(7), 1528–1538. https://doi.org/10.1007/s10439-012-0733-y
116. Zemánek, M., Burša, J., & Děták, M. (2009). Biaxial tension tests with soft tissues of arterial wall. *Engineering Mechanics, 16*(1), 3–11.
117. Bass, C. R., et al. (2001, November). SAE technical material properties for modeling traumatic aortic rupture (Vol. 45).
118. Vorp, D. A., Schiro, B. J., Ehrlich, M. P., Juvonen, T. S., Ergin, M. A., & Griffith, B. P. (2003). Effect of aneurysm on the tensile strength and biomechanical behavior of the ascending thoracic aorta. *Annals of Thoracic Surgery, 75*(4), 1210–1214.
119. Deng, Y.-C., Kong, W., & Ho, H. (1999). Development of a finite element human thorax model for impact injury studies. *SAE International, 108*, 1193–1199. https://doi.org/10.4271/1999-01-0715
120. Abé, H., Hayashi, K., & Sato, M. (1996) *Data book on mechanical properties of living cells, tissues, and organs*. Springer.
121. Richens, D., Field, M., Hashim, S., Neale, M., & Oakley, C. (2004). A finite element model of blunt traumatic aortic rupture. *European Journal of Cardio-Thoracic Surgery, 25*(6), 1039–1047. https://doi.org/10.1016/j.ejcts.2004.01.059
122. Darvish, K., Lobovsky, L., & Lee, S.-H. (2004). Analysis and modeling of aortic tissue material properties (Vol. 47039, pp. 335–336).

123. Holzapfel, G. A., Gasser, T. C., & Ogden, R. W. (2000). A new constitutive framework for arterial wall mechanics and a comparative study of material models. *Journal of Elasticity and the Physical Science of Solids, 61*(1), 1–48. https://doi.org/10.1023/A:1010835316564
124. Doyle, B. J., et al. (2009). Experimental modelling of aortic aneurysms: Novel applications of silicone rubbers. *Medical Engineering & Physics, 31*(8), 1002–1012. https://doi.org/10.1016/j.medengphy.2009.06.002
125. Rodríguez, J. F., Martufi, G., Doblaré, M., & Finol, E. A. (2009). The effect of material model formulation in the stress analysis of abdominal aortic aneurysms. *Annals of Biomedical Engineering, 37*(11), 2218. https://doi.org/10.1007/s10439-009-9767-1
126. Elad, D., Foux, A., & Kivity, Y. (1988). A model for the nonlinear elastic response of large arteries. *Journal of Biomechanical Engineering, 110*(3), 185–189.
127. Fung, Y. C. (1975). On mathematical models of stress-strain relationship for living soft tissues. *Polymer Mechanics, 11*(5), 726–740. https://doi.org/10.1007/BF00859649
128. Bosanquet, D., Farboud, A., & Luckraz, H. (2009). A review diaphragmatic injury. *Respiratory Medicine CME, 2*(1), 1–6. https://doi.org/10.1016/j.rmedc.2009.01.002
129. Boriek, A. M., Kelly, N. G., Rodarte, J. R., & Wilson, T. A. (2000). Biaxial constitutive relations for the passive canine diaphragm. *Journal of Applied Physiology Bethesda Md 1985, 89*(6), 2187–2190.
130. Asensio, J. A., et al. (2000). Diaphragmatic injuries. *Operative Techniques in General Surgery, 2*(3), 176–191. https://doi.org/10.1053/otgn.2000.17746
131. Petrone, P., Asensio, J. A., & Marini, C. P. (2017). Diaphragmatic injuries and post-traumatic diaphragmatic hernias. *Current Problems in Surgery, 54*(1), 11–32. https://doi.org/10.1067/j.cpsurg.2016.11.001
132. Scharff, J. R., & Naunheim, K. S. (2007). Traumatic diaphragmatic injuries. *Thoracic Surgery Clinics, 17*(1), 81–85. https://doi.org/10.1016/j.thorsurg.2007.03.006
133. Favre, J.-P., Cheynel, N., Benoit, L., & Favoulet, P. (2005). Traitement chirurgical des ruptures traumatiques du diaphragme. *EMC—Chirurgie, 2*(3), 242–251. https://doi.org/10.1016/j.emcchi.2005.04.004
134. Reber, P. U., Schmied, B., Seiler, C. A., Baer, H. U., Patel, A. G., & Buchler, M. W. (1998). Missed diaphragmatic injuries and their long-term sequelae. *Journal of Trauma, 44*(1), 183–188.
135. Feliciano, D. V., et al. (1984). Civilian trauma in the 1980s. A 1-year experience with 456 vascular and cardiac injuries. *Annals of Surgery, 199*(6), 717–724.
136. Mihos, P., et al. (2003). Traumatic rupture of the diaphragm: experience with 65 patients. *Injury, 34*(3), 169–172.
137. Sangster, G., Ventura, V. P., Carbo, A., Gates, T., Garayburu, J., & D'Agostino, H. (2007). Diaphragmatic rupture: a frequently missed injury in blunt thoracoabdominal trauma patients. *Emergency Radiology, 13*(5), 225–230. https://doi.org/10.1007/s10140-006-0548-y
138. McElwee, T. B., Myers, R. T., & Pennell, T. C. (1984). Diaphragmatic rupture from blunt trauma. *The American Surgeon, 50*(3), 143–149.
139. Murray, J. A., et al. (1998). Occult injuries to the diaphragm: prospective evaluation of laparoscopy in penetrating injuries to the left lower chest. *Journal of the American College of Surgeons, 187*(6), 626–630.
140. Ozpolat, B., Kaya, O., Yazkan, R., & Osmanoglu, G. (2009). Diaphragmatic injuries: a surgical challenge. Report of forty-one cases. *The Thoracic and Cardiovascular Surgeon, 57*(6), 358–362. https://doi.org/10.1055/s-0029-1185579
141. Bisgaard, C., Rodenberg, J. C., & Lundgaard, J. (1985). Spontaneous rupture of the diaphragm. *Scandinavian The Journal of Thoracic and Cardiovascular Surgery, 19*(2), 177–180.
142. Hanna, W. C., Ferri, L. E., Fata, P., Razek, T., & Mulder, D. S. (2008). The current status of traumatic diaphragmatic injury: Lessons learned from 105 patients over 13 years. *The Annals of thoracic surgery, 85*(3), 1044–1048. https://doi.org/10.1016/j.athoracsur.2007.10.084
143. Strumpf, R. K., Humphrey, J. D., & Yin, F. C. (1993). Biaxial mechanical properties of passive and tetanized canine diaphragm. *American Journal of Physiology-Heart and Circulatory Physiology, 265*(2 Pt 2), H469–H475.

144. Angelillo, M., Boriek, A. M., Rodarte, J. R., & Wilson, T. A. (1997). Theory of diaphragm structure and shape. *Journal of Applied Physiology Bethesda Md 1985, 83*(5), 1486–1491.
145. Kim, M. J., Druz, W. S., Danon, J., Machnach, W., & Sharp, J. T. (1976). Mechanics of the canine diaphragm. *Journal of Applied Physiology, 41*(3), 369–382.
146. Margulies, S. S., Lei, G. T., Farkas, G. A., & Rodarte, J. R. (1994). Finite-element analysis of stress in the canine diaphragm. *Journal of Applied Physiology Bethesda Md 1985, 76*(5), 2070–2075.
147. Behr, M., et al. (2006). 3D reconstruction of the diaphragm for virtual traumatology. *Surgical and Radiologic Anatomy, 28*(3), 235–240. https://doi.org/10.1007/s00276-006-0080-5
148. Condurache, A., Aach, T., Eck, K., Bredno, J., & Stehle, T. (2005). Fast and robust diaphragm detection and tracking in cardiac x-ray projection images. In *Medical imaging* (pp. 1766–1775). International Society for Optics and Photonics.
149. Segars, W. P., Lalush, D. S., & Tsui, B. M. W. (2001). Modeling respiratory mechanics in the MCAT and spline-based MCAT phantoms. *IEEE Transactions on Nuclear Science, 48*(1), 89–97.
150. Timc-imag, L., & Gaillard, L. (2003). Modèle fonctionnel du diaphragme pour l ' acquisition et le diagnostic en imagerie médicale.
151. Martins, J. A. C., Pires, E. B., Salvado, R., & Dinis, P. B. (1998). A numerical model of passive and active behavior of skeletal muscles. *Computer Methods in Applied Mechanics and Engineering, 151*(3), 419–433. https://doi.org/10.1016/S0045-7825(97)00162-X
152. Aulignac, D., Martins, J. A. C., & Pires, E. B. (2004). Physical modeling of the pelvic floor muscles using shell elements (pp. 1–13).
153. d'Aulignac, D., Martins, J. A. C., Pires, E. B., Mascarenhas, T., & Jorge, R. M. N. (2005). A shell finite element model of the pelvic floor muscles. *Computer Methods in Biomechanics and Biomedical Engineering, 8*(5), 339–347. https://doi.org/10.1080/10255840500405378
154. Humphrey, J. D., & Yin, F. C. (1987). On constitutive relations and finite deformations of passive cardiac tissue: I. A pseudostrain-energy function. *Journal of Biomechanical Engineering, 109*(4), 298–304.
155. Blümel, M., Guschlbauer, C., Daun-Gruhn, S., Hooper, S. L., & Büschges, A. (2012). Hill-type muscle model parameters determined from experiments on single muscles show large animal-to-animal variation. *Biological Cybernetics, 106*(10), 559–571. https://doi.org/10.1007/s00422-012-0530-6
156. Stuhmiller, J. H., Ho, K. H., Vander Vorst, M. J., Dodd, K. T., Fitzpatrick, T., & Mayorga, M. (1996). A model of blast overpressure injury to the lung. *Journal of Biomechanics, 29*(2), 227–234.
157. Roberts, S. B., & Chen, P. H. (1970). Elastostatic analysis of the human thoracic skeleton. *Journal of Biomechanics, 3*(6), 527–545. https://doi.org/10.1016/0021-9290(70)90037-0
158. Plank, G. R., & Eppinger, R. H. (1993). An improved finite element model of the human thorax. In *Proceedings: International TECHNICAL CONFERENCE ON THE ENHANCED SAFETY OF VEHICLES* (Vol. 1993, pp. 902–907).
159. Lee, J. B., & Yang, K. H. (2001). Development of a finite element model of the human abdomen. *Stapp Car Crash Journal, 45*, 79–100.
160. Sundaram, S. H., & Feng, C. C. (1977). Finite element analysis of the human thorax. *Journal of Biomechanics, 10*(8), 505–516. https://doi.org/10.1016/0021-9290(77)90104-X
161. Neathery, R. F. (1974). Analysis of chest impact response data and scaled performance recommendations. *SAE Transactions, 83*, 3763–3779.
162. Plank, G. R., Kleinberger, M., & Eppinger, R. H. (1998). Analytical investigation of driver thoracic response to out of position airbag deployment. *SAE Transactions, 107*, 2941–2954.
163. Y. Huang, A. I. King, J. M. Cavanaugh, Y. Huang, A. I. King, and J. M. Cavanaugh. (1994, November). Finite element modeling of gross motion of human cadavers in side impact. In *Presented at the stapp car crash conference*. SAE International. https://doi.org/10.4271/942207
164. Cavanaugh, J., & Walilko, T. (1996). Abdominal injury and response in side impact. In *Proceedings of stapp car*. https://doi.org/10.4271/962410

165. Happee, R., et al. (2000). Mathematical human body models representing a mid size male and a small female for frontal, lateral and rearward impact loading. *Proceedings of International Research Council on Biomechanics of Injury Conference, 28*, 67–84.
166. Jost, R., & Nurick, G. N. (1999). Finite element modelling of the human body in vehicle side impact. *International Journal of Crashworthiness*. Retrieved November 04, 2024, from https://www.tandfonline.com/doi/abs/10.1533/cras.1999.0090
167. Jost, R., & Nurick, G. N. (2001). Finite element simulation of biomechanical response of the human body subjected to lateral impact. *International Journal of Crashworthiness, 6*(1), 123–134. https://doi.org/10.1533/cras.2001.0167
168. Iwamoto, M., Kisanuki, Y., Watanabe, I., Furusu, K., & Miki, K. (2022). Development of a finite element model of the total human model for the human shoulder. In *IRCOBI Conference* (pp. 31–42).
169. Ruan, J., El-Jawahri, R., Li, C., Barbat, S., & Prasad, P. (2003). Prediction and analysis of human thoracic impact responses and injuries in cadaver impacts using a full human body finite element model. *Stapp Car Crash Journal, 47*, 299.
170. Lasry, D., Hoffmann, R., Yang, K. H., Pan, H. (1991, October). Mathematical modeling of the hybrid III dummy chest with chest foam. SAE International, Warrendale, PA, SAE Technical Paper 912892. https://doi.org/10.4271/912892
171. Yang, K. H., & Jialiang, L. (1992, November). Finite element modeling of hybrid III head-neck complex. SAE International, Warrendale, PA, SAE Technical Paper 922526. https://doi.org/10.4271/922526
172. Shah, C. S., Yang, K. H., Hardy, W., Wang, H. K., & King, A. I. (2001). Development of a computer model to predict aortic rupture due to impact loading. *Stapp Car Crash Journal, 45*, 161–182.
173. Maddali, M., Shah, C. S., & Yang, K. H. (2005). Finite element modeling of aortic tissue using high speed experimental data (Vol. 42312, pp. 97–102).
174. Gaur, P., et al. (2020). Inverse material characterisation of human aortic tissue for traumatic injury in motor vehicle crashes. *International Journal of Crashworthiness, 27*(2), 1–20. https://doi.org/10.1080/13588265.2020.1807678
175. Jansova, M., Hync, L., Toczyski, J., & Baudrit, P. (2015). Evaluation of human thorax FE model in various impact scenarios ˇ echova *9*, 5–20.
176. Shaw, J. M., Herriott, R. G., McFadden, J. D., Donnelly, B. R., & Bolte, J. H. (2006). Oblique and lateral impact response of the PMHS thorax. *Stapp Car Crash Journal, 50*, 147–167.
177. Kroell, C. K., Schneider, D. C., & Nahum, A. M. (1971) Impact tolerance and response of the human thorax. SAE International. https://doi.org/10.4271/710851

Computational Insights into Bone Mechanics: Molecular-Level Approaches

Piyush Uniyal, Krashn Kumar Dwivedi, and Akarsh Verma

Abstract Bone is a remarkable natural material known for its ability to adapt its structure in response to physiological demands. It is also considered a composite material, composed of a tough organic phase, a brittle mineral phase, and water molecules. These components organize at multiple length scales, creating a structure that is both extremely tough and lightweight. Additionally, bone exhibits a remarkable property called mechanotransduction, where it responds to external mechanical loads by dynamically adjusting its morphology through targeted bone deposition or resorption. However, due to disease or aging, a decline in bone quality substantially impairs the skeleton's load-bearing capacity, leading to fractures. Bone fractures are very common in the elderly population suffering from osteoporosis or other metabolic conditions such as diabetes. To prevent fractures associated with poor bone quality, it is crucial to understand the deformation and failure mechanisms of bone at every hierarchical level. Therefore, this review mainly focuses on the molecular mechanisms underlying bone tissue deformation and failure. In-depth understanding of these mechanisms holds promise for the design of targeted treatment regimens and therapeutic interventions tailored to address specific bone pathologies.

Keywords Bone · Mineralized collagen fibril · Molecular dynamics simulation · Mechanical properties

P. Uniyal (✉)
Department of Mechanical Engineering, KU Leuven, Leuven, Belgium
e-mail: piyu.uniyal@gmail.com

K. K. Dwivedi
Department of Mechanical Engineering and Materials Science, Washington University in St. Louis, St. Louis, MO, USA

A. Verma
Department of Mechanical Engineering, University of Petroleum and Energy Studies (UPES), Dehradun, India

K. K. Dwivedi et al. (eds.), *Mechanics and Materials Science of Biological Materials*,
Materials Horizons: From Nature to Nanomaterials,
https://doi.org/10.1007/978-981-96-5366-9_5

1 Introduction

Bone is a load-bearing, mineralized connective tissue with a complex hierarchical structure. It fulfills a multitude of biomechanical functions, including skeletal support, vital organ protection, and mineral homeostasis. Additionally, bone acts as a lever system in conjunction with the muscular system to facilitate locomotion. Bone also exhibits a remarkable property known as mechanotransduction. This process enables the tissue to dynamically remodel its structure by selectively depositing or resorbing bone matrix in response to the prevailing mechanical loading environment. This capability aids in repairing microcracks and fractures that occur during daily activities. This tissue's intricate arrangement across various length scales contributes to its high mechanical performance. Factors contributing to this performance, aside from bone mineral density, are known as bone quality. Compromised bone quality leads to a significant reduction in skeletal load-bearing capacity. This manifests as increased fracture risk, diminished joint mobility, and functional limitations, ultimately contributing to a decline in health-related quality of life. Furthermore, it poses a substantial economic burden on healthcare systems. Fracture events are a significant clinical concern, particularly among the elderly population with osteoporosis and co-morbidities such as diabetes. Bone research is directed toward several key areas: developing predictive models for fracture risk assessment, elucidating the mechanisms by which diseases alter bone quality, evaluating the impact of therapeutic interventions on bone health, and designing biocompatible bone graft substitutes and optimized implants for skeletal repair. Therefore, understanding the mechanisms of bone failure is crucial for developing treatments for bone disorders and designing effective artificial grafts and implants. The intricate hierarchical organization of bone tissue, spanning from the molecular to the macroscopic level, provides it remarkable toughness. Deciphering the mechanisms by which bone deformation and failure occur at the molecular scale holds immense potential for the development of targeted therapeutic strategies for specific skeletal pathologies. This review delves into prior research on bone tissue, with a particular focus on elucidating the molecular mechanisms governing its deformation under diverse loading conditions.

2 Bone Microstructure

Bone tissue exhibits a remarkable hierarchical structure, characterized by a composite composition at the molecular level. This composite consists of three primary components: a ductile organic matrix (32–44% volume), a brittle mineral phase (33–43% volume), and water (15–20% volume) [1]. The organic matrix is mostly composed of collagen I (about 90%), while the remaining fraction encompasses various other collagen types and non-collagenous proteins including osteocalcin and osteopontin [2]. Notably, the hydroxyapatite (HA) crystals within the mineral phase possess a relatively low degree of crystallinity [1]. Type 1 collagen, a fibrous protein, forms the main component of the matrix. It is composed of three polypeptide chains arranged in a triple helix structure, consisting of two α1 chains and one α2 chain. The triple helix has a diameter of 1.5 nm and a length of 300 nm, and it assembles into cylindrical fibrils with a diameter of about 30–60 nm through crosslinks and hydrogen bonds [3]. In addition to type 1 collagen, the organic matter in bone comprises approximately 10% of other collagens and a variety of non-collagenous proteins such as osteopontin, osteonectin, and osteocalcin, which play crucial roles in bone formation and are particularly associated with growth regions in developing bone.

3 Molecular Scale Mechanics of Bone Under Various Conditions

3.1 Mineralization

Bone, a naturally occurring biocomposite, is comprised of a complex interplay between collagen, mineralized components, and water. This hierarchical organization, spanning from the molecular to the macroscopic level, provides bone with its remarkable combination of strength as well as toughness. While the mechanical properties and deformation behavior of bone have been extensively documented across various length scales, a gap exists in our understanding of these processes at the molecular level. At the nanoscale, collagen fibrils—the fundamental building blocks of bone—interact with mineralized components. The interplay between these constituents, along with their relative abundance, significantly influences the overall mechanical behavior of bone tissue.

A previous study [4] employed molecular dynamics simulations to investigate the deformation behavior of collagen microfibrils with varying mineral densities (0–40%). The CHARMM force field [5] was adapted to accurately represent collagen and hydroxyapatite structures. The simulations revealed preferential mineral deposition within gap regions, with minimal presence in overlap regions. Furthermore, uniaxial tensile tests demonstrated a positive correlation between mineral content and the modulus of collagen fibrils. Additionally, significant differences in deformation mechanisms were observed between non-mineralized and mineralized fibrils. In non-mineralized samples, initial strain deformation was localized within the gap regions. However, increasing mineralization stiffened these areas, shifting the primary deformation zone to the overlap regions.

Another study [6] explored the large-scale deformation mechanisms of single collagen fibrils under tensile load. At low strain (around 5%), the initial elastic deformation is primarily due to uncrimping of the collagen fibril. In mineralized fibrils, the presence of minerals restricts this uncrimping and deformation primarily occurs within the overlap regions. As strain increases to 20%, slippage between minerals and collagen facilitates further mineral displacement and stress transfer from the gap to the overlap region. This is followed by collagen backbone stretching and ultimately catastrophic failure of the fibril. Notably, the study found that mineralization enhances collagen fibril strength by tenfold and toughness by thirty-fivefold. These results highlight the significant contribution of mineralization to the overall strength and toughness of collagen fibrils, and how mineral content directly influences bone's deformation behavior.

A recent coarse-grained molecular dynamics framework was used to investigate the impact of mineral arrangement on the mechanical behavior of mineralized collagen fibrils [7]. The study found that hydroxyapatite (HAp) restricts collagen sliding at large deformations, causing collagen molecules to experience greater tensile strain. Furthermore, it was noted that within mineralized collagen fibrils, mineralized regions contribute to stiffness, while non-mineralized regions play a key role in crack deflection.

3.2 Role of Hydration

Water, making up 10–20% of bone volume, is distributed in two main compartments: unbound within pore spaces and bound to the organic matrix. While its influence on bone mechanics across various scales is established, its role at the molecular level remains less explored. A previous study [8] investigated the impact of hydration on deformation behavior in mineralized and non-mineralized collagen fibrils. Tensile tests were conducted at varying water content (0, 2, and 4 wt%). They observed a decrease in Young's modulus for both fibril types with increasing hydration. At higher hydration levels (around 4 wt%), water preferentially accumulated within the gap region, leading to increased deformation there. Conversely, lower hydration levels resulted in more pronounced deformation within the overlap region. Notably, the gap/overlap deformation ratio increased by 40% for non-mineralized and 16% for mineralized fibrils with increasing hydration. These findings offer valuable insights into understanding bone disorders associated with water content alterations.

3.3 Effect of Crosslinking

The stability of collagen's intricate fibrillar structure is reinforced by post-translational modifications that form both intermolecular and interfibrillar crosslinks. There are two primary types of crosslinks in bone: enzymatic crosslinks and non-enzymatic crosslinks formed by advanced glycation end products (AGEs). Enzymatic crosslinks are formed by lysyl oxidase, an enzyme targeting specific lysine residues in the non-helical collagen ends, primarily in terminal regions. In contrast, AGEs can form anywhere along the collagen molecule. A recent study [9] investigated the impact of crosslink maturity on collagen deformation using steered molecular dynamics simulations in LAMMPS [10]. The ReaxFF force field was employed to evaluate the crosslinks' mechanical response. Collagen fibril deformation exhibits three stages: initial elastic deformation due to uncoiling, a linear region dominated by molecular sliding, and a final stage involving tropocollagen backbone stretching until rupture. Crosslink density influences fibril stiffness by altering interconnected molecules, while their mechanical properties impact failure strength and strain. The study revealed that mature crosslinks enhance mechanical properties, offering double the strength and triple the toughness compared to immature ones. Additionally, trivalent crosslinks resulted in brittle behavior. Understanding deformation mechanics of mineralized collagen fibrils with consideration of crosslinks can provide insights into bone diseases associated with altered crosslink profiles, such as Diabetes (T2D) [11].

In a study [12], the geometry and atomic partial charges of arginine-lysine crosslinks were introduced. Additionally, AGE-based intra crosslinks were incorporated into the collagen fibril to examine structural changes. The findings indicated that AGE crosslinks do not promote water diffusion through the collagen triple helix.

3.4 Viscoelasticity and Rate of Deformation

The viscoelastic behavior of bone, characterized by its ability to exhibit both elastic and viscous responses to stress, plays a critical role in mitigating fracture risk. This property allows for energy dissipation through mechanisms like stress relaxation and viscous damping, effectively acting as a biological shock absorber. Extensive research has been conducted at the micro and macro scales to elucidate the contributions of water and the organic matrix to bone viscoelasticity. However, a gap exists in our understanding at the fibrillar level. A previous molecular dynamics simulation [13] investigated the viscoelastic properties of mineralized collagen fibrils under desiccated and hydrated conditions, applying both shear and longitudinal deformation modes. The study revealed a positive correlation between mineralization and the dynamic modulus, indicating increased stiffness. Interestingly, the presence of intrafibrillar water resulted in a decrease in the elastic component, likely due to enhanced molecular mobility within the fibril. Conversely, a significant increase in viscosity was observed under hydrated conditions. These findings underscore the critical role of water in the mechanical response of mineralized collagen fibrils. By promoting energy dissipation through viscous mechanisms, water contributes significantly to the fracture resistance of bone tissue.

The viscoelastic nature of bone also manifests as creep behavior, a time-dependent deformation under constant stress. This phenomenon can lead to permanent bone deformation if sustained loads exceed physiological limits. Understanding creep behavior is crucial in various clinical scenarios, including spinal deformities and implant loosening. While extensive research has explored creep at the macroscopic level, knowledge at the molecular level remains scarce. A previous study [14] employed steered molecular dynamics simulations with the CHARMM force field [15] to investigate the creep response of collagen molecules. Two collagen molecules were covalently linked via lysine-lysine crosslinks and subjected to an axial pulling force ranging from 10 to 2000 pN. The results revealed a direct correlation between applied force and steady-state strain, with higher forces inducing greater deformation. Conversely, larger molecular size exhibited lower strain susceptibility. Interestingly, the observed force wave speed aligned with the theoretical one-dimensional elastic wave speed. This finding suggests a potential link between molecular-level creep and macroscopic wave propagation phenomena in bone. The significant creep behavior observed in collagen fibrils at the molecular level underscores the importance of this hierarchical phenomenon in understanding the overall viscoelasticity of bone tissue.

The viscoelastic nature of bone necessitates consideration of the loading rate when analyzing its deformation characteristics. A previous study [16] investigated the underlying molecular mechanisms governing the deformation response of tropocollagen molecules under varying loading regimes. The study specifically focused on the collagen peptide sequence [(Gly-Pro-Hyp)10]3, subjected to a wide range of pulling speeds (from 0.01 to 100 m/s). The mechanical properties and deformation behavior were then meticulously analyzed. The computational findings revealed a loading rate dependence for the Young's modulus of tropocollagen. At low loading rates (< 0.5 m/s), the mechanical properties remained relatively insensitive to the pulling speed. However, a significant increase in Young's modulus was found beyond this threshold. Interestingly, distinct deformation mechanisms were identified at different loading rates. Under slow loading conditions, the triple helix exhibited complete uncoiling followed by partial recoiling and ultimately straightening. At intermediate rates, the molecule displayed a linear uncoiling behavior with increasing strain. Conversely, high loading rates prevented complete uncoiling during the stretching process. These observations provide valuable insights into the strain rate dependency of mechanical responses observed at the macroscopic level [17].

3.5 Interfaces Within Bone

The hierarchical architecture of bone is characterized by the presence of numerous mineral-protein interfaces across various length scales. These interfaces play a crucial role in enhancing the overall toughness of bone tissue by facilitating efficient energy dissipation mechanisms. A study [18] employed molecular dynamics simulations to investigate the mechanical behavior at the interface between osteopontin (OPN) and hydroxyapatite (HA). The study specifically examined four distinct 18-residue OPN peptide sequences, each featuring variations in the number of acidic amino acids. The GROMOS force field was utilized to model the OPN peptides, followed by the application of pulling forces in both the thickness and interfacial directions. The findings revealed that the mechanical properties at this interface are primarily governed by attractive electrostatic interactions between acidic residues within the OPN peptide and calcium ions present in the hydroxyapatite lattice. Notably, a higher number of acidic amino acids within the OPN sequence resulted in increased energy dissipation during the deformation process. This phenomenon can be attributed to the formation of transient bonds between the acidic residues and the hydroxyapatite surface.

These transient bonds induce a stick–slip motion between the OPN peptide and the HA, leading to significant energy dissipation—a key mechanism for enhancing the fracture resistance of bone.

Another study investigated the interfacial water adsorption behavior between hydroxyapatite (HA) and collagen [19]. The study compared two distinct collagen configurations: a tightly packed triple helix, representative of healthy bone, and a partially unfolded triple helix, mimicking diseased bone. The presence of collagen was found to significantly enhance water adsorption onto the HA surface. Interestingly, the density of adsorbed water differed considerably between the folded and partially unfolded collagen structures. Collagen disrupts the bulk water's hydrogen bond network, facilitating a more favorable adsorption process onto the HA surface. The study quantified that the tightly packed triple helix structure enables a 30% reduction in the occupied area per molecule. This translates to a higher water binding capacity for healthy bone compared to diseased bone, where the unfolded collagen structure offers less surface area for interaction. Additionally, the study suggests an age-dependent weakening of water adsorption strength on the HA surface, potentially leading to a decrease in bound water content over time.

A study [20] investigated the effect of the HAp crystal structure and protein interactions on mechanical properties. The findings revealed changes in stability during loading, highlighting the crucial interaction between -COO and Ca, as well as the role of amino groups in aspartic acid's structural deformation. Additionally, calcium vacancies were found to affect carboxyl group adsorption, while hydrogen phosphate contributed to stabilizing the process.

4 Bone Diseases and Disorders

Bone quality is known to deteriorate with age and in diseases such as osteoporosis and diabetes [21, 22], leading to increased fracture risk and significant healthcare burdens. Developing targeted treatments for these bone disorders requires a comprehensive understanding of the structural alterations occurring at various hierarchical levels within bone tissue. This section specifically focuses on the molecular-level changes observed in bone structure across different diseased states.

4.1 Osteogenesis Imperfecta (OI)

Osteogenesis imperfecta (OI), or brittle bone disease, is a genetic disorder characterized by heightened bone fragility, reduced bone mineral density (BMD), and potential complications in other connective tissues. Clinical presentation allows for the classification of OI into four primary types. Mutations within the COL1A1 and COL1A2 genes, encoding type I collagen—the most abundant human protein—contribute to roughly 80–90% of OI cases. Previous studies utilizing cortical bone samples from OI patients have revealed a disorganized collagen architecture, elevated mineral content, diminished mineral heterogeneity, and increased vascular canal density [23–26]. Missense mutations, particularly those altering glycine codons, are the most frequent cause of OI. In silico investigations at the molecular level suggest these mutations disrupt collagen fibril structure, consequently impacting its mechanical properties [27]. This translates to poor fibril packaging, decreased crosslinking density, and impaired mineral crystal growth. The presence of glycine substitutions not only hinders proper fibril packing but also diminishes lateral crystallinity, the crystal arrangement within the fibrils. Furthermore, inadequate lateral packing may impede mineral crystal growth. Notably, with increasing disease severity, a molecular softening and weakening of intermolecular adhesion are observed. Collectively, these factors contribute to a significant reduction in mechanical strength at the collagen fibril level.

4.2 Osteoporosis

Osteoporosis is a skeletal disorder characterized by increased bone fragility and a heightened susceptibility to fractures, even with minimal trauma or during routine activities. This condition often progresses silently, with patients remaining asymptomatic until a fracture event. Such fractures can inflict substantial pain, potentially lead to debilitating complications, and significantly compromise an individual's functional independence [28]. To elucidate the specific bone abnormalities and underlying pathophysiology of osteoporosis, a traditional disease model is typically employed [29]. Primary osteoporosis manifests in two distinct subtypes: one linked to estrogen deficiency following menopause and another associated with the aging process itself [30]. Riggs et al. [31] proposed that type I osteoporosis, primarily affecting postmenopausal women, is characterized by predominant loss of trabecular bone. In contrast, type II osteoporosis, observed in both sexes, involves deterioration of both cortical and trabecular bone compartments. The progressive increase in bone porosity associated with osteoporosis has a detrimental effect on its mechanical properties and overall load-bearing capacity [21, 32, 33].

4.3 Type 2 Diabetes Mellitus (T2DM)

Diabetes is a prevalent chronic metabolic disorder that imposes a significant burden on public health systems worldwide. Currently, T2D affects over 387 million adults globally, accounting for approximately 10% of the population, with projections indicating this number could rise to 700 million by 2045 [34]. Until recently, the skeleton was not considered a target organ affected by T2D. However, it is now well-established that T2D independently increases the risk of bone fractures, irrespective of factors such as body mass index (BMI) and other osteoporosis-related factors [35].

Recent epidemiological and pathophysiological studies, coupled with advancements in investigative techniques, have enhanced our understanding of how T2D alters bone metabolism and increases fracture risk. Epidemiological data indicate that older adults with T2D have a significantly higher likelihood of fractures, including a 50%–80% increased risk of extremity fractures and accounting for approximately 20% of all clinical fractures in both men and women [36]. Interestingly, despite slightly higher bone mineral density observed in patients with T2D compared to nondiabetics [11, 37], the risk of fractures remains elevated.

While the mechanical response of diabetic bone has been explored across various scales (micro to macro) [11, 21, 38–40], a gap remains in our understanding of the molecular mechanisms underlying T2D-induced bone alterations. A recent study employed steered molecular dynamics simulations to investigate the impact of T2D on the energy dissipation behavior of bone at the molecular level [41]. The study focused on the role of osteocalcin (OC) at the hydroxyapatite (HA) interface, highlighting the crucial α-helical conformation of the protein for energy dissipation. In T2D, osteocalcin undergoes glycation, a process involving the formation of covalent crosslinks between arginine residues and the N-terminus. This disrupts the α-helical structure, significantly compromising the bone's capacity for energy dissipation. Additionally, glycation weakens the affinity of osteocalcin for HA crystals. The loss of the α-helix due to glycation diminishes the stick–slip mechanism and sacrificial bonding—key energy dissipation mechanisms in healthy bone.

4.4 Aging

Age-related decline in bone fracture resistance is well documented. A recent study employed a mouse model to investigate the impact of aging on post-translational modifications (PTMs) of collagen molecules [42]. Specifically, the study focused on deamidation, a process affecting the collagen triple helix with advancing age. Deamidation involves a rearrangement of hydrogen bonding with water molecules, leading to increased hydration around the triple helix. This occurs via a reaction between backbone nitrogen and a carbonyl sidechain, forming an intermediate ring that alters sidechain functionalities. To elucidate the effects of deamidation on Type I collagen at the molecular level, the study utilized molecular dynamics simulations. These simulations examined how deamidation disrupts hydrogen bonding patterns and influences overall collagen stability. Increased water binding by deamidated sidechains disrupts localized hydrogen bonding along the collagen backbone, critical for maintaining helical conformation. This alteration in optimal hydrogen bonding compromises the stability of the triple helix, potentially leading to localized unraveling at deamidation sites. This suggests a redistribution of stability within the deamidated triple helix compared to its native form. Furthermore, the increased interaction of water molecules with deamidated sidechains may induce swelling (diameter increase) of mineralized collagen fibrils (MCFs). Such swelling can impact the distribution of interstitial water between minerals and collagen fibrils. The potential migration of water from interstitial spaces to the extrafibrillar matrix could enhance the propensity for microcrack formation, ultimately reducing bone's fracture resistance at the macroscopic level.

4.5 Collagen Denaturation

The intricate triple helical conformation of bone collagen is essential for its structural integrity. Various pathological conditions and environmental factors, including mechanical stress and temperature variations, can disrupt this native structure. Such disturbances translate to a reduction in bone tissue's mechanical strength at the macroscopic level. Consequently, a comprehensive understanding of alterations in collagen's secondary structure holds significant promise for developing targeted treatments for bone disorders associated with collagen denaturation.

Fourier-Transform Infrared (FTIR) spectroscopy is a powerful tool for investigating the secondary structure of proteins. A recent study [17] utilized FTIR to quantify thermally induced changes in collagen's secondary structure. The Amide I band is particularly sensitive to protein secondary structure, primarily reflecting C=O stretching vibrations with some contribution from the polypeptide backbone. In untreated bone, the prominent peak at 1660 cm^{-1} in the Amide I band signifies collagen in its native, triple helical state. However, thermal treatment leads to a decrease in the intensity of this component, accompanied by an increase in contributions from other components. Moreover, the increase in the peak around 1645 cm^{-1} is associated with random unordered structures; following heat treatment suggests a transition from helical to coil conformations within collagen.

Previous studies have attributed the peak centered around 1630 cm^{-1} to left-handed α-helices in denatured collagen. Thermal treatment likely disrupts the hydrogen bonds stabilizing the triple helix, leading to partial unfolding and conversion into individual helices and random coils. The observed increase in collagen's unordered structures with heat treatment, as revealed by FTIR analysis, corresponds to the decline in bone's mechanical response at the macroscopic scale. The organic matrix phase of bone, primarily composed of type I collagen, plays a crucial role in bone mechanics, particularly in the post-yield regime. Alterations in the collagen phase are observed in various diseases such as osteogenesis imperfecta [26, 43] and type 2 diabetes [11], contributing to reduced bone fracture resistance.

5 Conclusion and Future Perspectives

Skeletal fractures are a significant global health concern, particularly among the aging population. These fractures contribute to a substantial loss of mobility and independence, increased mortality rates, and a diminished quality of life. Consequently, they pose a significant burden on healthcare systems, economies, and societies. The primary objective of bone research lies in preventing these fractures and promoting overall bone health.

Fragility fractures, a defining characteristic of aging and various diseases such as osteoporosis and diabetes, are a major clinical challenge. They arise from a decline in bone quality, characterized by reduced bone mineral density (BMD) and compromised mechanical properties. Previous studies have established that alterations in the bone's hierarchical structure and composition at the molecular level significantly influence its mechanical response. These changes include: variations in the degree of mineralization, modifications in hydration levels, alterations in collagen crosslinking and disruptions in collagen conformation. These molecular-level changes can be linked to the development and progression of various bone disorders. Investigating bone mechanics at this fundamental level offers valuable insights for the development of novel therapeutic interventions for bone diseases.

Bone possesses a unique hierarchical structure, encompassing various length scales from the intricate arrangement of collagen molecules to the entire bone organ. Each level within this hierarchy exhibits distinct deformation and damage mechanisms. A comprehensive understanding of the mechanical behavior across all relevant length scales, from the molecular level to the macroscopic level, is essential for effective fracture prevention and treatment development. While extensive literature exists on the mechanics of healthy and diseased bone at the macroscopic scale (e.g., whole bone bending or tensile testing), research on the molecular mechanics of bone under varying conditions remains limited.

To bridge this knowledge gap, further investigations at the molecular level are warranted. These investigations should be coupled with experimental validation using diseased bone tissue models that accurately represent the pathological state of interest. Additionally, multi-scale modeling approaches that integrate experimental data from various length scales hold significant promise. These models can elucidate the complex interplay between deformation mechanisms across different hierarchical levels, providing valuable insights for the development of targeted therapies that enhance bone health and reduce fractures associated with compromised bone mechanics.

Conflict of Interest The authors declare no conflict of interest.

References

1. Sabet, F. A., Najafi, A. R., Hamed, E., & Jasiuk, I. (2016). Modelling of bone fracture and strength at different length scales: A review. *The Royal Society Interface Focus, 6*(1), 20150055.
2. Ural, A., & Vashishth, D. (2014). Hierarchical perspective of bone toughness–from molecules to fracture. *International Materials Reviews, 59*(5), 245–263.
3. Stock, S. R. (2015). The mineral—Collagen interface in Bone. *Calcif Tissue International [Internet]* 262–280. https://doi.org/10.1007/s00223-015-9984-6

4. Nair, A. K., Gautieri, A., Chang, S. W., & Buehler, M. J. (2013). Molecular mechanics of mineralized collagen fibrils in bone. *Nature Communication [Internet], 4*, 1724–1729. https://doi.org/10.1038/ncomms2720
5. Brooks, B. R., Bruccoleri, R. E., Olafson, B. D., States, D. J., Swaminathan, S. A., & Karplus, M. (1983). CHARMM: A program for macromolecular energy, minimization, and dynamics calculations. *Journal of Computational Chemistry, 4*(2), 187–217.
6. Depalle, B., Qin, Z., Shefelbine, S. J., & Buehler, M. J. (2016). Large deformation mechanisms, plasticity, and failure of an individual collagen fibril with different mineral content. *Journal of Bone and Mineral Research, 31*(2), 380–390.
7. Tavakol, M., & Vaughan, T. J. (2023). A coarse-grained molecular dynamics investigation of the role of mineral arrangement on the mechanical properties of mineralized collagen fibrils. *Journal of the Royal Society Interface, 20*(198), 20220803.
8. Fielder, M., & Nair, A. K. (2019). Effects of hydration and mineralization on the deformation mechanisms of collagen fibrils in bone at the nanoscale. *Biomechanics and Modeling in Mechanobiology [Internet], 18*(1), 57–68. https://doi.org/10.1007/s10237-018-1067-y
9. Depalle, B., Qin, Z., Shefelbine, S. J., & Buehler, M. J. (2015). Influence of cross-link structure, density and mechanical properties in the mesoscale deformation mechanisms of collagen fibrils. *Journal of the Mechanical Behavior of Biomedical Materials, 1*(52), 1–13.
10. Plimpton, S. (1995). Fast parallel algorithms for short-range molecular dynamics. *Journal of Computational Physics, 117*(1), 1–19.
11. Sihota, P., Yadav, R. N., Dhaliwal, R., Bose, J. C., Dhiman, V., Neradi, D., et al. (2021). Investigation of mechanical, material, and compositional determinants of human trabecular bone quality in type 2 diabetes. *The Journal of Clinical Endocrinology & Metabolism, 106*(5), e2271–e2289.
12. Nash, A., Noh, S. Y., Birch, H. L., & de Leeuw, N. H. (2021). Lysine–arginine advanced glycation end-product cross-links and the effect on collagen structure: A molecular dynamics study. *Proteins: Structure, Function and Bioinformatics, 89*(5), 521–530.
13. Milazzo, M., David, A., & Buehler, M. J. (2021). Molecular origin of viscoelasticity in mineralized collagen fibrils 3390–3400.
14. Ghodsi, H., & Darvish, K. (2016). Characterization of the viscoelastic behavior of a simplified collagen micro-fibril based on molecular dynamics simulations. *Journal of the Mechanical Behavior of Biomedical Materials, 63*, 26–34. https://doi.org/10.1016/j.jmbbm.2016.06.006
15. MacKerell, A. D., Jr., Bashford, D., Bellott, M., Dunbrack, R. L., Jr., Evanseck, J. D., Field, M. J., et al. (1998). All-atom empirical potential for molecular modeling and dynamics studies of proteins. *The Journal of Physical Chemistry B, 102*(18), 3586–3616.
16. Gautieri, A., Buehler, M. J., & Redaelli, A. (2009). Deformation rate controls elasticity and unfolding pathway of single tropocollagen molecules. *Journal of the Mechanical Behavior of Biomedical Materials, 2*(2), 130–137.
17. Uniyal, P., Sihota, P., & Kumar, N. (2022). Effect of organic matrix alteration on strain rate dependent mechanical behaviour of cortical bone. *Journal of the Mechanical Behavior of Biomedical Materials, 1*(125), 104910.
18. Lai, Z. B., Wang, M., Yan, C., & Oloyede, A. (2014). Molecular dynamics simulation of mechanical behavior of osteopontin-hydroxyapatite interfaces. *Journal of the Mechanical Behavior of Biomedical Materials [Internet], 36*, 12–20. https://doi.org/10.1016/j.jmbbm.2014.04.002
19. Welborn, V. V. (2021). Environment-controlled water adsorption at hydroxyapatite-collagen interfaces. RSC.
20. Sun, Y., Wang, Y., Ji, C., Ma, J., & He, B. (2024). The impact of hydroxyapatite crystal structures and protein interactions on bone's mechanical properties. *Scientific Reports, 14*(1), 9786.
21. Yadav, R. N., Sihota, P., Uniyal, P., Neradi, D., Bose, J. C., Dhiman, V., et al. (2021). Prediction of mechanical properties of trabecular bone in patients with type 2 diabetes using damage based finite element method. *Journal of Biomechanics, 123*, 110495.
22. Yadav, R. N., Uniyal, P., Sihota, P., Kumar, S., Dhiman, V., Goni, V. G., et al. (2021). Effect of ageing on microstructure and fracture behavior of cortical bone as determined by experiment

and Extended Finite Element Method (XFEM). *Medical Engineering & Physics, 1*(93), 100–112.
23. Carriero, A., Zimmermann, E. A., Paluszny, A., Tang, S. Y., Bale, H., Busse, B., et al. (2014). How tough is brittle bone? Investigating osteogenesis imperfecta in mouse bone. *Journal of Bone and Mineral Research, 29*(6), 1392–1401.
24. Carriero, A., Doube, M., Vogt, M., Busse, B., Zustin, J., Levchuk, A., et al. (2014). Altered lacunar and vascular porosity in osteogenesis imperfecta mouse bone as revealed by synchrotron tomography contributes to bone fragility. *Bone [Internet], 61*, 116–124. https://doi.org/10.1016/j.bone.2013.12.020
25. Dong, X. N., Zoghi, M., Ran, Q., & Wang, X. (2010). Collagen mutation causes changes of the microdamage morphology in bone of an OI mouse model. *Bone [Internet], 47*(6), 1071–1075. https://doi.org/10.1016/j.bone.2010.08.013
26. Andriotis, O. G., Chang, S. W., Vanleene, M., Howarth, P. H., Davies, D. E., & Shefelbine, S. J., et al. (2015). Structure—Mechanics relationships of collagen fibrils in the osteogenesis imperfecta mouse model.
27. Gautieri, A., Uzel, S., Vesentini, S., Redaelli, A., & Buehler, M. J. (2009). Molecular and mesoscale mechanisms of osteogenesis imperfecta disease in collagen fibrils. *Biophysical Journal [Internet], 97*(3), 857–865. https://doi.org/10.1016/j.bpj.2009.04.059
28. Looker, A. C., Fan, B., & Shepherd, J. A., et al. (2017). FRAX-based estimates of 10-year Probability of hip and major osteoporotic fracture among adults aged 40 and over: United States, 2013 and 2014. *National Health Statistics Report, 103*, 1–16.
29. Ettinger, M. P. (2003). Aging bone and osteoporosis: Strategies for preventing fractures in the elderly. *Archives of Internal Medicine, 163*(18), 2237–2246.
30. Marcus, R., & Majumder, S. (2001). The nature of osteoporosis. Osteoporosis 3–17
31. Riggs, B. L., Wahner, H. W., Seeman, E., Offord, K. P., Dunn, W. L., Mazess, R. B., et al. (1982). Changes in bone mineral density of the proximal femur and spine with aging: Differences between the postmenopausal and senile osteoporosis syndromes. *Journal of Clinical Investigation, 70*(4), 716–723.
32. Kumar, N., Kumar, A., Uniyal, P., Ramalingaiah, B., Sharma, S., Goni, V. G., et al. (2020). Mimicking high strength lightweight novel structures inspired from the trabecular bone microarchitecture. *Philosophical Transactions of the Royal Society A: Mathematical, Physical and Engineering Sciences., 378*(2167), 20190448.
33. Uniyal, P., Sihota, P., Tikoo, K., & Kumar, N. (2021). Anatomical variation in intracortical canal network microarchitecture and its influence on bone fracture risk. *Journal of the Mechanical Behavior of Biomedical Materials, 123*, 104770.
34. Idf Diabetes Atlas [Internet]. (2017). file:///C:/Users/CeciliaTorqueti/Downloads/IDF_DA_8e-EN-final (1).pdf
35. Schwartz, A. V. (2003). Diabetes mellitus: Does it affect bone? *Calcified Tissue International, 73*(6), 515–519.
36. Rubin, M. R., & Patsch, J. M. (2016). Assessment of bone turnover and bone quality in type 2 diabetic bone disease: Current concepts and future directions. *Bone Research, 4*(1), 1–9.
37. Hofbauer, L. C., Brueck, C. C., Singh, S. K., & Dobnig, H. (2007). Osteoporosis in patients with diabetes mellitus. *Journal of Bone and Mineral Research, 22*(9), 1317–1328.
38. Karim, L., & Bouxsein, M. L. (2016). Effect of type 2 diabetes-related non-enzymatic glycation on bone biomechanical properties. *Bone [Internet], 82*, 21–27. https://doi.org/10.1016/j.bone.2015.07.028
39. Karim, L., Moulton, J., Van Vliet, M., Velie, K., Robbins, A., & Malekipour, F., et al. (2018). Bone microarchitecture, biomechanical properties, and advanced glycation end-products in the proximal femur of adults with type 2 diabetes. *Bone [Internet]*. https://www.sciencedirect.com/science/article/pii/S8756328218302205
40. Farr, J. N., & Khosla, S. (2016). Determinants of bone strength and quality in diabetes mellitus in humans. *Bone [Internet], 82*, 28–34. https://doi.org/10.1016/j.bone.2015.07.027
41. Tavakol, M., & Vaughan, T. J. (2020). The structural role of osteocalcin in bone biomechanics and its alteration in type - 2 diabetes. *Scientific Reports [Internet]* 1–10. https://doi.org/10.1038/s41598-020-73141-w

42. Creecy, A., Brown, K. L., Rose, K. L., Voziyan, P., & Nyman, S. (2020). Post-translational modifications in collagen type I of bone in a mouse model of aging. *Bone [Internet]., 143*, 115763. https://doi.org/10.1016/j.bone.2020.115763
43. Wallace, J. M., Orr, B. G., Marini, J. C., & Banaszak, M. M. (2011). Nanoscale morphology of type I collagen is altered in the Brtl mouse model of Osteogenesis Imperfecta. *Journal of Structural Biology [Internet], 173*(1), 146–152. https://doi.org/10.1016/j.jsb.2010.08.003

Ultimate Biomaterial: Titanium a Gold Standard Choice for Current Orthopaedic Implants

Manjit Sandhu, Navin Kumar, Ravinder Singh Sawhney, Jeevan Jyoti, Kuldeep Singh, and Pawandeep Kaur

Abstract Biomaterials are integral to numerous medical applications, encompassing orthopaedics, dentistry, medication delivery, tissue engineering, and cardiovascular devices. Ensuring biocompatibility, these materials must not harm the human body. They fall into four main classes: metals, polymers, ceramics, and natural materials, each with distinct properties and applications. Titanium alloys have emerged as preferred orthopaedic implant materials due to exceptional biocompatibility and mechanical properties, despite challenges like stress shielding and toxic elements. Advancements in beta titanium alloys offer improved biocompatibility, while shape-memory alloys present unique functionalities. Challenges such as nickel allergy persist, but ongoing research enhances biocompatibility and performance. In orthopaedics, implants are crucial for bone and joint repair, yet material-related failures prompt revisions. Factors like prosthesis design and surgical technique influence implant success, with titanium alloys often preferred for their fatigue strength. Strategies like surface coatings and ion implantation mitigate wear and friction issues. Biomaterials have evolved over four generations to address therapeutic needs, from bio-inert materials to bioactive and biodegradable substances. Fourth-generation biomaterials mimic natural tissues, coupling with electronic systems for advanced diagnostic and therapeutic applications. Classification into metallic, ceramic, polymeric, and composite categories aids understanding and application. Continued development holds promise for enhancing patient outcomes across medical disciplines.

Keywords Biomaterials · Orthopaedic implants · Titanium alloys · Biocompatibility · Material-related failures · Surface modifications · Fourth-generation biomaterials · Metallic biomaterials · Ceramic biomaterials · Polymeric biomaterials · Composite biomaterials

M. Sandhu (✉) · R. S. Sawhney · K. Singh · P. Kaur
Department of Electronics Technology, Guru Nanak Dev University, Amritsar, India
e-mail: manjit.ece@gndu.ac.in

N. Kumar · J. Jyoti
Department of Mechanical Engineering, Indian Institute of Engineering Ropar, Rupnagar, India

K. K. Dwivedi et al. (eds.), *Mechanics and Materials Science of Biological Materials*, Materials Horizons: From Nature to Nanomaterials,
https://doi.org/10.1007/978-981-96-5366-9_6

1 Introduction

In today's world, biomaterials are extremely important. In the last few decades, numerous types of biomaterials have been created as a result of advancements in material science and medical technology with characteristics appropriate for a range of applications [1]. Biomaterials are primarily used in the following domains: orthopaedics, dentistry, medication delivery, skin tissue engineering, and cardiovascular devices [2]. Making sure a material is safe and biocompatible with the human body is crucial before designating it as an implant. According to Williams' definition, a biomaterial is considered biocompatible if it can function as intended in conjunction with a medical therapy without causing any unanticipated side effects for the patient or other therapy recipients. Rather, the goal should be to maximize the therapeutic effect in a manner that is pertinent to the clinical setting while also eliciting the most advantageous cellular or tissue response possible [3]. This means that since harmful substances can harm a patient's entire body, they shouldn't be released into the body by the implant [4, 5]. A human body will react in some way to any material inserted into it. The material's reactivity increases at the point where it comes into contact with damaged tissue and the implant, which explains why this response is more noticeable there. The number of neighbours between atoms at the interface is thought to be infinite, making it essentially a two-dimensional imperfection. They therefore have more energy than the materials' core atoms [6].

Generally speaking, biomaterials can be classified as either natural materials, polymers, metals and alloys, or ceramics. They are extensively employed in many different medical specialties, including drug delivery systems, skin replacement, cardiovascular biomaterials [7], and hyper thermic cancer treatment [8]. To enhance their performance, research is being done in large quantities. [9]. The characteristics and uses of metallic implants [10], polymers [11], and bioactive glasses and glass–ceramics [12, 13] are discussed in a number of reviews.

Injury-related conditions that impair joint, muscle, or tendon pain and cause tissue degradation are referred to as musculoskeletal disorders. Millions of people worldwide have been impacted by musculoskeletal disorders, which can lead to physical disabilities and persistent pain. It is anticipated that as lifestyles change, the number of individuals afflicted with bone and joint disorders will rise in the upcoming years. As per a study conducted by Kurtz and colleagues [14], 572,000 hip replacement procedures are anticipated to be performed overall by 2030. Additionally, it is anticipated that 3.48 million knee replacement procedures will be performed overall. It follows that a greater need for novel and enhanced implants will drive the need for better implant materials and surgical techniques. For many musculoskeletal problems, including total joint replacement, surgery is required. [15] Here is the area where bone regrowth occurs and repair are greatly aided by biomaterials. Biomaterial-based surgical implants are used to repair damaged parts and restore their functionality [16]. There are currently several implant material types in use, each with their own set of characteristics. For example, cobalt chromium alloys, titanium-based alloys, and 316L stainless steel are the top metals for orthopaedic applications such as hip joint

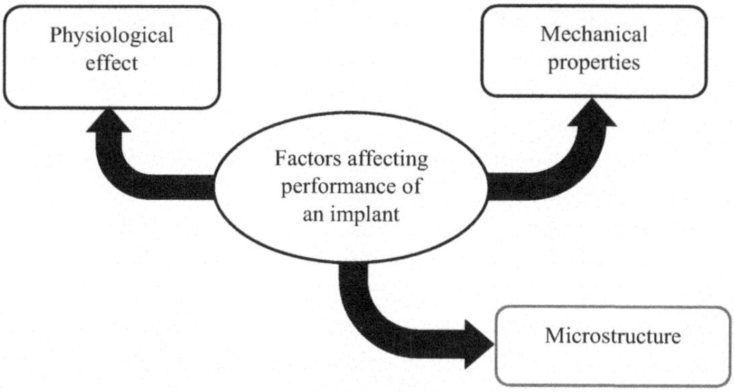

Fig. 1 It's critical to understand that the mechanical characteristics, physiological aspects, and microstructure of a material will impact how well an implant performs when choosing one for a given application

replacement, fracture fixation, and dentistry [17–19]. Orthopaedic surgery has made tremendous progress in recent years with the introduction of new and enhanced techniques for bone repair; however, certain aspects still need to be refined. Long-term failure of alloys can occur for a number of reasons, including low strength, wear, corrosion, and mismatches in Young's modulus between body parts. Consequently, a second surgery is required. These cosmetic procedures are pricy, uncomfortable, and occasionally even unsuccessful. The problems with current implant materials are being addressed by the creation of novel metal alloys. It is critical to create materials that are resilient, resistant to wear and corrosion, biocompatible, bioactive, and able to withstand extended periods of time without failing. Therefore, it is crucial to have accurate understanding of the variables influencing implant material performance [20] (Fig. 1).

The most popular material option for orthopaedic implants is now titanium and alloys based on titanium. With regard to this review, orthopaedic implant titanium and titanium alloys will be considered in the main. Learn about the changes that occur throughout the process, as well as the microstructure, mechanical properties, biocompatibility, bone bonding, resistance to corrosion and wear, and how processing parameters impact these metrics.

2 Titanium and Its Biocompatibility

The transition metal titanium has been demonstrated to be extremely valuable in the medical field, particularly in the area of implant technology. It is shiny and has remarkable properties. Titanium alloys are frequently used in the following dental implant applications:

1. **Orthopaedic Implants**: Materials like titanium are frequently used in knee and hip replacements. Because it is extremely strong and corrosion-resistant, it is particularly helpful in load-bearing implants.
2. **Dental Implants**: The preferred material for dental implants is titanium. These implants not only replace missing teeth, but they also provide a strong base for future teeth, whether they are permanent or removable.
3. **Bone Plates and Screws**: The production of screws and other fixation devices requires the use of titanium. When doing joint and bone fusion surgeries, these are utilized.
4. **Spinal Fusion Devices**: In order to treat a variety of spinal disorders, spinal fusion surgeries employ titanium alloys as spinal fusion devices. Rods, screws, and cages in the spinal column are made of this material.
5. **Heart Valves and Pacemakers**: Because it does not react with other substances, titanium finds use in medical devices such as pacemakers and heart valves. It is a component of the housings of various cardiac implant devices.
6. **Craniofacial Plates**: Titanium screws and plates are used to stabilize the bone structure during reconstructive surgery, especially for the face bones.
7. **Cochlear Implants**: Cochlear implants are made of titanium because it is biocompatible and can withstand the conditions found in the human body. This technology helps people who have lost their hearing.

Because of its exceptional biocompatibility, titanium alloys are preferred in these medical applications. A substance is considered biocompatible if it can be introduced into the human body and does not produce any adverse reactions, either locally or systemically. A series of biological reactions take place in the biological microenvironment of biomaterials once they are incorporated into the human body. These responses depend on a number of variables, including surface characteristics, the specific human body part affected, and the biomaterial's composition. These factors include interactions with various cell and tissue types, proteins, and physiological fluids. First, blood coagulation is triggered by blood proteins adhering quickly to the surface of the biomaterial. Acute inflammation ensues, in which white blood cells aggregate at the location in an effort to break down what they understand to be an alien object. If this problem is not fixed, there may be a persistent inflammatory response in which macrophages surround the biomaterial and possibly form giant cells that resemble a foreign body. As new blood vessels and fibroblasts surround the area, granulation tissue may eventually form. If the biomaterial can't break down naturally, it could cause an adverse reaction where a fibrous tissue capsule grows densely around the orthopaedic implant. It all depends on the biomaterial; some reactions go away and the body integrates the material, while others cause inflammation that lasts for a long time. Keep in mind that the surface of the biomaterial is very important in creating this unfavourable reaction from the body, as it interferes with the initial point of contact, which affects how quickly the wound heals and how well the implant integrates over time. Implants designed to make contact with bone have specific surface topography and chemical compositions that control the nature and degree of these interactions. These interactions can be seen in the formation of calcium phosphate layers, interactions with various cell types, such as osteoblasts and macrophages, and ion and biomolecule adsorption [21]. To ensure optimal performance and long-term viability, the orthopaedic implant must be compatible with the surrounding tissue.

Since titanium has a low ion release, is non-magnetic, mimics bone structure well, has a strength-to-weight ratio, is inert, and promotes osseointegration through protein adsorption, it is considered biocompatible. It also lacks immunogenicity. Due to their superior compatibility, longevity, and ability to work with diagnostic imaging methods such as magnetic resonance imaging scans, titanium alloys are chosen over stainless steel and Cr-Co alloys for implant applications. Titanium is frequently tested for cytotoxicity using particular cell types, such as L929 and MC3T3 E1 cells, in vitro in order to determine its safety and efficacy. These tests consistently show titanium to be a viable and reliable implant material.

Despite all of titanium's benefits, there is still much to learn about its complete biocompatibility. The two most adaptable Ti alloys used in medical practice today are commercially pure titanium and Ti6Al4V. Conversely, CP-Ti exhibits low wear resistance, low strength, and polishing difficulty. Furthermore, there are certain restrictions on the Ti6Al4V alloy. Approximately 110 GPa is its Young modulus, and the Ti6Al4V alloy exhibits stress-shielding potential compared to the 10–30 GPa of human cortical bone. Stress shielding can lead to implant failure since it promotes bone resorption, which occurs when the porous bone loses its adherence to the implant during use. The presence of the hazardous elements aluminium and vanadium makes it a potential long-term risk factor for conditions such as peripheral neuropathy, Alzheimer's disease, osteomalacia, and other neurological disorders [22]. Consequently, novel materials free of such cytotoxic components and possessing a low elastic modulus are urgently required.

The development of novel beta titanium alloys containing metal concentrations that are not harmful including Ta, Nb, and Zr has resulted from concerns over these elements. As an illustration, it has been noted that the concentration of metal release is significantly decreased when Ta is added [23]. Research carried out by Okazaki and associates [24] have exhibited that beta alloys (Ti–15Zr–4Nb–4Ta) exhibit higher cell growth rates than Ti6Al4V. The beta titanium alloy Ti-35Nb-7Zr-6Ta-2Fe-0.5S has a remarkable similarity to the elastic modulus of bone, is very strong, and is biocompatible, making it a potential option for orthopaedic implants [25]. During a 21-day culture of human osteoblasts, this alloy showed promise for load-bearing implants by outperforming the traditional Ti-6Al-4 V alloy in terms of collagen I synthesis and cell population density. Furthermore, in contrast to the Extra Low Interstitial Ti-6Al-4 V alloy, Ti-xTa alloys have demonstrated greater biocompatibility in cell viability studies [22].

There are two primary methods to improve titanium's and its alloys' biocompatibility. First off, the creation of titanium alloys of type β, which show better growth rates and cell viability than traditional Ti–6Al–4 V alloys—has resulted from the reduction of toxic elements. Second, methods for surface modification—which are covered in more detail in Sect. 5—have the potential to enhance osseointegration and antibacterial qualities.

3 Types of Titanium-Based Alloys

The α phase (98.9–99.6% Ti) pure titanium crystal structure emerges as the metal cools to ambient temperature. It is hexagonal in shape. At 882 °C, the hexagonal close-packed structure of the α-phase transforms into a cubic system with a central body [26]. This is another example of an allotropic phase change. This phase transformation temperature is affected significantly by auxiliary components. Both β-stabilizers and β-stabilizers are groups of elements that either raise or lower the transient temperature [18–20]. β-stabilizers include elements like Al, O, N, and C. Transition metals, known as β-stabilizers, impart strength to titanium metal based on their content addition [27]. To ensure that the β phase stays at room temperature, the alloying elements can be adjusted. As a result, several Ti alloy varieties are recognized, such as α, near-α, α + β, and β. Shape-memory alloys based on titanium also have unique characteristics, such as high elasticity.

3.1 α-Type and Near-α Ti Alloys

Titanium alloys that include only the α phase are referred to as "α-type alloys," and these include Ti and CP-Ti alloys. For CP-Ti, there are four standard grades: those with an oxygen content between 0.18 and 0.40 weight per cent and those with a ferrous content between 0.20 and 0.50 weight per cent [28]. The β phase is typically found in fewer than 5 vol.% in near-α or super-α titanium alloys, where the α phase is the most prevalent. An insignificant quantity of β-stabilizers (1–2 weight per cent) in them sets them apart from Ti alloys of the α type [27]. There hasn't been much research done on near-α Ti alloys for use in biomedicine yet [27]. With no loss of load-bearing capacity, CP-Ti shows better corrosion resistance than near-α Ti alloys [29].

Similar qualities that make both α-type and near-α Ti alloys ideal for high-temperature applications include outstanding resistance to corrosion, excellent weldability, and high creep resistance [30]. Nonetheless, heat treatment typically has no effect on the mechanical strength of HPP structures because of their intrinsic stability, which makes them relatively weak at room temperature [31]. CP–Ti, an α-type Ti alloy, has been utilized as the initial generation of Ti biomaterials for implants for almost half a century. Initially, Co-Cr alloys and stainless steel were replaced with CP-Ti to make implants because these materials contained undesirable elements like Ni, Co, and Cr. But, due to its moderate strength, CP-Ti might not be able to completely satisfy the more stringent mechanical needs of certain connective or hard tissues that support loads. The advent of (α + β)-type Ti alloys, which typically offer greater strength than α-type Ti alloys, was a suitable way to get around this restriction.

3.2 $(\alpha + \beta)$-Type Ti Alloys

In [α+β]-Type Ti alloys, the percentage of β-stabilizer is higher (5–30 vol.%) than in near-α Ti alloys. The ageing process (which takes place between 480 and 650°C) and α-β alloys' microstructure are influenced by various aspects of the alloy composition. This results in a finely mixed α and β phase, which are subsequently strengthened through ageing and solution treating [32]. These alloys stand out for having strong room temperature strength, a moderate melting point, and good formability [32]. For the purpose of maximizing their qualities, they can be heated [33], which keeps their elastic modulus [34] at a level that is comparable, but with the correct treatments, their strengths can rise by 30–50%. Through the development of a TiO2/OH film on the bone surfaces, they also exhibit strong corrosion resistance and promote rapid osteointegration [35].

Except for CP-Ti, 50% of all Ti production is made up of the dominant (α+β)-type Ti alloy Ti–6Al–4V (Ti64). Although it was created for the aerospace industry, its strength and ability to withstand corrosion have made it useful in a range of biomedical applications, including orthopaedic implants [21, 27, 36]. With fewer O, N, C, and H impurities, Ti-6Al-4V ELI is an enhanced version of the original material. It is utilized in eleven bone implants and numerous prosthetic hip joints. Of the four alloys that are ASTM-standardized for use in medical devices, the most widely used α–β alloys are Ti-6Al-4V and Ti-6Al-4V ELI (Table 2). Because Ti-6Al-4V contains toxic vanadium, alternatives have been created, such as Ti-6Al-7Nb, which has better wear resistance and is utilized in medical devices like wires, screws, and fasteners as well as prosthetic hips and femur hip stems [37, 38]. These alloys still include aluminium, which raises health risks and may contribute to conditions like Alzheimer's [39]. It no longer contains vanadium. Biomedical implants are not the best application for (α+β)-type Ti alloys, which came into being between the years 1950 and 1990, due to their higher moduli than bone (Table 1). There are potential complications, such as the implant becoming loose and the adjacent bone resorbing. To improve biocompatibility and achieve an appropriate elastic modulus, it has been possible to create β-type Ti alloys with non-toxic β-stabilizers like Mo, Ta, and Zr [40].

3.3 β-Type Ti Alloys

The second wave of titanium biomaterials is composed of titanium alloys due of their greater Young's moduli, which may result in stress shielding in implants. Usually, they are present in these alloys to some degree since their moduli are closer to those of bone.

β alloys possess exceptional forgeability and cold formability due to their ability at room temperature, maintain the β phase completely, through rapid cooling post-solution treatment [34]. Moreover, alloys with β have a high hardness. Fine α-phase

Table 1 Classes of metallic biomaterials and their primary application as implants

Materials	Applications	Primary utilizations	Applications status
Stainless steels	Orthopaedic Orthodontic cardiovascular	Temporary devices (screws, plates and hip nails), total hip replacements	Routinely applied
Co-based alloys	Orthopaedic Orthodontic cardiovascular	Total joints replacements, dental implants, removable partial dentures, orthodontic wire leads, femoral stems, bone implant applications, load-bearing implants, bearing surface implant	Routinely applied
Ti-based alloys	Orthopaedic Orthodontic cardiovascular	Dental implants, orthodontic wire leads, cardiovascular, vascular stents, heart valve parts, stem, total hip replacements	Routinely applied
Miscellaneous NiTi	Orthodontic cardiovascular	Vascular stents, Vena cava filter, Intracranial aneurysm clips, catheter guide wires, orthopaedic staples, orthodontic dental arch wires	FDA approved
Mg	Orthopaedic craniofacial	Biodegradable orthopaedic implants	Animal test
Ta		A radiographic marker, wire structures for neurosurgery and plastic surgery	FDA approved

particles precipitate within the β-matrix following additional ageing at temperatures between 450 and 650 °C. This adds to the material's strength, yielding increased fatigue, ultimate tensile strength, and fracture toughness. Nevertheless, in comparison to α–β alloys, their higher density may lessen their creep strength and tensile ductility [41]. Alloys with β can partially transform to a martensitic state at room temperature when heated to α or other ordered phases, suggesting that they are metastable despite exhibiting minimal to no work hardening [41]. Compared to Ti-6Al-4V α-β alloys, they are more biocompatible and have a lower elastic modulus, and their use for orthopaedic implants increased dramatically in the 1990s [42]. The notable biocompatibility of these alloys without vanadium is mainly due to their alloying with elements such as Nb, Zr, Mo, and Ta.

The enhanced densities observed in β-type Ti alloys are ascribed to the presence of Mo, Ta, Nb, and Zr [43]. Due to their high melting points, scarcity, and cost, these elements complicate the preparation of alloys [44]. Recent research is looking into less expensive alternatives that use elements like Fe, Mn, Sn, and Cr to provide the best qualities possible for future biomedical applications [45]. Despite having lower ultimate strength values than cobalt alloys, alloys with β and (α+β) are similar to cobalt alloys and 316L stainless steel. Their yield strengths also show that, in contrast to cobalt alloys, 316L stainless steels are not as far from the lower bound of the range.

Heat-induced solid-state, diffusion-less martensitic phase transformations provide titanium-based shape-memory alloys with unique functionality. Alternatively, orthopaedic implant biomechanical advantages can be obtained with β-type titanium alloys.

3.4 *Ti-Based Shape-Memory Alloys*

One noteworthy class of smart materials is shape-memory alloys based on titanium that demonstrate the shape-memory effect, or the capacity to reverse deformation and regain original shape. Since there is a phase transition between the austenite and martensite phases at different temperatures, temperature affects this phenomenon. Owing to their exceptional mechanical properties, Ti-based shape-memory alloys such as TiNi [46], TiNiAg [47], TiZr [48], and TiNbSn [49] are highly valuable in biomedical and aerospace applications. Out of every shape-memory alloy based on titanium, TiNi (Nitinol) is the most researched and used because of its unique combination of properties and economic viability.

It has been observed that in artificial saliva, the TiNi alloy is slightly more corrosion-resistant than Co-Cr-based alloys, while maintaining biocompatibility with stainless steel and other titanium implants [50]. Furthermore, a lower number of thrombus formations were noted in TiNi stents than in stainless steel stents [50]. Owing to its remarkable strength, minimal stiffness, and shape-recovery characteristics, it finds application in various orthopaedic settings such as bone implants, tissue engineering, joint replacements, stents, and other settings where osseointegration is desired [51].

A titanium stent is used. However, TiNi alloys' distinct characteristics, such as their transformation hysteresis, phase composition, and general structure, are mostly determined by their chemical makeup. This includes their super elasticity and shape-memory effects. In one instance, TiNi films were produced by co-sputtering at 723K with various Ti:Ni ratios by Fu et al. [52]. At 51.3% Ti, the material underwent a two-stage phase transition from martensite to the R-phase, also known as the rod-like eutectic phase, and had minimal residual stress at room temperature. Nevertheless, a significantly increased residual stress resulted from increased internal stress and limited relaxation during the R-phase shift when this composition was deviated from (47.3% or 53.0%). However, incorporating alloying components with various compositions, such as Zr [53], Nb [54], Cu [55], Fe [56], and Ag [57], has been used in some studies to modify TiNi alloy, which has an impact on its transformation behaviour and mechanical properties. Shape-memory alloy (SMA) based on titanium properties are influenced by element additions, as summarized in Table 3. Citation [58] indicated that the transformation temperature can be considerably raised by the Zr addition. It is possible to slightly increase the transformation hysteresis of the TiNi alloy by adding Cu or Fe. Alternatively, the TiNiNb alloy, which has an additional Nb added to it, exhibits better shape memory and superelastic rebound.

Nickel, which is present in TiNi shape-memory alloys, has the potential to release hazardous ions, which can cause Ni allergy and hypersensitivity. Non-toxic Ti-Zr alloys of the β type have shown great promise in the last decade and can now be used to produce biocompatible shape-memory alloys. High biocompatibility alloying elements including Nb, Mo, and Sn are added to improve these alloys, leading to regulated mechanical characteristics and behaviours of martensitic transformation [59]. For more information on how different alloying elements affect the superelasticity and martensitic transformation behaviours of these o-type Ti alloys, see Endoh et al. [60]. They found that Ti-Zr-Mo-Sn alloys have better superelasticity at room temperature than other alloys, which means that they could be used instead of bone tissue. β-Ti alloys have great biocompatibility and are excellent shape-memory alloys without nickel, which makes them very attractive as biomaterials.

4 Orthopaedic Implants

According to Williams, there are two main purposes for which implants are used in orthopaedic surgery: either to replace or fix broken bones and joints. Currently in widespread use, the artificial hip joint is a legitimate example of a material used successfully in medicine. On the other hand, considering that there are a few but notable number of material-related errors, orthopaedic implant failures remain a cause for concern [61]. Revision surgery is necessary for these kinds of failures, and it is expensive for the health system and painful for the patient.

Numerous elements, including prosthesis design, surgical technique, and material selection, influence the mechanical efficacy of prosthetic joints. Early and ongoing usage of wrought stainless steel and cobalt-chromium alloy has kept these metals at the forefront of automotive technology. Since they've become stronger, titanium alloys Ti-318 and Ti-550 have supplanted commercially pure titanium, i.e. Ti-160, which was initially released in the 1930s [62]. The hip prosthesis may fail to function properly in one of three ways. A fracture is an obvious example. Problems that really matter, though, include implant loosening and articulating surface wear. The most common cause of hip joint prosthesis metallic component fracture is femoral stem fatigue failure.

This issue should be lessened by the exceptional fatigue strength of titanium and titanium alloy. Even though there aren't many reports of titanium fractures, it's still vital to understand why they happen so that remedies can be found. Ti-160 is susceptible to surface imperfections that act as stress concentrations, as demonstrated by Hughes and Jordan [63]. Upon closer inspection of an implant's fracture surface, surface defects were observed, and the fracture appeared more suggestive of an impact breakdown than weariness. The resolution of this issue can be achieved by placing greater emphasis on remembering the surgeons about the final product's surface finish of the importance of exercising extreme caution to prevent scratching the implant surface during their procedures.

Given that the fatigue strength of Ti-160 is much lower than that of Ti-318, these surface imperfections may also be prime locations for the onset of fatigue. In order to address this issue, Ti-318 was introduced, and according to Dobbs' report [62], it seems to perform better than Ti-160. However, he acknowledges that this conclusion is based on a limited amount of clinical data. He did document a single case of a fatigue-related Ti-318 stem prosthesis rupture, but he believed this to be the result of a flaw in the design. It seems as though the issues related to the hip prosthesis' early failure have been fixed. Nonetheless, given the tendency towards implants with porous surface coatings, which significantly impact fatigue strength, keeping a careful watch out for any signs of potential problems is essential.

When using implant materials for joint replacement, wear and friction are also significant facets of their behaviour. For a smooth gliding action to occur, it is crucial that there be as little friction as possible between the articulating surfaces. Furthermore, it's critical to reduce the likelihood of articulating surface damage, which could show up as wear, creep, or surface loss. By causing a local tissue reaction, the discharge of wear particles into the tissues around them may have a negative impact on the implant's overall biocompatibility. When tested at low loads, the coefficient of friction of titanium moving over itself is low, but it quickly rises as the load is increased. For reasons related to titanium's extreme reactivity, disruption of the oxide coating happens at high loads, leading to local welding. Titanium is therefore not a good bearing surface due to its propensity for galling and seizing.

Because titanium is unable to form a layer that the lubricant can physically or chemically absorb, applying lubricants does not solve this issue. An alternate strategy is to apply surface coatings, but this could negatively impact titanium's excellent biocompatibility as well as its fatigue strength and corrosion resistance. Thus, it was initially advised against using titanium for articulating surfaces. The solution to this issue was to employ a coupled prosthesis, in which the bearing surface would be composed of either alumina or cobalt-chromium alloy and the prosthesis's main body would be constructed of titanium alloy (Fig. 2). The bearing surface material that is now utilized most frequently is ultra-high molecular weight high density polyethylene. However, study conducted by Miller et al. demonstrated that stainless steel, titanium, and cobalt-chromium alloy all performed similarly to UHMWPE. This suggests that titanium should not wear in situations where it articulates against a UHMWHD polyethylene bearing surface. It has also encouraged titanium to be used more widely as a direct articulating surface, which has resulted in significantly less complex implant designs.

The recent development of nitrogen is deposited by ions onto the articulating surfaces is a response to ongoing concerns regarding the wear potential of titanium alloy [64]. It is a high energy process where a large number of atoms are moved by collisions from their initial positions to consistent replacement locations within the crystal lattice where the implanted atoms are located. Hard second-phase nitride, with a size of less than 0.1/~m, is created within titanium and its composition due to nitrogen implantation. According to some claims, the rate at which nitrogen implants cause titanium alloy to erode against UHMWPE is dramatically reduced because these tiny nitride particles stop dislocations from moving, increasing the surface layer's microhardness by two times. When compared to thermal diffusion processes, one benefit of the ion bombardment technique is its speed.

Revision surgery has been attributed primarily to hip prosthesis loosening. The femoral stem or the acetabular cup was found to be loose in 50% of cases, according

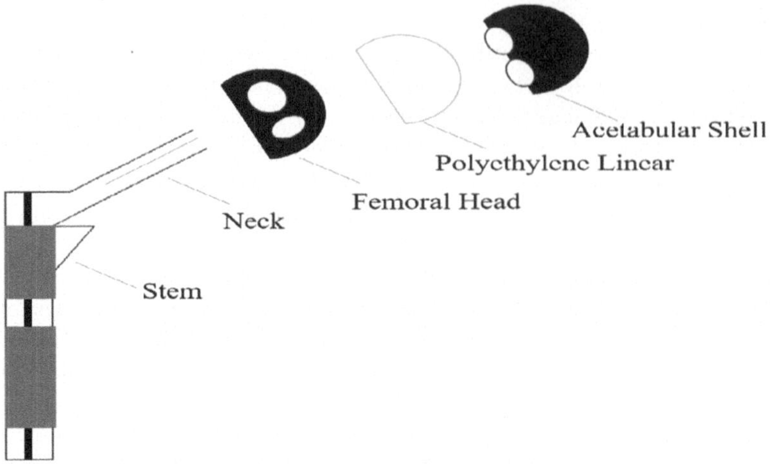

Fig. 2 Complete hip replacement system that uses titanium wire mesh sintered onto the acetabular cup and stem surfaces to allow for bone ingrowth fixation

to a recent study by August et al. [65] of 230 McKeeFarrar hip arthroplasties. Eight % of cases were stem fractures, and 4.8% were infections, and loosening for 78% of the 64 that required revision. Many have also noted the high rate of loosening in long-term patients [66], which is extremely concerning because young adults who want to live active, fulfilling lives are more likely to experience this issue [65]. A great deal of effort has been put into improving the cement technique [67], which has improved the quality of prosthesis fixation. Traditionally, implant fixation is accomplished by using a poly (methyl methacrylate) bone cement. In spite of this, the concept of cementless fixation is currently under extensive investigation due to the remarkable outcomes of bony ingrowth into the porous surfaces of certain materials. Additionally, a considerable proportion of patients—roughly 0.8% of all patients—have a cement allergy, for which cementless fixation would be the most effective treatment option [65]. Due to its exceptional bone biocompatibility, titanium is a clear choice for this strategy. Though not novel, the idea of cementless fixation was first documented in 1943 when Moore and Bohlman successfully fixed a joint prosthesis.

Since cobalt-chromium alloys are easily transformed into porous surface coatings, a great deal of research on coated implants for orthopaedic uses has concentrated on these materials [68]. Still, efforts to create a titanium alloy implant with a porous surface are underway because the metal shares many advantages with alloys made of cobalt and chromium, that are lightweight, have a high fatigue strength, etc. [68, 69]. It is crucial that the femoral component be designed carefully. To compensate for the decreased fatigue strength in the example, the porous mesh is only placed where the stem has a thick cross-section and away from areas of high stresses. These novel prostheses haven't yet gone through a lot of clinical testing. Given that the goal is to improve long-term performance—that is, performance lasting longer than five years— it will take some time to make a firm assessment of their clinical performance when compared to cemented prostheses.

5 Failure of Implants

These painful revision procedures have a relatively low success rate in addition to being highly costly. Between 2005 and 2030, overall hip revision surgery rates are predicted to rise by 137%, while knee revision surgery rates are predicted to rise by 607% [14]. As a result, enormous growth in the production of implants is predicted for the next few years. The need for implants is growing, so it's critical that biomaterials development efforts pick up speed. Orthopaedic implant materials ought to possess superior fatigue and wear resistance, elevated ductility, an optimal blend of high strength and low modulus, and negligible cytotoxicity, particularly in applications that bear weight [70]. In the environment of the body, they ought to be more resistant to corrosion. At the moment, the component materials used for these applications include cobalt-chromium alloys, titanium-based alloys, and 316L stainless steel. Sadly, these materials have demonstrated a propensity to fail after prolonged use due to a number of factors, such as having a modulus that is higher than bone, being less resistant to wear and corrosion, and not being biocompatible. Figure 3 outlines the various reasons that revision surgery might be required. The increased likelihood of revision surgery can also be explained by the longer life expectancy. Orthopaedic implants were only thought to last for roughly 15 years because THR was previously only performed on patients under the age of 65 [71]. The situation has changed, though, as a result of medical technology advancements that have extended life expectancy and improved prognosis for people who have experienced physical trauma as a result of mishaps, sports, inappropriate or excessive exercise habits, or automobile accidents. As a result, it is now anticipated that the implants will last a lifetime without experiencing failure or requiring revision surgery. Thus, it is crucial to develop suitable materials with great longevity and biocompatibility. Titanium alloys are swiftly overtaking other materials as biomaterials, despite the fact that they are presently employed in a variety of settings. An overview of titanium alloys' many benefits that make them perfect for bio applications is provided in this paper. Ten sections make up this article: requirements for biomaterials; current biomedical materials and their status; classification of titanium alloys; how structure and properties relate to each other; modulus and heat treatment; biomedical alloys' wear and corrosion properties and solutions; surface modifications needed for improved osseointegration and high resistance to wear and corrosion; and biocompatibility concerns with titanium alloys and future biomaterials..

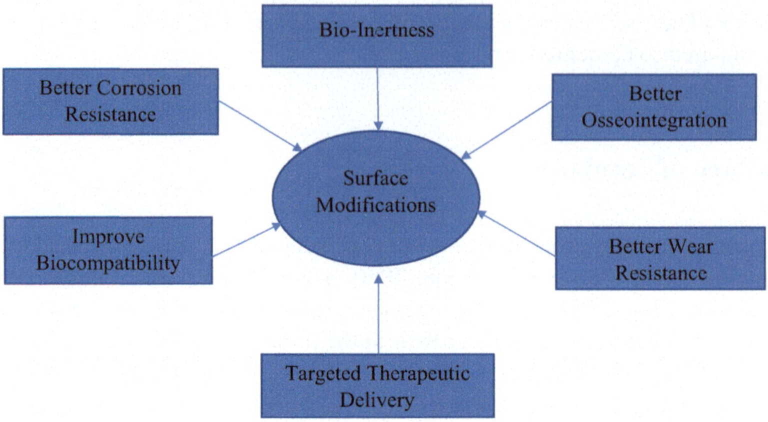

Fig. 3 Titanium (Ti) surface modification benefits, which are essential for improving the biomedical device industry's use of titanium materials by improving their performance and characteristics

6 Requirements of a Biomaterial

The intended medical application influences the design and choice of biomaterials. The interdisciplinary process of creating novel biomaterials frequently necessitates cooperation between pathologists, clinicians, biomedical engineers, material scientists, and engineers. For an implant to function longer without experiencing rejection, it needs to have the following qualities:

6.1 Mechanical Properties

A material's mechanical properties determine what kind of material is best for a particular use. Hardness, modulus, elongation, and tensile strength are a few of the characteristics that are crucial. When cyclic loading is applied to an implant, its fatigue strength determines how the material will react to repeated cyclic loads or strains. This property ensures the implant's long-term viability. "Biomechanical incompatibility" refers to the state in which an implant fails due to either inadequate strength or an imbalance in the mechanical characteristics between the implant and the bone. An equivalent modulus to that of bone is anticipated for the material used in its place. A bone's type and the direction of measurement can affect its bone modulus, which can range in magnitude from 4 to 30 Gpa [72]. Because the materials used to make current implants are more rigid than bone, they do not allow the necessary stress to be transferred to neighbouring bone. This leads to bone resorption around the implant, which in turn causes implant shifting. We refer to this biomechanical incompatibility as the "stress shielding effect" [73] that it results in the death of bone cells. Therefore, it is imperative to use a material with an outstanding combination of low modulus and high strength close to the bone during implantation in order to avoid implant loosening and to reduce the frequency of revision surgery.

6.2 Biocompatibility

Ideally, the materials utilized to create implants would be completely harmless, causing no allergic reactions or inflammatory reactions in humans. The success of biomaterials is heavily dependent on how the human body reacts to implants, which is why biocompatibility is so important [74]. Both the material's degradation in the body environment and the host response it elicits are major factors that influence a material's biocompatibility, cataloguing biological substances in accordance with their physiological responses. While bio-tolerant implants can also be manufactured, bioactive materials are preferred due to their ability to promote high bone integration. Implant materials undergo various reactions with human tissues and fluids, and our system determines whether or not these materials are acceptable. One issue with biocompatibility is thrombosis, which occurs when blood clots and platelets adhere to biomaterial surfaces. Another issue is fibrous tissue encapsulation, which can happen when biomaterials are implanted in soft tissues.

6.3 High Corrosion and Wear Resistance

Implants release non-compatible metal ions into the body due to their poor resistance to wear and corrosion in bodily fluids. New research has shown that the released ions have the potential to cause allergic and toxic responses [75]. The primary factors influencing a material's service life are its resistance to wear and abrasion. Another consequence of poor wear resistance is implant loosening, and research has demonstrated that the tissue where wear debris is deposited undergoes a cascade of reactions [76]. This is why it's so important to make implants that are resistant to wear and corrosion so they last as long as possible in the human body.

6.4 Osseointegration

Implant loosening is caused by micromotions that prevent an implant's surface from integrating with surrounding bone and tissues [77]. A fibrous tissue developed between the implant and the bone indicates improper integration [77]. Surface-appropriate materials are therefore essential for the implant to properly integrate with the surrounding bone. A healthy osseointegration is largely dependent on surface topography, surface chemistry, and surface roughness.

7 Surface Modifications for Enhanced Biocompatibility

Despite being widely used in cardiovascular and hard tissue replacement applications, titanium and its alloys have long-term performance issues [78]. First off, titanium alloys are bio-inert. They also differ greatly in composition from surrounding bones, even with their exceptional biocompatibility. Because of this, the Ti alloy orthopaedic implants undergo mechanical osseointegration with the bone rather than a strong chemical bond, known as fibride capsule formation failure [79]. Second, the sub-surface of Ti alloys is still susceptible because of insufficient resistance to plastic shear, inadequate surface oxide film adherence, and work hardening properties [80]. As a result, the proportionate motion of implants in relation to bones causes adhesive and abrasive wear that unadulterated titanium alloys are unable to sustain, which eventually causes implant failure and loosening [81, 82]. Thirdly, complicated surroundings within the human body can lead to the stable passive layer that Ti alloys naturally form on their surface to peel and dissolve, despite the Ti alloys' excellent corrosion resistance [83, 84]. The interplay of external forces and body fluids causes this process, and it may cause the body to release contaminants or hazardous materials that may be toxic, inflamed, or cause blood clotting [27, 85, 86]. Therefore, a range of techniques are often used to modify the implants' surface properties in order to enhance their resistance to wear, corrosion, and biocompatibility. Some advantages of surface modification of titanium are shown in Fig. 3.

8 Evolution of Biomaterials

Four generations and four categories have been identified in the evolution of biomaterials, based on different therapeutic requirements. Different biomaterial generations are depicted [87, 90].

8.1 First-Generation Biomaterials

Since 1950, efforts to replace or repair host tissue without endangering the biological system have resulted in the selection and development of bio-inert materials. Due to their bio-inert nature, these materials cause fibrous tissues to encapsulate the implanted materials, isolating the implants from the surrounding bone tissues and ultimately leading to implant failure [88]. First-generation biomaterials, which include Co-Cr alloys, 316 L stainless steel, titanium and its alloys, alumina, zirconia, and polyurethane, were important components of implant materials [89].

8.2 Second-Generation Biomaterials

These bioactive, second-generation biomaterials can directly bond with bone tissue and can interact with biological systems to trigger biological responses. Bone formation can be facilitated by the potential chemical reaction that occurs between the bioactive implant and the surrounding tissue. Material that is bio-inert can become bioactive or bioresorbable with a small compositional alteration. The interaction between cells is encouraged by appropriate surface treatments or coatings of bioactive agents on metallic implant material. Biogas, calcium phosphates, and bioactive ceramics are a few types of bioactive materials [34].

8.3 Third-Generation Biomaterials

Biomaterials of the third generation have been designed and developed since 1990. These substances possess the capacity for both biodegradability and bioactivity [90]. Once inserted, these implants have the ability to gradually dissolve and replace damaged bone tissue by growing new bone [91]. These substances are sometimes referred to as biodegradable or bioresorbable implant materials. Examples are alloys based on magnesium and biodegradable polymers like PLA and PLGA [92].

8.4 Fourth-Generation Biomaterials

To give cells the right kind of stimulation, these are reproducing the metabolic environment and molecular architecture of natural tissues, emulating the extracellular matrix [93]. The integration of bodily functions through electronic systems has the potential to yield valuable insights for both basic and clinical medicine, including diagnostic and therapeutic applications.

The fourth-generation biomaterials are designed to achieve the following objectives:

- Tracking cellular reactions and enabling the exchange of bioelectric signals with host tissues.
- Regulating cellular bioelectric signals for tissue regeneration.

According to Ning (2016), these bioelectric signals can give additional details about how a host material reacts to outside stimuli and can also be used to respond appropriately to the host and outside environment in order to control cell behaviour [94].

9 Classification of Biomaterials on the Basis of Materials

9.1 Metallic Biomaterials

The materials most commonly used for load-bearing implants are metals.

Certain orthopaedic replacement procedures are among the most common ones that use metallic implants. A few instances of metallic implants were screws, wires, fracture fixation plates, and entire joint prostheses for the shoulders, knees, ankles, hips, and other joints [95]. In addition, metallic implants are utilized as dental materials in orthopaedics, maxillofacial surgery, and cardiovascular medicine [96]. In biomedical applications, the most commonly used metals and alloys are tantalum-based alloys, cobalt-chromium alloys, titanium and its alloys, and 316 L stainless steel [97]. The classes of metallic biomaterials are shown in Table 1 along with their main use as implant materials [98].

9.2 Ceramic Biomaterials

Ceramic materials have long been used in dentistry as restorative or replacement materials. Dental cements, crowns, and dentures are a few of the materials used in dentistry. A portion of the ceramic scaffolds have been utilized to augment bone and joint replacements in order to restore normal function. The bio ceramics' low fracture toughness, however, severely limits their application in load-bearing situations. The biocompatibility and corrosion resistance of implant materials are enhanced by the use of a variety of ceramic biomaterials, such as hydroxyapatite, zirconia, and alumina [99].

9.3 Polymeric Biomaterials

Repeatedly bonded small units of monomer molecules combine to form larger molecules known as polymers. The Greek words "polys-many" and "meros-units" are the source of the word "polymer," which describes a macromolecule that is created when numerous monomer units are bonded together to form longer chain structures. One chain structure's bonding between distinct monomer molecules determines the polymer's mechanical, chemical, and physical characteristics. The different physical and chemical characteristics of polymers allow them to be categorized as natural, synthetic, inorganic, and organic polymers.

Biomedical materials made of different polymers are used to replace or restore bodily functions. Joint prostheses are made of polyethylene, and bioresorbable screws, plates, and sutures for fracture repair are made of polycaprolactone. Polymers with highly comparable physical properties to soft tissues are utilized in the production of skins, vessel walls, and joint prosthesis cups. Polymeric biomaterials include PLA, polycaprolactone, polyethylene, polyurethane, and natural biopolymers such as collagen, alginates, chitosan, and so on. [100]

9.4 Composites Biomaterials

It is possible to create heterogeneous materials with superior mechanical, chemical, and physical properties by combining two or more components to create composite biomaterials. They are employed in many different fields because of their advantageous qualities. Because they have a lower elastic modulus, carbon-based and carbon-reinforced polymer composite materials could be used for bone repair and replacement. Since they are lightweight, low density, and highly durable, composite materials are used to make prosthetic limbs. For biomedical applications, these materials make excellent scaffolds [101].

10 Conclusion

Overall, the conclusion drawn from the comprehensive discussion on biomaterials underscores their indispensable role in modern medicine. Biomaterials, particularly titanium and its alloys, have revolutionized various medical applications, ranging from orthopaedics to dentistry, cardiovascular devices, and tissue engineering. They are excellent choices for implant materials because of their remarkable mechanical qualities, resistance to corrosion, and biocompatibility, which are necessary for replacing or mending damaged tissues.

However, despite the significant advancements made in biomaterials, challenges persist, particularly concerning material-related failures in orthopaedic implants. Strategies to address these challenges include improving surface finish, enhancing fatigue strength, and exploring innovative techniques like ion implantation of nitrogen. Moreover, the stringent requirements for orthopaedic implants necessitate continuous research and development efforts to optimize alloy compositions, surface modifications, and processing techniques, ultimately improving patient outcomes and reducing the need for revision surgeries.

The evolution of biomaterials through four generations reflects the dynamic nature of medical technology, with each generation offering distinct characteristics and therapeutic applications. From bio-inert materials to bioactive and biodegradable substances, biomaterials have evolved to meet the evolving needs of healthcare, enabling safer and more effective treatments for various medical conditions. Additionally, the classification of biomaterials into metallic, ceramic, polymeric, and composite categories highlights the diverse range of materials available for medical applications, each with its unique properties and advantages.

In conclusion, the continuous advancement of biomaterials holds great promise for addressing healthcare challenges and improving patient outcomes. Collaboration between researchers, engineers, clinicians, and other stakeholders is essential in driving innovation and translating research findings into practical solutions that benefit patients worldwide. By harnessing the potential of biomaterials technology, the field of medicine can continue to advance, offering safer, more durable, and more effective treatments for the benefit of humanity.

Conflict of Interest The authors declare no conflict of interest.

References

1. Raghavendra, G. M., Varaprasad, K., & Jayaramudu, T. (2015). Biomaterials: design, development and biomedical applications. In *Nanotechnology applications for tissue engineering* (pp. 21–44). William Andrew Publishing.
2. Narayan, R. J. (2010). The next generation of biomaterial development. *Philosophical Transactions of the Royal Society A: Mathematical, Physical and Engineering Sciences, 368*(1917), 1831–1837.
3. Williams, D. F. (2008). On the mechanisms of biocompatibility. *Biomaterials, 29*(20), 2941–2953.
4. Smith, A. J., Dieppe, P., Vernon, K., Porter, M., & Blom, A. W. (2012). Failure rates of stemmed metal-on-metal hip replacements: Analysis of data from the National Joint Registry of England and Wales. *The Lancet, 379*(9822), 1199–1204.
5. Bradberry, S. M., Wilkinson, J. M., & Ferner, R. E. (2014). Systemic toxicity related to metal hip prostheses. *Clinical Toxicology, 52*(8), 837–847.
6. Li, X., Tao, L., Chen, Z., Fang, H., Li, X., Wang, X., ... & Zhu, H. (2017). Graphene and related two-dimensional materials: Structure-property relationships for electronics and optoelectronics. *Applied Physics Reviews, 4*(2).
7. Jaganathan, S. K., Supriyanto, E., Murugesan, S., Balaji, A., & Asokan, M. K. (2014). Biomaterials in cardiovascular research: Applications and clinical implications. *BioMed Research International.*
8. Chatterjee, D. K., Diagaradjane, P., & Krishnan, S. (2011). Nanoparticle-mediated hyperthermia in cancer therapy. *Therapeutic Delivery, 2*(8), 1001–1014.
9. Hu, R., Ma, S., Li, H. U., Ke, X., Wang, G., Wei, D., & Wang, W. (2011). Effect of magnetic fluid hyperthermia on lung cancer nodules in a murine model. *Oncology Letters, 2*(6), 1161–1164.
10. Eliaz, N., & Metoki, N. (2017). Calcium phosphate bioceramics: A review of their history, structure, properties, coating technologies and biomedical applications. *Materials, 10*(4), 334.
11. Prasad, K., Bazaka, O., Chua, M., Rochford, M., Fedrick, L., Spoor, J., ... & Bazaka, K. (2017). Metallic biomaterials: current challenges and opportunities. *Materials, 10*(8), 884.
12. Cabal, B., Alou, L., Cafini, F., Couceiro, R., Sevillano, D., Esteban-Tejeda, L., … & Moya, J. S. (2014). A new biocompatible and antibacterial phosphate free glass-ceramic for medical applications. *Scientific Reports, 4*(1), 5440.
13. Kaur, G., Pandey, O. P., Singh, K., Homa, D., Scott, B., & Pickrell, G. (2014). A review of bioactive glasses: Their structure, properties, fabrication and apatite formation. *Journal of Biomedical Materials Research Part A: An Official Journal of The Society for Biomaterials, The Japanese Society for Biomaterials, and The Australian Society for Biomaterials and the Korean Society for Biomaterials, 102*(1), 254–274.

14. Kurtz, S., Ong, K., Lau, E., Mowat, F., & Halpern, M. (2007). Projections of primary and revision hip and knee arthroplasty in the United States from 2005 to 2030. *JBJS, 89*(4), 780–785.
15. Fayaz, H. C., Jupiter, J. B., Pape, H. C., Smith, R. M., Giannoudis, P. V., Moran, C. G., ... & Parvizi, J. (2011). Challenges and barriers to improving care of the musculoskeletal patient of the future-a debate article and global perspective. *Patient Safety in Surgery, 5*, 1–10.
16. Szczęsny, G., Kopec, M., Politis, D. J., Kowalewski, Z. L., Łazarski, A., & Szolc, T. (2022). A review on biomaterials for orthopaedic surgery and traumatology: From past to present. *Materials, 15*(10), 3622.
17. Aherwar, A., Singh, A. K., & Patnaik, A. (2016). Cobalt Based Alloy: A Better Choice Biomaterial for Hip Implants. Trends in Biomaterials & Artificial Organs, 30(1).
18. Uhthoff, H. K., Bardos, D. I., & Liskova-Kiar, M. A. R. I. A. (1981). The advantages of titanium alloy over stainless steel plates for the internal fixation of fractures. An experimental study in dogs. *The Journal of Bone & Joint Surgery British, 63*(3), 427–484.
19. Cordeiro, J. M., Beline, T., Ribeiro, A. L. R., Rangel, E. C., da Cruz, N. C., Landers, R., ... & Barão, V. A. (2017). Development of binary and ternary titanium alloys for dental implants. *Dental Materials, 33*(11), 1244–1257.
20. Elias, C. N. (2011). *Factors affecting the success of dental implants. Implant dentistry: a rapidly evolving practice* (pp. 319–364). Rijeka: InTech.
21. Hanawa, T. (2022). Biocompatibility of titanium from the viewpoint of its surface. *Science and Technology of Advanced Materials, 23*(1), 457–472.
22. Raganya, M. L., Moshokoa, N. M., Obadele, B., Olubambi, P. A., & Machaka, R. (2019). The microstructural and mechanical characterization of the β-type Ti-11.1 Mo-10.8 Nb alloy for biomedical applications. *IOP Conference Series: Materials Science and Engineering, 655*(1), 012025). IOP Publishing.
23. Zhou, Y. L., Niinomi, M., Akahori, T., Fukui, H., & Toda, H. (2005). Corrosion resistance and biocompatibility of Ti–Ta alloys for biomedical applications. *Materials Science and Engineering: A, 398*(1–2), 28–36.
24. Okazaki, Y. (2001). A new Ti–15Zr–4Nb–4Ta alloy for medical applications. *Current Opinion in Solid State and Materials Science, 5*(1), 45–53.
25. Kopova, I., Stráský, J., Harcuba, P., Landa, M., Janeček, M., & Bačákova, L. (2016). Newly developed Ti–Nb–Zr–Ta–Si–Fe biomedical beta titanium alloys with increased strength and enhanced biocompatibility. *Materials Science and Engineering: C, 60*, 230–238.
26. Jeong, H. I., Lee, C. M., & Kim, D. H. (2022). Manufacturing of Ti-Nb-Cr-V-Ni-Al refractory high-entropy alloys using direct energy deposition. *Materials, 15*(19), 6570.
27. Zhang, L. C., & Chen, L. Y. (2019). A review on biomedical titanium alloys: Recent progress and prospect. *Advanced Engineering Materials, 21*(4), 1801215.
28. Bhat, B. N., Pandey, A. B., Tamirisakandala, S., Nathal, M. V., Ellis, D. L., McGill, P. B., ... & Zhu, D. (2018). *Aerospace materials characteristics (no. M17–6171)*. American Institute of Aeronautics and Astronautics.
29. Zhang, Y., Sun, D., Cheng, J., Tsoi, J. K. H., & Chen, J. (2020). Mechanical and biological properties of Ti–(0–25 wt%) Nb alloys for biomedical implants application. *Regenerative Biomaterials, 7*(1), 119–127.
30. Moshokoa, N. A. (2020). *The design and development of a novel Beta Ti alloys with low elastic modulus for biomedical application*. University of Johannesburg (South Africa).
31. Grover, T., Pandey, A., Kumari, S. T., Awasthi, A., Singh, B., Dixit, P., ... & Saxena, K. K. (2020). Role of titanium in bio implants and additive manufacturing: An overview. *Materials Today: Proceedings, 26*, 3071–3080.
32. McAndrew, A. R., Colegrove, P. A., Bühr, C., Flipo, B. C., & Vairis, A. (2018). A literature review of Ti-6Al-4V linear friction welding. *Progress in Materials Science, 92*, 225–257.
33. Santhosh, R., Geetha, M., & Nageswara Rao, M. (2017). Recent developments in heat treatment of beta titanium alloys for aerospace applications. *Transactions of the Indian Institute of Metals, 70*, 1681–1688.

34. Chen, Q., & Thouas, G. A. (2015). Metallic implant biomaterials. *Materials Science and Engineering: R: Reports, 87*, 1–57.
35. Sidambe, A. T. (2014). Biocompatibility of advanced manufactured titanium implants—A review. *Materials, 7*(12), 8168–8188.
36. Dai, N., Zhang, L. C., Zhang, J., Zhang, X., Ni, Q., Chen, Y., ... & Yang, C. (2016). Distinction in corrosion resistance of selective laser melted Ti-6Al-4V alloy on different planes. *Corrosion Science, 111*, 703–710.
37. Iijima, D., Yoneyama, T., Doi, H., Hamanaka, H., & Kurosaki, N. (2003). Wear properties of Ti and Ti–6Al–7Nb castings for dental prostheses. *Biomaterials, 24*(8), 1519–1524.
38. Tamilselvi, S., Raman, V., & Rajendran, N. (2006). Corrosion behaviour of Ti–6Al–7Nb and Ti–6Al–4V ELI alloys in the simulated body fluid solution by electrochemical impedance spectroscopy. *Electrochimica Acta, 52*(3), 839–846.
39. Li, Y., Yang, C., Zhao, H., Qu, S., Li, X., & Li, Y. (2014). New developments of Ti-based alloys for biomedical applications. *Materials, 7*(3), 1709–1800.
40. Niinomi, M. (2003). Recent research and development in titanium alloys for biomedical applications and healthcare goods. *Science and Technology of Advanced Materials, 4*(5), 445.
41. Niinomi, M. (2002). Recent metallic materials for biomedical applications. *Metallurgical and Materials Transactions A, 33*, 477–486.
42. Niinomi, M. (1999). Recent titanium R&D for biomedical applications in Japan. *JOM Journal of the Minerals Metals and Materials Society, 51*, 32–34.
43. Yang, Y., Li, G. P., Wang, H., Wu, S. Q., Zhang, L. C., Li, Y. L., & Yang, K. (2012). Formation of zigzag-shaped $\{1\,1\,2\}<1\,1\,1>$ β mechanical twins in Ti–24.5 Nb–0.7 Ta–2 Zr–1.4 O alloy. *Scripta Materialia, 66*(5), 211–214.
44. Narita, K., Niinomi, M., Nakai, M., Hieda, J., & Oribe, K. (2012). Development of thermo-mechanical processing for fabricating highly durable β-type Ti–Nb–Ta–Zr rod for use in spinal fixation devices. *Journal of the Mechanical Behavior of Biomedical Materials, 9*, 207–216.
45. Niinomi, M., Nakai, M., Hieda, J., Cho, K., Akahori, T., Hattori, T., & Ikeda, M. (2013). Research and development of low-cost titanium alloys for biomedical applications. *Key Engineering Materials, 551*, 133–139.
46. Fashu, S., Lototskyy, M., Davids, M. W., Pickering, L., Linkov, V., Tai, S., ... & Tarasov, B. P. (2020). A review on crucibles for induction melting of titanium alloys. *Materials & Design, 186*, 108295.
47. Yıldırım, A., Abboosh, O., Soyer, P., Kılıç, G., Savur, A., Dağdelen, F., ... & Tunalı, Y. (2022). Investigation of thermal and antimicrobial properties of NiTiX (X= Ta, Ag, and Nb) shape memory alloys (Master's thesis). Kırşehir Ahi Evran Üniversitesi-Fen Bilimleri Enstitüsü).
48. Li, S., Choi, M. S., & Nam, T. H. (2021). Phase stability of the amorphous phase and non-equilibrium phase in a β Ti-Zr-based shape memory alloy. *Scripta Materialia, 195*, 113721.
49. Li, S., Lee, W. T., Yeom, J. T., Kim, J. G., Oh, J. S., Lee, T., ... & Nam, T. H. (2022). Towards bone-like elastic modulus in TiNbSn alloys with large recovery strain for biomedical applications. *Journal of Alloys and Compounds, 925*, 166724.
50. Thierry, B., Merhi, Y., Bilodeau, L., Trepanier, C., & Tabrizian, M. (2002). Nitinol versus stainless steel stents: Acute thrombogenicity study in an ex vivo porcine model. *Biomaterials, 23*(14), 2997–3005.
51. Sharma, N., Jangra, K. K., & Raj, T. (2018). Fabrication of NiTi alloy: A review. *Proceedings of the Institution of Mechanical Engineers, Part L: Journal of Materials: Design and Applications, 232*(3), 250–269.
52. Fu, Y., & Du, H. (2003). Effects of film composition and annealing on residual stress evolution for shape memory TiNi film. *Materials Science and Engineering: A, 342*(1–2), 236–244.
53. Zhou, J., Wu, Y., Wang, H., Liu, X., Jiang, S., Wang, X., & Lu, Z. (2019). Work-hardenable Zr-based bulk metallic glass composites reinforced with ex-situ TiNi fibers. *Journal of Alloys and Compounds, 806*, 1497–1508.
54. Yan, W., Wu, M., Huang, X., Zhang, L., & Qu, X. (2023). Investigations of the microstructure and mechanical properties of the Nb-Ti/Nb-Ti-Ni brazed joints. *Materials Today Communications, 35*, 106190.

55. Kaftaranova, M., Hodorenko, V., Anikeev, S., Artyukhova, N., Shabalina, A. V., & Gunther, V. (2022). Investigation of the effect of copper addition on physical and mechanical properties of TiNi-Cu porous alloy. *Metals, 12*, 1696.
56. Li, P., Jia, Y., Wang, Y., Li, Q., Meng, F., & He, Z. (2019). Effect of Fe addition on microstructure and mechanical properties of as-cast Ti49Ni51 alloy. *Materials, 12*(19), 3114.
57. Baigonakova, G., Marchenko, E., Chekalkin, T., Kang, J. H., Weiss, S., & Obrosov, A. (2020). Influence of silver addition on structure, martensite transformations and mechanical properties of TiNi–Ag alloy wires for biomedical application. *Materials, 13*(21), 4721.
58. Hong, S. H., Kim, J. T., Park, H. J., Kim, Y. S., Suh, J. Y., Na, Y. S., ... & Kim, K. B. (2017). Influence of Zr content on phase formation, transition and mechanical behavior of Ni-Ti-Hf-Zr high temperature shape memory alloys. *Journal of Alloys and Compounds, 692*, 77–85.
59. Biesiekierski, A., Wang, J., Gepreel, M. A. H., & Wen, C. (2012). A new look at biomedical Ti-based shape memory alloys. *Acta Biomaterialia, 8*(5), 1661–1669.
60. Endoh, K., Tahara, M., Inamura, T., & Hosoda, H. (2017). Effect of Sn and Zr content on superelastic properties of Ti-Mo-Sn-Zr biomedical alloys. *Materials Science and Engineering: A, 704*, 72–76.
61. Dobbs, H. S., & Robertson, J. L. M. (1982). Alloys for orthopaedic implant use. *Engineering in Medicine, 11*(4), 175–182.
62. Dobbs, H. S. (1982). Fracture of titanium orthopaedic implants. *Journal of Materials Science, 17*, 2398–2404.
63. Hughes, A. N., & Jordan, B. A. (1972). Metallurgical observations on some metallic surgical implants which failed in vivo. *Journal of Biomedical Materials Research, 6*(2), 33–48.
64. Van Noort, R. (1987). Titanium: The implant material of today. *Journal of Materials Science, 22*, 3801–3811.
65. August, A. C., Aldam, C. H., & Pynsent, P. B. (1986). The McKee-Farrar hip arthroplasty. A long-term study. *The Journal of Bone & Joint Surgery British, 68*(4), 520–527.
66. Wang, C. T., & Pilliar, R. M. (1989). Fracture toughness of acrylic bone cements. *Journal of Materials Science, 24*, 3725–3738.
67. Reigstad, A., & Hetland, K. R. (1984). Rearthroplasty after conventional total hip prosthesis and double-cup prosthesis: A comparative study. *Archives of Orthopaedic and Traumatic Surgery, 103*, 152–155.
68. Kilner, T., Dempsey, A. J., Pilliar, R. M., & Weatherly, G. C. (1987). The effects of nitrogen additions to a cobalt-chromium surgical implant alloy: Part 1 processing and microstructure. *Journal of Materials Science, 22*, 565–574.
69. Ronningen, H., Lereim, P., Galante, J., Rostoker, W., Turner, T., & Urban, R. (1983). Total surface hip arthroplasty in dogs using a fiber metal composite as a fixation method. *Journal of Biomedical Materials Research, 17*(4), 643–653.
70. Long, M., & Rack, H. J. (1998). Titanium alloys in total joint replacement—A materials science perspective. *Biomaterials, 19*(18), 1621–1639.
71. AlYafei, H. F. S. (2020). Tribo-mechanical, biocompatibility, and corrosion properties investigation of zirconium and SST 304 by application of hydroxyapatite/reduced graphene oxide/palladium-platinum nanocomposite coatings [Doctoral dissertation].
72. Katz, J. L. (1980). Anisotropy of Young's modulus of bone. *Nature, 283*(5742), 106–107.
73. Sumner, D. R., Turner, T. M., Igloria, R., Urban, R. M., & Galante, J. O. (1998). Functional adaptation and ingrowth of bone vary as a function of hip implant stiffness. *Journal of Biomechanics, 31*(10), 909–917.
74. Williams, D. F., Bezuidenhout, D., De Villiers, J., Human, P., & Zilla, P. (2021). Long-term stability and biocompatibility of pericardial bioprosthetic heart valves. *Frontiers in Cardiovascular Medicine, 8*, 728577.
75. Hallab, N. J., Anderson, S., Stafford, T., Glant, T., & Jacobs, J. J. (2005). Lymphocyte responses in patients with total hip arthroplasty. *Journal of Orthopaedic Research, 23*(2), 384–391.
76. Sargeant, A., & Goswami, T. (2006). Hip implants: Paper V. *Physiological Effects. Materials & Design, 27*(4), 287–307.

77. Viceconti, M., Muccini, R., Bernakiewicz, M., Baleani, M., & Cristofolini, L. (2000). Large-sliding contact elements accurately predict levels of bone–implant micromotion relevant to osseointegration. *Journal of Biomechanics, 33*(12), 1611–1618.
78. Fukuda, A., Takemoto, M., Saito, T., Fujibayashi, S., Neo, M., Yamaguchi, S., ... & Nakamura, T. (2011). Bone bonding bioactivity of Ti metal and Ti–Zr–Nb–Ta alloys with Ca ions incorporated on their surfaces by simple chemical and heat treatments. *Acta Biomaterialia, 7*(3), 1379–1386.
79. Liu, X., Chu, P. K., & Ding, C. (2004). Surface modification of titanium, titanium alloys, and related materials for biomedical applications. *Materials Science and Engineering: R: Reports, 47*(3–4), 49–121.
80. Oladokun, A. O. (2017). Mechanism of fretting corrosion at the modular taper interface of hip prosthesis [Doctoral dissertation]. University of Leeds.
81. Yang, X., & Hutchinson, C. R. (2016). Corrosion-wear of β-Ti alloy TMZF (Ti-12Mo-6Zr-2Fe) in simulated body fluid. *Acta biomaterialia, 42*, 429–439.
82. Bahraminasab, M., Sahari, B. B., Edwards, K. L., Farahmand, F., & Arumugam, M. (2013). Aseptic loosening of femoral components–materials engineering and design considerations. *Materials & Design, 44*, 155–163.
83. Zhang, Y., Li, J., Che, S., Yang, Z., & Tian, Y. (2019). Chemical leveling mechanism and oxide film properties of additively manufactured Ti–6Al–4V alloy. *Journal of Materials Science, 54*, 13753–13766.
84. Hua, N., Wang, W., Wang, Q., Ye, Y., Lin, S., Zhang, L., ... & Liaw, P. K. (2021). Mechanical, corrosion, and wear properties of biomedical Ti–Zr–Nb–Ta–Mo high entropy alloys. *Journal of Alloys and Compounds, 861*, 157997.
85. Richter, M., & Matusiewicz, H. (2021). Review of the local tissue reaction to metallic spinal implant debris: Ions and nanoparticles. *World Journal of Advanced Research and Reviews, 9*(3), 167–187.
86. Gobbi, S. J., Reinke, G., Gobbi, V. J., Rocha, Y., Sousa, T. P., & Coutinho, M. M. (2020). Biomaterial: Concepts and basics properties. *European International Journal of Science and Technology, 9*(2), 23–42.
87. Hernández-Montes, V., Betancur-Henao, C. P., & Santa-Marín, J. F. (2017). Titanium dioxide coatings on magnesium alloys for biomaterials: A review. *Dyna, 84*(200), 261–270.
88. Patel, N., Best, S. M., Bonfield, W., Gibson, I. R., Hing, K. A., Damien, E., & Revell, P. A. (2002). A comparative study on the in vivo behavior of hydroxyapatite and silicon substituted hydroxyapatite granules. *Journal of Materials Science: Materials in Medicine, 13*, 1199–1206.
89. Hench, L. L. (1980). Biomaterials. *Science, 208*, 826–831. https://doi.org/10.1126/science.6246576
90. Hench, L. L., & Polak, J. M. (2002). Third-generation biomedical materials. *Science, 295*(5557), 1014–1017.
91. Spaans, C. J., Belgraver, V. W., Rienstra, O., De Groot, J. H., Veth, R. P. H., & Pennings, A. J. (2000). Solvent-free fabrication of micro-porous polyurethane amide and polyurethane-urea scaffolds for repair and replacement of the knee-joint meniscus. *Biomaterials, 21*(23), 2453–2460.
92. Prakasam, M., Locs, J., Salma-Ancane, K., Loca, D., Largeteau, A., & Berzina-Cimdina, L. (2017). Biodegradable materials and metallic implants—A review. *Journal of Functional Biomaterials, 8*(4), 44.
93. Sachot, N., Mateos-Timoneda, M. A., Planell, J. A., Velders, A. H., Lewandowska, M., Engel, E., & Castano, O. (2015). Towards 4th generation biomaterials: A covalent hybrid polymer–ormoglass architecture. *Nanoscale, 7*(37), 15349–15361.
94. Ning, C., Zhou, L., & Tan, G. (2016). Fourth-generation biomedical materials. *Materials Today, 19*(1), 2–3.
95. Pilliar, R. M. (2021). Metallic biomaterials. *Biomedical Materials* 1–47.
96. Pandey, A., Awasthi, A., & Saxena, K. K. (2020). Metallic implants with properties and latest production techniques: A review. *Advances in Materials and Processing Technologies, 6*(2), 405–440.

97. Morsiya, C. (2022). A review on parameters affecting properties of biomaterial SS 316L. *Australian Journal of Mechanical Engineering, 20*(3), 803–813.
98. Asri, R. I. M., Harun, W. S. W., Samykano, M., Lah, N. A. C., Ghani, S. A. C., Tarlochan, F., & Raza, M. R. (2017). Corrosion and surface modification on biocompatible metals: A review. *Materials Science and Engineering: C, 77*, 1261–1274.
99. Paqué, P. N., & Özcan, M. (2024). A review on biocompatibility of dental restorative and reconstruction materials. *Current Oral Health Reports* 1–10.
100. Griffith, L. G. (2000). Polymeric biomaterials. *Acta Materialia, 48*(1), 263–277.
101. Lakes, R. S. (2007). Composite biomaterials. In *Biomaterials* (pp. 4–1). CRC Press.

Biomechanics and Material Properties of Cancer Cells with Implications for Therapeutic Approach

Priya and Dhruv Kumar

Abstract As compared to cancer cells to their normal counterparts explain distinctive biomechanical and material properties that are essential to interpreting the behaviour, course, and responsiveness of cancer cells to medication. Cytoskeleton and extracellular matrix modification impact the distinct mechanical properties of cancerous cells, with enhanced deformability, changed viscoelasticity, and changing adhesion property. Changes affect the total hallmarks of cancer. Furthermore, cancer cells' mechanical characteristics can be useful for therapeutic and diagnostic purpose. In order to know the mechanobiology of cancerous cells, many techniques help to understand their biomechanical features. These techniques are optical tweezers, microfluidics, and atomic force microscopy (AFM). The innovation of diagnostic techniques which differentiate cancerous cells from normal cells by using mechanical process can be facilitated by knowing these characteristics. Furthermore, novel therapeutic approaches can be attainable if therapies work on the biomechanical characteristics of cancerous cells, like those that affect adhesion and stiffness. Here, possible uses in cancer detection and treatment, biomechanics and material characteristics play a crucial part in the biology of cancer.

Keywords Biomechanics · Biomaterial · Cancer · Therapeutic approach

1 Introduction

For several decades, researchers have been examining the relationships between human diseases and biomechanics. By applying conventional "solid" and "fluid" mechanics concepts from physics and engineering to the study of biological and physiological problems, the mechanics of organ function, the mechanical response of tissues, joints, and articulating surfaces, and the physical and mechanical processes

Priya · D. Kumar (✉)
Department of Allied Sciences, School of Health Sciences and Technology, UPES, Dehradun, Uttarakhand, India
e-mail: dhruv.kumar@ddn.upes.ac.in

K. K. Dwivedi et al. (eds.), *Mechanics and Materials Science of Biological Materials*, Materials Horizons: From Nature to Nanomaterials,
https://doi.org/10.1007/978-981-96-5366-9_7

related to blood flow through the vasculature have all enhanced significantly [1]. The study of cellular and molecular processes that contribute to the onset and spread of cancer, a class of diseases marked by unchecked cell proliferation, tissue invasion, and occasionally metastasis (spread to other parts of the body), is known as cancer biology [2]. Genetic mutations and epigenetic changes that interfere with normal cellular processes can cause cancer by causing aberrant cell proliferation, evading apoptosis (programmed cell death), sustained angiogenesis (formation of new blood vessels), and other characteristics that promote tumour growth and survival [3]. Genetic Mutations and Oncogenes are also considerable Mutations in certain genes, known as oncogenes, can drive the uncontrolled growth of cells [4]. Oncogenes are often mutated forms of normal genes (proto-oncogenes) that promote cell growth and division. Mutations that activate oncogenes can lead to continuous cell proliferation [5]. Tumour Suppressor Genes: These are genes that normally inhibit cell growth and promote apoptosis. When tumour suppressor genes, such as TP53 (which codes for the p53 protein) or RB1, are inactivated by mutations, cells lose regulatory control overgrowth and division, contributing to cancer development [5]. Cell cycle dysregulation tightly regulated by various checkpoints. Dysregulation of the cell cycle, often due to mutations in genes that control these checkpoints, can lead to uncontrolled cell proliferation, a hallmark of cancer [5]. In apoptosis evasion cancer cells often acquire the ability to evade apoptosis, allowing them to survive and proliferate despite signals that would normally trigger cell death in damaged or abnormal cells [6]. Angiogenesis sustains their growth, tumours require a blood supply that provides oxygen and nutrients. Cancer cells can secrete factors like VEGF (vascular endothelial growth factor) that promote angiogenesis, the growth of new blood vessels into the tumour [7]. The process by which cancer cells migrate to distant organs from the main tumour to create secondary tumours is known as metastasis. The steps involved in this process are following detachment from the initial site, invasion of neighbouring tissues, invasion of the lymphatic system, and colonisation of other areas [2]. Response to therapeutic approach significantly influenced by this microenvironment of cancer cells [8]. Malignant tissue usually displays high risk of mutations, chromosomal abbreviation, and DNA damage because of complex genomic instability, which helps to enhance to the genetic diversity in tumour and its resistance to treatment [9]. Without affecting the sequence of DNA, genetic changes like methylation of DNA and histone modification can modify in gene expression pattern. Tumour suppressor gene can be silence or activation of oncogene may result from these different modifications [10]. They could generate an immunosuppressive or downregulate antigen presentation or express immunological checkpoint proteins like PD-L1 [11]. In the area of research, role the biomechanics of malignant cells is becoming more crucial. The mechanical characteristics and behaviours of cancer cells are essential for their ability to divide, infiltrate neighbouring tissues, and travel to far-off locations inside the body [2]. Biomechanics is a multidisciplinary approach that combines biology, physics, and engineering, offers a distinct viewpoint on cancer.

1.1 Mechanotransduction of Cancer Cell Invasion and Metastasis

Mechanotransduction pathways helps to maintain cellular reactions in to mechanical stress and strain in normal cells, helps to maintain tissue homeostasis [12]. As an instance, high ECM stiffness can start oncogenic processes in cells by activating different pathways like focal adhesion kinase (FAK) and Hippo-YAP/TAZ. Cancer cells move through the extracellular matrix (ECM) because of their frequently changed mechanical feature, such as decreased stiffness and greater deformability [13]. For invasion process, cancer cells' interaction of mechanical property with the extracellular matrix (ECM) is necessary, particularly their approach to attach is essential [14]. Cancer cells detect mechanical signals from the outer extracellular matrix (ECM) to modify the ECM to develop pathways for invasion. Cells have potential to detect and provide response to mechanical stimuli from surroundings by a process called mechanotransduction [15]. Angiogenesis and metastasis can be aided by the intricate mechanical stimulus so cancer cells and the stromal cells that surround them through such as fibroblasts and immune cells exchange [16]. Within the cancerous cell microenvironment, modifications in tissue stiffness and mechanical stress can trigger signalling pathways that support the survival, invasion, and multiplication of cancer cells.

1.2 Cancer Cell Plasticity and Adaptation

Cancer cells' mechanical features are dynamic and can alter in response to many different external stimuli or medical interventions. Cancer cells are able to endure a variety of circumstances due to their flexibility, or plasticity, such as mechanical stress, hypoxia, exposure to radiation or chemotherapy, and more [17]. Knowing how cancer cells adapt biomechanically may help us understand how tumours become resistant to treatments and help us find new therapeutic targets that can be used to avoid or overcome resistance. Cancer diagnosis and prognosis can be made with the use of biophysical markers: Biophysical markers that can be used to differentiate between malignant and normal cells include the mechanical characteristics of cells, such as adhesion, viscoelasticity, and stiffness [18]. Those features can be measured, and the results can be used for diagnostic and prognostic purposes, using methods like optical tweezers and atomic force microscopy (AFM). To help stratify patients and customise treatment plans, alterations in the biomechanics of cancer cells and their microenvironment can also reveal information about the aggressiveness of the tumour and the chance of metastasis [19]. New method for treatment of cancer is to focus on the mechanical characteristics of cancer cells. Among the possible therapeutic target are modifying the stiffness of the tumour microenvironment to prevent cancer from progression and interfering with the pathways which involves mechanical process that allow cancerous cells to invade and growth [20]. The aim of cell biomechanics is to find a mechanical characteristic which interacts with cells with their surroundings [21]. As compared to normal cells, the biomechanical characteristics of cancerous cells are significantly altered. These alterations are required in malignancy and primary biomechanical distinctions among healthy and malignant cells can be divided into several categories, including cell stiffness, deformability, motility and adhesion.

2 Biomechanic Properties of Cancer Cells and Analysis Approach

For a cell to stay intact, to communicate with the external environment, and to control the flow of materials into and out of the cell, its membranes are essential [22]. These changes helps in growth, invasion, and metastasis cells. The main changes in the structure and cancer cell membranes are as follows (Fig. 1).

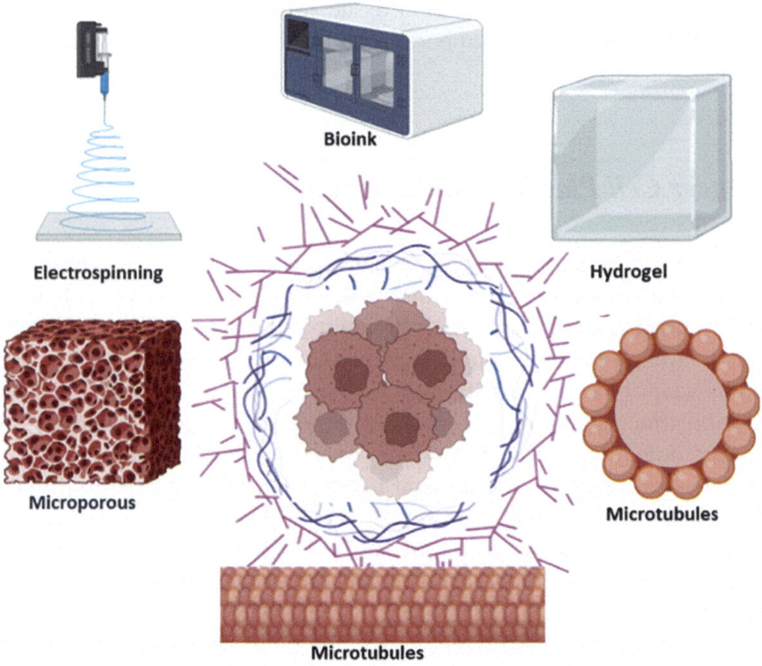

Fig. 1 Different component of cancerous Biomaterial treatment approach

2.1 Lipid Composition

Cancer cells have ability to modify lipid compositions in their cell membranes, which influence signalling and interactions with the external environment [23]. Phosphatidylserine, which is normally present in to the inner leaflet of the plasma membrane in healthy cells, which is increased on the outside leaflet of the membrane. Depending on the cancer, the level of cholesterol in cancerous cell membranes can change [24]. High cholesterol can cause lipid rafts formation and microdomains rich in sphingolipids and cholesterol. These lipid rafts used as a platforms for signalling, promoting proliferation in cancer cells [25]. Increased level of sphingosine-1-phosphate promote cell growth but these alterations can influence apoptosis and cell survival. The processes of cell invasion and migration are made easier by this increased mobility.

2.2 Protein Composition

Cell signalling and immune recognition may all be impacted by changes in the protein composition of the cell membrane in cancerous cells. Epidermal growth factor receptor (EGFR) and human epidermal growth factor receptor 2 (HER2) are examples of receptor tyrosine kinases (RTKs) that are overexpressed or have altered versions in cancer cell membranes [26]. High level of glycolipids and glycoproteins on the membranes of cells are observed. Excessive secretion of mucins are glycosylated proteins that shield cancer tissue [27]. Due to alterations in composition of lipid, including high level of unsaturated fatty acids these cells show more fluid membranes [28]. The dynamics of the malignant cell membrane, increase the production lipid rafts [29]. The clumping of signalling molecules is possible by these dynamic alterations, which also improve signal transduction pathways that support the growth, survival, and migration of cancer cells.

3 Altered Glycosylation Patterns and Cell to ECM

The process of attaching sugar molecules to proteins and lipids, known as glycosylation, is frequently dysregulated in cancerous cells. Aberrant Glycosylation in Cancer cells frequently exhibit aberrant glycosylation patterns, such as increased sialylation (addition of sialic acid) and altered glycan branching. These changes can affect cell–cell and cell–ECM interactions, immune recognition, and signalling [30]. The expression of some glycan structures, such as sialyl-Tn antigen and Tn antigen is upregulated in cancer and may serve as markers for cancer treatment [31]. Cancer cell membranes are adapted to help altered interactions with other cells and the ECM [32]. The enzymes have the ability to be attracted to and activated on the cell membrane, to show composition of the membrane.

4 Cytoskeletal Alterations

Actin filaments intermediate filaments and microtubules are made up of majority of the cytoskeleton and responsible for structural integrity [34]. Cytoskeleton is significantly rearranged which increases motility and deformability of cells [35]. The epithelial-mesenchymal transition (EMT) is determined by the high expression of intermediate filaments like vimentin [36]. These cytoskeletal rearrangment allow cancer cells to resist the mechanical stresses that arise during metastasis.

4.1 Actin Filaments (Microfilaments)

Actin filaments promote an essential role in intracellular transport, motility, and shape maintenance. Actin polymerisation is usually increased in malignant cells. Cancer cells can move by tissues and also pierce the extracellular matrix (ECM) [37]. Cancer cells often exhibit altered dynamics of filament formation and breakdown [38].

4.2 Microtubules

Microtubules are essential for cell division, tubular structure formation and intracellular transport [39]. Increased microtubule stability in certain cancer cells can lead to resistance to medications that target microtubules, such as paclitaxel [40]. Microtubule-associated proteins (MAPs), which maintain or destabilise microtubules, may be generated differently or there may be mutations in tubulin, the protein that makes up microtubules. For mitosis to occur, microtubules are essential, especially for the establishment of the mitotic spindle, which separates the chromosomes during cell division [41]. Aneuploidy, or an abnormal number of chromosomes, is a characteristic of many malignancies and is generally caused by aberrant spindle formation and chromosomal segregation in cancer cells. These aberrant mitotic processes are facilitated by modifications in microtubule dynamics, which also encourage genetic instability and the development of tumours.

5 Nucleus Mechanics and Its Role in Cancer Progression

Nuclear architecture refers to the organisation and structural features of the nucleus, including the nuclear envelope, chromatin arrangement, and nuclear matrix. In cancer cells, significant alterations in nuclear architecture are commonly observed. These changes are not merely a consequence of malignant transformation but play an active role in promoting cancer progression, metastasis, and resistance to therapy [42]. Key changes in nuclear architecture in cancer cells include alterations in nuclear shape and size, chromatin organisation, nuclear envelope integrity, and nuclear matrix composition. One of the hallmark features of cancer cells is an altered nuclear morphology. This includes changes in nuclear size (often enlarged nuclei) and irregular nuclear shape (nuclear pleomorphism). Cancer cells frequently exhibit an increased nuclear-to-cytoplasmic ratio, characterised by larger nuclei compared to normal cells. This change is partly due to increased DNA content (aneuploidy) and altered nuclear envelope structure. Enlarged nuclei are often seen in rapidly proliferating cells, reflecting increased DNA synthesis and altered cell cycle regulation [43]. Cancer cells frequently feature aberrantly shaped lobulated nuclei with uneven shapes. The chromatin organisation and nuclear envelope structural disruptions are indicated by these morphological changes. Since more aggressive malignancies usually display higher nuclear irregularity, nuclear pleomorphism is a diagnostic sign used in histopathology to grade tumours. An essential component in controlling gene expression is chromatin, a combination of proteins and DNA [44]. Chromatin architecture is frequently significantly altered in cancer cells, which has an impact on gene regulation and genomic stability. Heterochromatin (hyper-condensed) and euchromatin (decondensed) chromatin regions are commonly observed in cancer cells [45]. Gene expression becomes aberrant as a result of this aberrant distribution, which impacts DNA accessibility to transcription factors and other regulatory proteins. Tumour suppressor genes may be silenced by hyper-condensation in some areas, whilst oncogenes may be activated by decondensation. DNA methylation and histone changes are frequently observed in cancerous cells and are a component of the wider modifications in chromatin structure [46]. Hypomethylation of DNA can lead to genomic instability, while hypermethylation of promoter regions of tumour suppressor genes can result in their silencing. Histone modifications, such as acetylation and methylation, are also altered, contributing to the dysregulation of gene expression profiles that drive cancer progression [47]. In cancer cells, there is a disruption of nuclear subdomains, such as nucleoli, nuclear speckles, and PML bodies (promyelocytic leukaemia bodies). The nucleolus, involved in ribosomal RNA synthesis, is often enlarged and more numerous in cancer cells, reflecting increased ribosome biogenesis required for rapid cell growth and proliferation [48]. The nuclear envelope consists of a double lipid bilayer and is supported by the nuclear lamina, a meshwork of intermediate filaments composed mainly of lamin proteins. In many cancers, there is a deregulation in the expression of nuclear lamins. Lamin A/C is often downregulated or abnormally localised, while Lamin B1 levels can be either increased or decreased depending on the cancer type [49]. These changes affect nuclear stability, chromatin organisation, and DNA repair mechanisms. Loss of lamin A/C has been associated with increased nuclear deformability, facilitating cancer cell migration through dense tissues [50]. Nuclear envelope integrity shortcomings such as nuclear envelope rupture and blebbing, are frequently seen in cancer cells. These flaws may result in the nuclear contents being released into the cytoplasm, which leads to genomic instability and incite inflammatory reactions that may accelerate the growth of tumours. The movement of molecules between the cytoplasm and nucleus is controlled by the altered function of the nuclear pore complex (NPC). Proteins involved in cell cycle regulation, DNA repair, and apoptosis can have their localisation and activity affected by altered nucleocytoplasmic transport (nucleocytoplasmic transport) in cancer cells due to alterations in NPC composition and function [51]. In in addition to helping to preserve nuclear shape and mechanical stability, the nuclear matrix acts as a scaffold for the arrangement of chromatin and nuclear bodies. Nuclear stiffness and elasticity are frequently altered in cancer cells due to modifications in the nuclear matrix protein composition. Cancer cell invasion and migration may be facilitated by these changes, which may also have an impact on nuclear mechanics, gene expression, and chromatin organisation. Increased nuclear flexibility, which enables cancer cells to more readily deform their nuclei and move through small crevices in tissues, is a common feature [52]. Cancer cells can adapt to various mechanical conditions during invasion and metastasis thanks to this enhanced deformability, which is linked to alterations in nuclear lamins and the nuclear matrix. Genomic instability, a defining feature of cancer, is intimately associated with altered nuclear architecture in cancer cells [52]. Chromosomal aberrations and the accumulation of mutations can result from modifications in the organisation of chromatin and nuclear structure, which might hinder DNA repair mechanisms. DNA repair accuracy and efficiency can be impacted by factors such as the spatial reorganisation of DNA repair foci and modified localisation of DNA repair proteins. Micronuclei, which are tiny, extranuclear structures containing entire or partial chromosomes, are frequently seen in cancer cells. One sign of genomic instability is micronuclei, which arise from incorrect chromosomal segregation. Inflammation and genomic instability can be further encouraged by their tendency to burst and leak their contents into the cytoplasm.

6 Implications of Cancer Cell Biomechanics in Diagnosis

The study of the physical characteristics and behaviours of cancer cells is known as cancer cell biomechanics, and it has a big impact on cancer detection. Cancer cells have different mechanical characteristics from healthy cells, including stiffness, deformability, and adhesion, because of modifications to their cytoskeletal organisation and cellular architecture. These biomechanical alterations may be used as biomarkers to identify malignancy. For example, compared to normal cells, cancer cells are frequently softer and more malleable, which facilitates their capacity to travel and infiltrate neighbouring tissues—a crucial aspect of metastasis [53]. These characteristics can be measured using methods like optical tweezers, microfluidics, and atomic force microscopy (AFM), which can provide important diagnostic data. Further, the prompt identification of alterations in cell mechanics may allow for an earlier diagnosis because it may occur before the obvious manifestation of standard biochemical indicators. Furthermore, knowing the biomechanics of cancer cells can aid in distinguishing between different cancer types and stages, resulting in more individualised therapy regimens. In conclusion, the biomechanics of cancer cells provide a novel non-invasive diagnostic instrument that facilitates the development of tailored therapies and increases accuracy and early being diagnosed.

7 Conclusion

The investigation of the biomechanics of cancer cells presents encouraging perspectives for improving cancer detection and therapy. The distinct mechanical characteristics of cancer cells—such as changed adhesion, deformability, and stiffness—set them apart from healthy cells and offer new biomarkers for early identification. Accurate and timely cancer diagnosis is made possible by methods such as optical tweezers, microfluidics, and atomic force microscopy (AFM), which allow for the precise monitoring of these biomechanical alterations. In addition, knowing these biomechanical traits enables the differentiation of cancer stages and types, opening the door to more specialised and focussed treatment approaches. Therefore, applying biomechanical analysis to clinical practice could greatly enhance patient outcomes, early intervention, and diagnostic accuracy in cancer therapy.

Data Availability Not applicable.

Declaration of Competing Interest The authors have no conflicts to disclose.

References

1. Suresh, S. (2007). Biomechanics and biophysics of cancer cells. *Acta Biomaterialia, 3*(4), 413–438.
2. Martin, T. A., Ye, L., Sanders, A. J., Lane, J., & Jiang, W. G. (2013). Cancer invasion and metastasis: molecular and cellular perspective. In *Madame curie bioscience database [internet]*. Landes Bioscience.
3. Ozyerli-Goknar, E., & Bagci-Onder, T. (2021). Epigenetic deregulation of apoptosis in cancers. *Cancers, 13*(13), 3210.
4. Brown, G. (2021). Oncogenes, proto-oncogenes, and lineage restriction of cancer stem cells. *International Journal of Molecular Sciences, 22*(18), 9667.
5. Botezatu, A., Iancu, I. V., Popa, O., Plesa, A., Manda, D., Huica, I., ... & Badiu, C. (2016). Mechanisms of oncogene activation. *New Aspects in Molecular and Cellular Mechanisms of Human Carcinogenesis, 9*(1).
6. Pfeffer, C. M., & Singh, A. T. (2018). Apoptosis: A target for anticancer herapy. *International Journal of Molecular Sciences, 19*(2), 448.

7. Lugano, R., Ramachandran, M., & Dimberg, A. (2020). Tumor angiogenesis: Causes, consequences, challenges and opportunities. *Cellular and Molecular Life Sciences, 77*, 1745–1770.
8. Wang, Q., Shao, X., Zhang, Y., Zhu, M., Wang, F. X., Mu, J., ... & Chen, K. (2023). Role of tumor microenvironment in cancer progression and therapeutic strategy. *Cancer Medicine, 12*(10), 11149–11165.
9. Tubbs, A., & Nussenzweig, A. (2017). Endogenous DNA damage as a source of genomic instability in cancer. *Cell, 168*(4), 644–656.
10. Al Aboud, N. M., Tupper, C., & Jialal, I. (2018). Genetics, epigenetic mechanism.
11. Kim, S. K., & Cho, S. W. (2022). The evasion mechanisms of cancer immunity and drug intervention in the tumor microenvironment. *Frontiers in Pharmacology, 13*, 868695.
12. Di, X., Gao, X., Peng, L., Ai, J., Jin, X., Qi, S., ... & Luo, D. (2023). Cellular mechanotransduction in health and diseases: from molecular mechanism to therapeutic targets. *Signal Transduction and Targeted Therapy, 8*(1), 282.
13. Mokhtari, R. B., Ashayeri, N., Baghaie, L., Sambi, M., Satari, K., Baluch, N., ... & Chakraborty, S. (2023). The hippo pathway effectors YAP/TAZ-TEAD oncoproteins as emerging therapeutic targets in the tumor microenvironment. *Cancers, 15*(13), 3468.
14. Fares, J., Fares, M. Y., Khachfe, H. H., Salhab, H. A., & Fares, Y. (2020). Molecular principles of metastasis: A hallmark of cancer revisited. *Signal Transduction and Targeted Therapy, 5*(1), 28.
15. Li, X., & Wang, J. (2020). Mechanical tumor microenvironment and transduction: Cytoskeleton mediates cancer cell invasion and metastasis. *International Journal of Biological Sciences, 16*(12), 2014.
16. De Visser, K. E., & Joyce, J. A. (2023). The evolving tumor microenvironment: From cancer initiation metastatic outgrowth. *Cancer Cell, 41*(3), 374–403.
17. Shuvalov, O., Daks, A., Fedorova, O., Petukhov, A., & Barlev, N. (2021). Linking metabolic reprogramming, plasticity and tumor progression. *Cancers (Basel).*
18. Kumari, A., Veena, S. M., Luha, R., & Tijore, A. (2023). Mechanobiological strategies to augment cancer treatment. *ACS Omega, 8*(45), 42072–42085.
19. Najera, J., Rosenberger, M. R., & Datta, M. (2023). Atomic force microscopy methods to measure Tumor mechanical properties. *Cancers, 15*(13), 3285.
20. Nguyen, L. T., Jacob, M. A. C., Parajón, E., & Robinson, D. N. (2022). Cancer as a biophysical disease: Targeting the mechanical-adaptability program. *Biophysical Journal, 121*(19), 3573–3585.
21. Kalli, M., Poskus, M. D., Stylianopoulos, T., & Zervantonakis, I. K. (2023). Beyond matrix stiffness: targeting force-induced cancer drug resistance. *Trends in Cancer.*
22. Choromańska, A., Chwiłkowska, A., Kulbacka, J., Baczyńska, D., Rembiałkowska, N., Szewczyk, A., ... & Saczko, J. (2021). Modifications of plasma membrane organization in cancer cells for targeted therapy. *Molecules, 26*(7), 1850.
23. Szlasa, W., Zendran, I., Zalesińska, A., Tarek, M., & Kulbacka, J. (2020). Lipid composition of the cancer cell membrane. *Journal of Bioenergetics and Biomembranes, 52*(5), 321–342.
24. Fadok, V. A., Bratton, D. L., Frasch, S. C., Warner, M. L., & Henson, P. M. (1998). The role of phosphatidylserine in recognition of apoptotic cells by phagocytes. *Cell Death & Differentiation, 5*(7), 551–562.
25. Codini, M., Garcia-Gil, M., & Albi, E. (2021). Cholesterol and sphingolipid enriched lipid rafts as therapeutic targets in cancer. *International Journal of Molecular Sciences, 22*(2), 726.
26. Paul, M. K., & Mukhopadhyay, A. K. (2004). Tyrosine kinase–role and significance in cancer. *International Journal of Medical Sciences, 1*(2), 101.
27. Koistinen, P., & Heino, J. (2013). Integrins in cancer cell invasion. In *madame curie bioscience database [internet].* Landes Bioscience.
28. Alves, A. C., Ribeiro, D., Nunes, C., & Reis, S. (2016). Biophysics in cancer: The relevance of drug-membrane interaction studies. *Biochimica et Biophysica Acta (BBA)-Biomembranes, 1858*(9), 2231–2244.

29. Maja, M., & Tyteca, D. (2022). Alteration of cholesterol distribution at the plasma membrane of cancer cells: From evidence to pathophysiological implication and promising therapy strategy. *Frontiers in Physiology, 13*, 999883.
30. Thomas, D., Rathinavel, A. K., & Radhakrishnan, P. (2021). Altered glycosylation in cancer: A promising target for biomarkers and therapeutics. *Biochimica et Biophysica Acta (BBA)-Reviews on Cancer, 1875*(1), 188464.
31. Huang, J., Huang, J., & Zhang, G. (2022). Insights into the role of sialylation in cancer metastasis, immunity, and therapeutic opportunity. *Cancers, 14*(23), 5840.
32. Janiszewska, M., Primi, M. C., & Izard, T. (2020). Cell adhesion in cancer: Beyond the migration of single cells. *Journal of Biological Chemistry, 295*(8), 2495–2505.
33. Fletcher, D. A., & Mullins, R. D. (2010). Cell mechanics and the cytoskeleton. *Nature, 463*(7280), 485–492.
34. Fife, C. M., McCarroll, J. A., & Kavallaris, M. (2014). Movers and shakers: Cell cytoskeleton in cancer metastasis. *British Journal of Pharmacology, 171*(24), 5507–5523.
35. Datta, A., Deng, S., Gopal, V., Yap, K. C. H., Halim, C. E., Lye, M. L., ... & Yap, C. T. (2021). Cytoskeletal dynamics in epithelial-mesenchymal transition: Insights into therapeutic targets for cancer metastasis. *Cancers, 13*(8), 1882.
36. Pratiwi, L., Elisa, E., & Sutanto, H. (2024). Probing the protrusions: lamellipodia and filopodia in cancer invasion and beyond. *Mechanobiology in Medicine* 100064.
37. Svitkina, T. (2018). The actin cytoskeleton and actin-based motility. *Cold Spring Harbor Perspectives in Biology, 10*(1), a018267.
38. Izdebska, M., Zielińska, W., Hałas-Wiśniewska, M., & Grzanka, A. (2020). Involvement of actin and actin-binding proteins in carcinogenesis. *Cells, 9*(10), 2245.
39. Huang, L., Peng, Y., Tao, X., Ding, X., Li, R., Jiang, Y., & Zuo, W. (2022). Microtubule organization is essential for maintaining cellular morphology and function. *Oxidative Medicine and Cellular Longevity, 2022*(1), 1623181.
40. Čermák, V., Dostál, V., Jelínek, M., Libusová, L., Kovář, J., Rösel, D., & Brábek, J. (2020). Microtubule-targeting agents and their impact on cancer treatment. *European Journal of Cell Biology, 99*(4), 151075.
41. Ferreira, L. T., Figueiredo, A. C., Orr, B., Lopes, D., & Maiato, H. (2018). Dissecting the role of the tubulin code in mitosis. In *Methods in cell biology* (Vol. 144, pp. 33–74). Academic.
42. Easwaran, H. P., & Baylin, S. B. (2011, March). Role of nuclear architecture in epigenetic alterations in cancer. In *Cold Spring Harbor symposia on quantitative biology* (pp. sqb-2010). Cold Spring Harbor Laboratory Press.
43. Fischer, E. G. (2020). Nuclear morphology and the biology of cancer cells. *Acta Cytologica, 64*(6), 511–519.
44. Singh, I., & Lele, T. P. (2022). Nuclear Morphological abnormalities in Cancer: A search for unifying mechanisms. *Nuclear, chromosomal, and genomic architecture in biology and medicine* (pp. 443–467). Springer International Publishing.
45. Nair, N., Shoaib, M., & Sørensen, C. S. (2017). Chromatin dynamics in genome stability: Roles in suppressing endogenous DNA damage and facilitating DNA repair. *International Journal of Molecular Sciences, 18*(7), 1486.
46. Shanmugam, M. K., Arfuso, F., Arumugam, S., Chinnathambi, A., Jinsong, B., Warrier, S., ... & Lakshmanan, M. (2018). Role of novel histone modifications in cancer. *Oncotarget, 9*(13), 11414.
47. Ehrlich, M. (2019). DNA hypermethylation in disease: Mechanisms and clinical relevance. *Epigenetics, 14*(12), 1141–1163.
48. Penzo, M., Montanaro, L., Treré, D., & Derenzini, M. (2019). The ribosome biogenesis—Cancer connection. *Cells, 8*(1), 55.
49. Chow, K. H., Factor, R. E., & Ullman, K. S. (2012). The nuclear envelope environment and its cancer connections. *Nature Reviews Cancer, 12*(3), 196–209.
50. Dos Santos, Á., & Toseland, C. P. (2021). Regulation of nuclear mechanics and the impact on DNA damage. *International Journal of Molecular Sciences, 22*(6), 3178.

51. Gauthier, B. R., & Comaills, V. (2021). Nuclear envelope integrity in health and disease: Consequences on genome instability and inflammation. *International Journal of Molecular Sciences, 22*(14), 7281.
52. Dahl, K. N., Ribeiro, A. J., & Lammerding, J. (2008). Nuclear shape, mechanics, and mechanotransduction. *Circulation Research, 102*(11), 1307–1318.
53. Bao, L., Kong, H., Ja, Y., Wang, C., Qin, L., Sun, H., & Dai, S. (2023). The relationship between cancer and biomechanics. *Frontiers in Oncology, 13*, 1273154.

A Review on Mechanisms of Single-Cell and Collective Cell Migration: Insights into Tissue Dynamics

Abhimanyu Kiran, Navin Kumar, and Vishwajeet Mehandia

Abstract Cell migration, whether occurring at the level of individual cells or as a collective effort, is a vital biological process shaped by the biochemical and biophysical characteristics of the surrounding microenvironment. In single-cell migration, movement is driven by actin-based protrusions, integrin-mediated focal adhesions, and actomyosin contractions, allowing cells to respond to environmental factors such as chemical gradients or substrate stiffness. On the other hand, collective cell migration involves the coordinated movement of interconnected cells within a monolayer, facilitated by structures like E-cadherin-based cell–cell junctions and the actin cytoskeleton. These intercellular interactions enhance the speed, directionality, and efficiency of migration compared to individual cells. Collective migration plays a pivotal role in maintaining tissue structure, remodeling, homeostasis, and in biological processes such as embryogenesis, morphogenesis, wound repair, and cancer metastasis. E-cadherin is crucial for maintaining mechanical stability, force distribution, and intercellular communication during collective migration, while its loss often triggers epithelial-to-mesenchymal transition (EMT), a critical step in cancer metastasis. Gaining a deeper understanding of the similarities and differences between single-cell and collective migration is key to unraveling the mechanisms behind tissue dynamics and disease development.

1 Cell Migration

The single-cell migration has been studied extensively [1, 2]. A typical cell expands by forming protrusion (driven mainly by actin polymerization) at the leading end. This protrusion may have small filopodia, large lamellipodia, or their combination. The cell form integrin-based focal adhesion at the leading end through which it

A. Kiran · N. Kumar
Department of Mechanical Engineering, IIT Ropar, Rupnagar, Punjab, India

V. Mehandia (✉)
Department of Chemical Engineering, IIT Ropar, Rupnagar, Punjab, India
e-mail: vishwajeet@iitrpr.ac.in

K. K. Dwivedi et al. (eds.), *Mechanics and Materials Science of Biological Materials*, Materials Horizons: From Nature to Nanomaterials,
https://doi.org/10.1007/978-981-96-5366-9_8

applies traction force on its substrate. Then the cell applies a pull to translocate its body forward toward the leading end and detaches the focal adhesion from the trailing end [3]. For efficient polarization and migration, cells require actin and myosin-based contractions [4]. The direction of cell migration is governed by the guidance signal (negative and positive), such as external mechanical and chemical cues present in its microenvironment. The cells move randomly in the absence of external guidance. The morphology, migration speed, degree of polarization, and dynamics may vary depending upon the cell type. The single cell's migration is affected by its external stimuli like substrate stiffness, roughness, and topography [5, 6].

Likewise, the cell monolayer is also affected by the biophysical and biochemical parameters of its environment. However, the collective cell migration cannot be explained by the large number of single cells migrating in a direction with a constant speed [7] because each cell of the monolayer is connected to its neighboring cells via E-cadherin (cell–cell junction protein) and actin cytoskeleton [8]. Therefore, the direction of cell migration in a monolayer is influenced by its neighboring cells [9, 10]. It is reported that the cells in a monolayer exhibit more directed and rapid migration compared to single cells. The cellular interactions via soluble and contact-mediated signals are responsible for highly coordinated and efficient collective cell migration [11]. Hence, collective cell motion results in a much more efficient mode of migration than the same number of single cells placed together without considering the intercellular interaction between them [8]. Therefore, intercellular interactions are vital for collective cell migration. E-cadherin is an intracellular protein present at the cell–cell junction of the epithelial cell monolayer that provides homeostasis and mechanical rigidity. It also helps in cell communication, force transmission, and collective cell migration. The loss of E-cadherin leads to the removal of cell–cell contact, thereby causing epithelial to mesenchymal (EMT) transition [12–14]. Many single cells or small cluster of cells gets detached from the monolayer due to incomplete EMT transition. These detached cells acquire the phenotype of the mesenchymal cells (that prefer to migrate) and move to various locations in the body. These cells can undergo mesenchymal to epithelial (MET) transition and settle there after reaching other parts. Thereby, this type of transition is observed during cancer metastasis. Therefore, it is essential to understand the connection and distinction between single and collective cell migration.

The cells may prefer to migrate together as compared to individually because of the following reasons: (a) To maintain the integrity of the tissue and its continuous remodeling, (b) displacement of immobile cells with their connected mobile cells, (c) maintain the homeostasis of the monolayer, (d) long-range communication to execute collective decision as a tissue [4]. Hence, collective cell migration sets as a fundamental example of how interdependent cells move together by forming an active tissue that can undergo continuous remodeling while maintaining its homeostasis and integrity of the monolayer.

In collective cell migration, cells move together as a cohesive group which is the hallmark of remodeling events in the tissue. The collective cell migration is observed in embryogenesis, morphogenesis, cancer metastasis, and wound healing [11]. The cells can collectively move as sheets, clusters, strands, or ducts by employing their cytoskeleton and molecular motors. Just like single cells, the mechanical and chemical cues present in the microenvironment could be responsible for long-range collective cell guidance [11]. Additionally, cadherins (junction proteins available between the cells) maintain intracellular properties such as force generation, collective polarization, decision making, and complex tissue organization [11].

1.1 Social Cell–Cell Interactions

The cells of the monolayer interact with its neighboring cell and its substrate. It is assumed that for epithelial cells, cell–cell interaction dominates the cell–substrate interactions. Therefore, in this section, we have discussed all the social cell–cell interactions affect cell polarity.

1.1.1 Local Alignment

Due to this social interaction, cells tend to align together. The aligned cell might deform or elongate together while migrating, but biochemical processes may regulate their migration. Most commonly, the Vicsek rule is utilized to explain the local alignment in discrete models. However, for continuum models, orientational stiffness in the polarity field is used.

1.1.2 Contact Regulation of Locomotion (CRL)

Due to this social interaction, cells tune their direction of migration upon coming in contact with other cells. Three types of CRL have been defined. First is contact inhibition of locomotion (CIL), where cells, upon head-to-head collision, retract and repolarize their lamellipodia in opposite direction [8, 15]. Thereby, move away from each other. The second is contact following of locomotion (CFL), where the head of a cell interacts with the tail of other cells, then the head of the cell starts to follow the tail of other cell [16, 17]. Third, contact enhancement of locomotion (CEL), in few cases, head-to-tail collision tend to enhance the locomotion [18].

1.1.3 Polarity-Flow Alignment

Just like liquid crystals, cells tend to migrate in the direction of lowest shear stress, and this phenomenon is also termed as plithotaxis [19–22]. However, unlike passive liquid crystals, the cell reorientation may result from an active mechanosensitive cellular response. The mechanism behind the phototaxis is unknown, but there may exist some feedback loops between polarity and cell migration.

1.2 Collective Cell Migration as a Sheet

There have been several categories of collective cell migration such as sheet migration in 2D, sprouting and branching in 3D, Streams in neural cells, posterior lateral line in zebrafish, and border cell migration in germ line cells. However, here we have only focused on the studies related to sheet migration. The sheet migration is most commonly observed in the epithelial cell monolayer. The cells of the monolayer maintain strong cell–cell adhesions and tissue integrity while moving forward as a sheet.

The monolayer used to study collective cell migration can be formed by using invasive or non-invasive methods. The invasive methods include stretch wound assay, where the wound is formed in the confluent monolayer by using pipette tip or blade or microneedle [23, 24], and laser ablation [25] (Fig. 1a). The non-invasive method includes artificial wound model [26], or undamaged gaps [27], where cells are seeded inside fixed confinement, and this confinement is removed after the formation of a fully confluent monolayer (Fig. 1b). The removal of the confinement or physical barrier provides the availability of free space, which is a sufficient condition to trigger the migration of the monolayer toward the freely available space [28, 29]. Thereby, cells at the leading site collectively migrate as a sheet toward the free space [28, 30–33]. In this type of non-invasive artificial wound system, no injury occurs, and it may not relate to the physiological scenario of wound healing. Cell proliferation is observed during wound healing, but cell proliferation inhibition by mitomycin C does not prevent the gap filling [33, 34]. This suggests that the absence of cell division may affect the rate by which wound close, but it cannot stop the healing mechanism. The stretch wound assay is generally performed on epithelial [35, 36] or endothelial cells [34], where cells move together as cohorts. Whereas confluent fibroblast cells also undergo wound healing after stretch test, many cells also move as independent cells [37].

The collective cell migration of the monolayer toward the freely available space has been extensively studied. Earlier it was assumed that few transformed marginal cells, also known as leader cells, show positive outgrowth in their phenotype, and they effectively pull their passive follower cells. Later, many studies, including individual cell tracking and the presence of cellular protrusions, indicated that both leader and follower cells of the monolayer are actively involved in collective cell migration [23, 34, 38]. The emergence of the leader cells is stochastic, but the availability of growth factors can induce their abundance.

It is observed that there exists coordination between the velocity vector of a cell with its neighboring cells. This coordinated motion between cells and their neighbors can occur due to cell–cell adhesion and mechanical coupling. Each cell may contribute to a small portion of net collective cell migration, or highly motile cells pull less motile cells with them by maintaining tissue integrity. Although this intracellular interaction between cell and its neighbor implicitly describes the collective cell dynamics. However, for large-scale coordinated collective cell motion, real-time signaling interactions are required. One such signaling interaction includes the

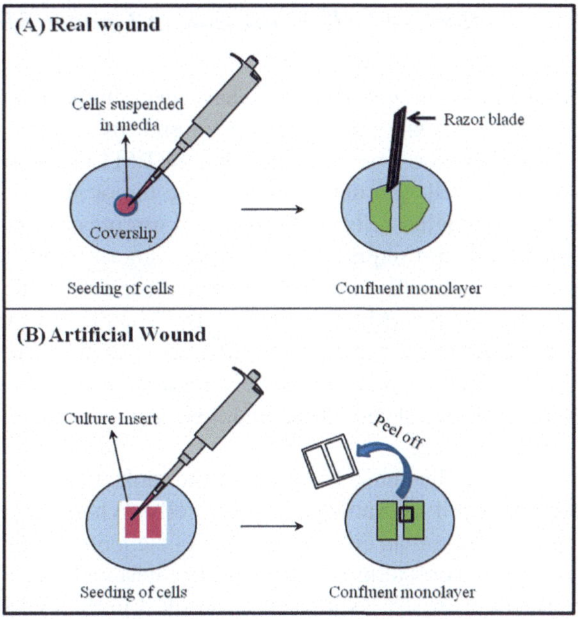

Fig. 1 The procedure of wound formation. **a** Real wound; **b** artificial wound

Ca^{2+} wave that propagates from one cell to other via extracellular ATP, and these waves can be modulated using growth factors [39, 40]. Another signaling interaction is the slower wave generated by coordinated ERK activation in the case of real wounds [29]. Since the ERK wave is affected by cell migration, it suggests that this wave is induced mechanically [37]. Additionally, a slow mechanical wave has been experimentally reported to propagate through the cell monolayer for long-range cell guidance [41, 42].

The sheet movement also occurs in developmental processes like dorsal closure in Drosophila and early morphogenetic changes in a few animals [43, 44]. For wounds, the active mechanism for healing may vary depending upon their shape. For smaller holes created by laser ablation, marginal cells exhibit a rich actin-ring around the gap. It leads to the development of a popular hypothesis that the Purse-string mechanism is responsible for the gap closure in the wounds [45]. Whereas in the scratch test assay, a straight cut is applied at the center of a confluent monolayer. In this case, forward sheet movements occurred without active purse-string mechanism [46–48]. For dorsal closure, the experimental and modeling of forces indicate that the sheet's translocation is generated by the combination of purse-string mechanism and active crawling of autonomous cells in the forward direction. In physiological scenarios, wound healing is a quite complex process compared to stretch wound assay performed in 2D culture dishes. Apart from epithelial cells, many other cells are also involved, and the formation of a fibrous clot can apply mechanical strain. It may be why intermediate filaments are required for sheet migration of keratinocytes cells in wound healing [49, 50].

1.3 Collective Cell Migration in Cancer Metastasis

In squamous carcinomas, which have an epithelial phenotype, cells get disseminated from the tumor site and collectively migrate by acquiring the mesenchymal phenotype [51–53]. Earlier, it was assumed that the EMT transition occurs in cancer metastasis, and single cells get detached from the tumor site. However, later, it is discovered that the epithelial cells in 3D spread in sprouts or groups. Therefore, complete EMT transition is not required for the spreading of tumor in its surrounding tissue [51]. For collective cell migration of the cancer cells is linked to the activity of matrix metalloproteases (MMPs) [54–56]. For MDCK cells, the forced expression of MMP is sufficient to trigger the invasion properties [55]. The availability of a small group of cancer cells in a lymph vessel suggests that cancer cells can reach the lymphatic system via metastasis. These findings suggest that complete EMT is not required for tumor spreading.

Many studies have highlighted the molecular and cellular routes to explain the collective cell migration of cancer cells. When cancer cells are cultured on a 3D matrix, they spread collectively due to the matrix degradation by MMPs [56]. Interestingly, the co-culture of stromal fibroblast and carcinoma cells showed that cancer cells follow the tracks made by fibroblast in the ECM. In other words, cancer cells recruit fibroblast for their migration [57]. This type of collaboration between two cell types can help in the effective spreading of individual cancer cells [58]. The knockdown of p120 in A321 cells helps the migration of cancer cells in 2D, but it impaired the migration in 3D [59]. The pancreatic tumor progression, an in vivo model, showed that the podoplanin protein expression could cause cancer invasion without EMT [60]. The presence of enriched Fascin (an actin regulator) has been reported in the front cells of colorectal carcinomas, and it promotes invasiveness without EMT [61]. These studies indicate the importance of investigating the migration of tumor cells in their physiological environment. In future, the proper understanding of tumor spreading can be utilized to stop it.

1.4 Models for Collective Cell Migration

The collective cell migration has been a topic of research for more than 60 years. There have been a lot of theoretical and experimental studies to understand the physical mechanism responsible for cell–cell communication to drive collective cell migration [9–17].

1.4.1 Theoretical Models

The theoretical models are divided into two classes, namely the discrete models and the continuum models [16]. The discrete models such as Vertex [62, 63], Voronoi

[64, 65], Cellular Potts model [66], and particle-based model [67, 68] consider each cell as an active particle with persistent motility. These models can incorporate individual cell motility [65, 69], contractibility [70], and cell–cell interaction [65, 68]. In contrast, the continuum models such as the active gel [71] model describe cell monolayer as a fluid or an elastic continuum with the freedom to incorporate active processes like contractibility and cellular polarization. These models can successfully capture the large-scale behavior observed in confined [72], expanding [42, 73], and wounded [73] monolayer.

1.4.2 Experimental Studies

There have been many experimental studies to understand the mechanism responsible for collective cell migration. The cells at the leading front of the monolayer have been categorized as leaders and followers [74]. The origin of the leader cell is stochastic [4], and only a few marginal cells get transformed into leader cells. These leader cells exhibit a distinct phenotype as they show positive outgrowth after their recruitment. These cells invade the freely available space as well as they pull their immediately physically connected follower cells. Earlier, it was assumed that the leader cells actively pull their passive follower cells [26, 34, 75, 76]. Later, it is found that the leader cells apply the pull of ~100 nN, which is not sufficient to pull the complete monolayer [7]. Therefore, leader cells do need some assistance from the follower cells to migrate effectively as a sheet. Since then, many studies have indicated that follower cells actively participate in collective cell migration. These studies include the presence of the significant amount of traction force up to ~10 cell rows [77], propagation of a slow mechanical wave through monolayer [41], the existence of correlated velocity field up to ~10–15 cell rows [78], strong alignment of cell's principal stress with velocity field [19], and mechanical interaction among the follower cells determine the emergence of leader cells [30]. Additionally, the presence of "cryptic" cell groups in the sub-marginal region has been reported [38]. The lamellipodia of these cryptic cells penetrate into the basal of their front cells, and it is assumed that these cryptic cell groups could be responsible for collective cell migration. All these studies suggest that leader cells and follower cells both actively participate in collective cell migration. However, the exact mechanism through which follower cells contribute to collective cell migration is unknown.

2 Material Properties of Epithelial Monolayer

The living cells are active materials with phenomenal mechanical properties. These properties are measured using rheology, which is the study of the flow of matter. Depending on the probing length scale, the rheology is classified into two categories: micro and macro-rheology.

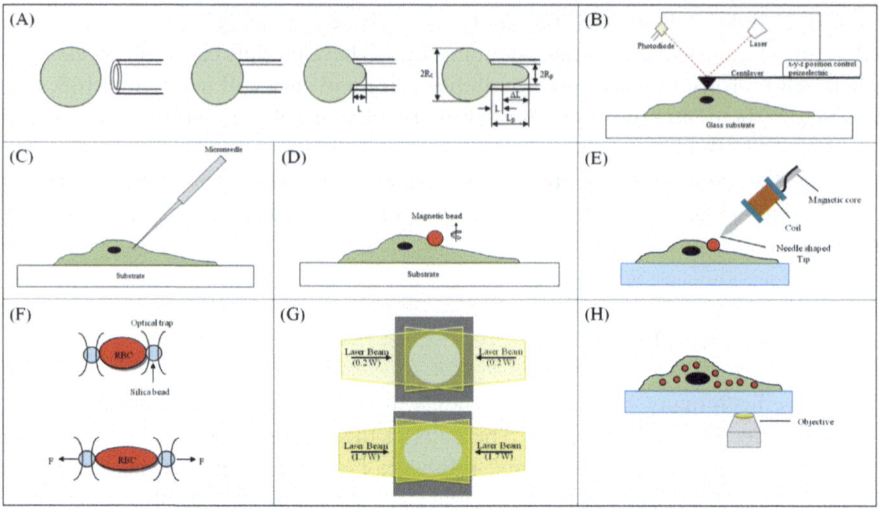

Fig. 2 Schematic representation of micro-rheometry techniques. **a** Micropipette aspiration; **b** micro-needle manipulation; **c** MTC; **d** optical stretcher; **e** optical stretchers

2.1 Micro-Rheology Techniques

2.1.1 Micropipette Aspiration

In this technique, the living cell's mechanical behavior is determined by aspirating the cell surface inside a glass tube by applying a negative pressure gradient (Fig. 2a). The leading end of the cell that went inside the glass tube is tracked. The living cell is viscoelastic in nature as it posses both viscous and elastic properties. If the viscosity of the cell is more than its elasticity, the cell behaves like a liquid drop. Whereas, if the elasticity is more than the viscosity, the cell behaves as a solid gel. The procedure of micropipette aspiration starts with suctioning of the cell surface inside the micropipette until a hemispherical projection is being formed. If the suction pressure is further increased, then the cell shows a liquid-like behavior with constant cortical tension and move inside the glass tube until a new equilibrium position is attained [79, 80].

2.1.2 Atomic Force Microscope (AFM)

It is a microscopic technique that can apply and measure the response simultaneously. In this technique, a probe is attached to a flexible cantilever beam that can smoothly scan the sample to record the electrostatic interaction forces between sample and probe. These forces cause the deflection of the cantilever beam. The cantilever beam behaves like a spring and follows Hook's law ($F = -Kx$), where K is the stiffness of the cantilever and x is the distance between probe and sample. The spherical probe is used for biological samples. The AFM can also perform the raster scan to measure the topology of the sample. At every step, the AFM tip senses the involved forces, and a feedback mechanism is used to regulate the position of the AFM tip using a piezoelectric scanner. The deflection of the AFM tip scanning a biological sample is measured by the change in the reflected laser beam position falling on a photodiode [81]. The AFM can operate in three modes: contact mode, tapping mode, and force volume mode. The movement of the cantilever tip on the sample surface is different for all these modes. The force volume mode is recommended for biological samples.

2.1.3 Micro-needle Manipulation

This technique is used to poke or tug the sub-cellular part (Fig. 2b). The micro-needles are called pokers [82], and they are formed by heating and pulling of glass fiber. The needles act as a mechanical spring, and hence, the magnitude of force applied by the spring is measured by the famous Hook's law ($F = Kx$), where K is the micro-needles bending stiffness, and x is the microneedle deflection measured by the optical [83] or electrical [84] method. The elasticity of the cell is determined from the slope of the force–displacement curve.

2.1.4 Magnetic Twisting Cytometry (MTC)

It is used to apply torque on cellular or sub-cellular components (Fig. 2c). This technique utilizes ECM coated magnetic beads (1–10 μm diameter), and when these beads are introduced to the cell, they only attach to specific ligand present at the surface or inside the cell body. Since the coated bead is magnetic in nature, therefore, torque can be applied on the cell–bead interface by applying magnetic field [85].

2.1.5 Magnetic Tweezers

This technique has the ability to apply and measure the force on ferromagnetic beads attached to the cells by introducing the gradient of the magnetic field. In brief, the ligands coated magnetic bead is attached to the integrin present at the targeted part of the cell membrane or sub-cellular structures. Therefore, forming the classical integrin-ligand catch bond. The motion of the magnetic bead movement can be precisely controlled by utilizing the magnetic field gradient generated by the electromagnetic coil [86]. The amount of force applied by the magnetic bead is proportional to the intensity of the magnetic field.

2.1.6 Optical Tweezers

This technique uses a laser light source and a microscope to trap small objects and control their movement via photons (Fig. 2d). When a focused light interacts with a small dielectric particle, then they experience scattering and gradient forces. The scattering force is produced when a photon travels through a medium enter inside an object, and it changes its direction due to the change of refractive index. This change in direction causes the change in momentum. Thereby, scattering force is being exerted on the object. The gradient force is produced by the gradient in the intensity of the light field. An optical trap is formed by focusing the laser to a diffraction-limited spot through a high-powered microscope. Any dielectric particle near the focus will experience a restoring force that tends to move the object toward the center of focus. In order to trap a particle, the gradient force should be greater than the scattering force. This condition can be attained using a highly focused laser.

The silicone beads used for creating the optical traps are coated with ECM proteins (or ligands). Therefore, integrin present at the surface of the cells makes a catch bond with the ligands coated on the silicon beads [87]. The position of the bead can be controlled precisely by adjusting the laser. The trapping force can be applied by varying the center position of the beam. If small displacement is made by using the trapping force, then the optical traps act as linear springs. That means that if any force is applied to pull the bead out of the trap, then the restoring force has a tendency to bring it back into the center of the trap. The restoring force is measured using the formula, $F = -Kx$, where K is the stiffness of the trap that can be measured by exerting a known force on the trapped bead and x is measured small displacement of the bead from the center of the trap. The displacement of the bead is measured either using the imaging technique or the photodiode technique.

2.1.7 Optical Stretchers

In this technique, a cell is trapped between two counter-propagating low-power laser divergent beams (Fig. 2e). If the intensity of laser power is increased, it does not displace the cell from the original position, but the cell stretching would occur [88].

2.1.8 Particle-Tracking

In this technique, small florescent beads (<1 μm) are inserted into the living cells that diffuse randomly in their cytosol [89]. Since the mass of the beads is infinitesimally small, their inertial effects can be neglected. Therefore, these beads move due to Brownian motion. The movement of these beads can be recorded using a fluorescent microscope. This technique has been used to evaluate the viscoelastic response of cells under serum starvation, pharmacological treatment, and wounds [90].

2.2 *Macro-Rheology Techniques*

The macro-rheology is generally performed by using oscillatory rheometers. Where a sinusoidal shear deformation is applied to a sample, and its response is measured [91]. In traditional rheometers, the sample is placed between two parallel plates, and a pre-normal force is maintained. Generally, the upper plate rotates to apply shear, and the bottom plate remains fixed. It is known as a single drive module, where the upper plate applies shear to the sample and measures its response. The macro-rheology has been performed on mucus [92–94], sputum [95], mesoglea [96], tumor tissues [97], lipid monolayer [98], synovial fluid [99], bone marrow [100], and various other biological samples. Recently, the traditional rheometer has been modified by adding a high-quality confocal microscope to perform the rheology of biological samples like collagen [101, 102]. Upon Comparing, micro and macro-rheological studies, it is found that micro-rheology underestimates the viscoelastic properties of biological samples at sub-cellular, cellular, and tissue scale [94, 103]. The probable reason for this may be that the viscoelastic properties are highly heterogeneous in the biological sample.

For monolayers, the CMR (Cell Monolayer Rheology) is a well-known macro-rheology technique used to investigate the material properties of the cell monolayer at a tissue scale. It is usually performed using an oscillatory rheometer [104–106]. The rheological properties may vary due to cytotoxicity, malignancy, and other abnormalities. The parallel plate configuration is used to perform the bulk rheology of the cell monolayer. These studies are performed on a discrete monolayer (Fig. 3) of fibroblast or HeLa, where a large number of single cells are attached to both the plates of the rheometer [105–107]. However, this discrete monolayer does not represent the actual monolayer having cell–cell contacts. This cell–cell contact provides mechanical strength and rigidity to the monolayer [12]. They are also responsible for force transmission among the monolayer [108].

3 Research Objectives

Many studies have implicitly reported the role of follower cells in collective cell migration. However, the mechanism by which follower cells contribute to collective cell migration is unknown. Therefore, our first objective is to identify the mechanism responsible for collective cell migration for expanding tissue. The majority of studies reported on collective cell migration use artificial model wounds that possess high cell number density. However, the cell number density is quite low in physiological scenarios like wounds (many cell damage due to applied cut) and cancer metastasis (a small group of cells get detached from the tumor site and begin to migrate collectively). Therefore, our second objective is to differentiate between artificial wound models and physiological scenarios like cancer metastasis, wound healing, and tissue expansion. Finally, considering the monolayer as an active matter, we assumed that

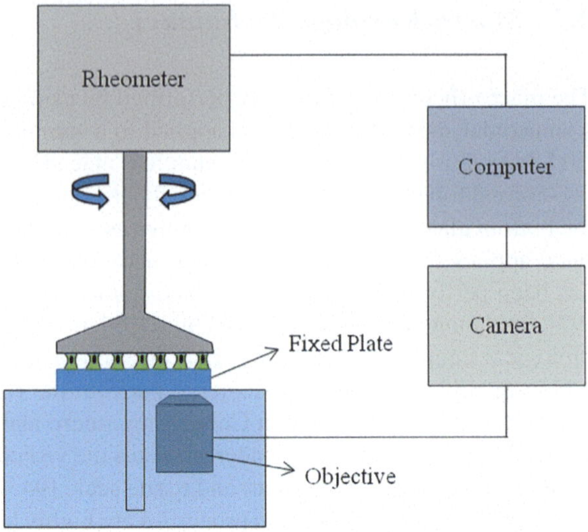

Fig. 3 Schematic representation of macro-rheology

its mechanical properties could be tuned by varying its activity. Therefore, our third objective is to measure the rheological properties of the monolayer with and without serum.

4 Dissertation Overview and Organization

The thesis consists of five chapters and is organized as follows:

Chapter 1: Introduction provides details about single-cell versus collective cell migration, 2D sheet migration, collective cell migration in cancer metastasis, theoretical models, and experimental studies on collective cell migration. A brief summary of experimental studies is provided, followed by the statement of research objectives for this work.

Chapter 2: Introduce a mechanism to explain the role of follower cells in the collective cell migration of front cell rows in freely expanding monolayers.

Chapter 3: Provides the comparison between artificial wound model and physiological scenarios like cancer metastasis, wound healing, and tissue expansion.

Chapters 4: Present the difference in the mechanical properties of the monolayer with and without serum.

Chapter 5: Summarizes and discusses the key results of Chaps. 2–4, a discussion on limitations of our work, and proposes directions for future research.

References

1. Ridley, A. J., Schwartz, M. A., & Burridge, K., et al. (2003). Cell migration: Integrating signals from front to back. *Science (80-), 302*(5651), 1704–1709.
2. Van Haastert, P. J. M., & Devreotes, P. N. (2004). Chemotaxis: Signalling the way forward. *Nature Reviews Molecular Cell Biology, 5*(8), 626–634.
3. Douglas, A., & Lauffenburger, A. F. H. (1996). Cell migration: Review a physically integrated molecular process. *Cell, 84*, 359–369. https://doi.org/10.1016/S0092-8674(00)81280-5
4. Rørth, P. (2009). Collective cell migration. *Annual Review of Cell and Developmental Biology, 25*(1), 407–429. https://doi.org/10.1146/annurev.cellbio.042308.113231
5. Nelson, C. M., & Tien, J. (2006). Microstructured extracellular matrices in tissue engineering and development. *Current Opinion in Biotechnology, 17*(5), 518–523.
6. Liu, Y.-J., Le Berre, M., Lautenschlaeger, F., et al. (2015). Confinement and low adhesion induce fast amoeboid migration of slow mesenchymal cells. *Cell, 160*(4), 659–672.
7. De Pascalis, C., & Etienne-Manneville, S. (2017). Single and collective cell migration: The mechanics of adhesions. *Molecular Biology of the Cell, 28*(14), 1833–1846.
8. Mayor, R., & Etienne-Manneville, S. (2016). The front and rear of collective cell migration. *Nature Reviews Molecular Cell Biology, 17*(2), 97.
9. Vicsek, T., Czirók, A., Ben-Jacob, E., Cohen, I., & Shochet, O. (1995). Novel type of phase transition in a system of self-driven particles. *Physical Review Letters, 75*(6), 1226.
10. Szabo, B., Szöllösi, G. J., Gönci, B., Jurányi, Z., Selmeczi, D., & Vicsek, T. (2006). Phase transition in the collective migration of tissue cells: Experiment and model. *Physical Review E, 74*(6), 61908.
11. Friedl, P., & Gilmour, D. (2009). Collective cell migration in morphogenesis, regeneration and cancer. *Nature Reviews Molecular Cell Biology, 10*(7), 445–457. https://doi.org/10.1038/nrm2720
12. Cooper, G. M., & Hausman, R. E. (2004). *The cell: Molecular approach*. Medicinska naklada.
13. Shook, D., & Keller, R. (2003). Mechanisms, mechanics and function of epithelial–mesenchymal transitions in early development. *Mechanisms of Development, 120*(11), 1351–1383.
14. Thiery, J. P., & Sleeman, J. P. (2006). Complex networks orchestrate epithelial–mesenchymal transitions. *Nature Reviews Molecular Cell Biology, 7*(2), 131–142.
15. Stramer, B., & Mayor, R. (2017). Mechanisms and in vivo functions of contact inhibition of locomotion. *Nature Reviews Molecular Cell Biology, 18*(1), 43–55.
16. Banerjee, S., & Marchetti, M. C. (2018). *Continuum models of collective cell migration*. http://arxiv.org/abs/1805.06531
17. Fujimori, T., Nakajima, A., Shimada, N., & Sawai, S. (2019). Tissue self-organization based on collective cell migration by contact activation of locomotion and chemotaxis. *Proceedings of the National Academy of Sciences, 116*(10), 4291–4296.
18. d'Alessandro, J., Solon, A. P., Hayakawa, Y., et al. (2017). Contact enhancement of locomotion in spreading cell colonies. *Nature Physics, 13*(10), 999–1005.
19. Tambe, D. T., Hardin, C. C., Angelini, T. E., et al. (2011). Collective cell guidance by cooperative intercellular forces. *Nature Materials, 10*(6), 469.
20. Trepat, X., & Fredberg, J. J. (2011). Plithotaxis and emergent dynamics in collective cellular migration. *Trends in Cell Biology, 21*(11), 638–646.
21. He, S., Liu, C., Li, X., Ma, S., Huo, B., & Ji, B. (2015). Dissecting collective cell behavior in polarization and alignment on micropatterned substrates. *Biophysical Journal, 109*(3), 489–500.
22. Zaritsky, A., Welf, E. S., Tseng, Y.-Y., et al. (2015). Seeds of locally aligned motion and stress coordinate a collective cell migration. *Biophysical Journal, 109*(12), 2492–2500.
23. Fenteany, G., Janmey, P. A., & Stossel, T. P. (2000). Signaling pathways and cell mechanics involved in wound closure by epithelial cell sheets. *Current Biology, 10*(14), 831–838.

24. Ammann, K. R., DeCook, K. J., Tran, P. L., Merkle, V. M., Wong, P. K., & Slepian, M. J. (2015). Collective cell migration of smooth muscle and endothelial cells: Impact of injury versus non-injury stimuli. *Journal of Biological Engineering, 9*(1), 19.
25. Brugués, A., Anon, E., Conte, V., et al. (2014). Forces driving epithelial wound healing. *Nature Physics, 10*(9), 683.
26. Poujade, M., Grasland-Mongrain, E., Hertzog, A., et al. (2007). Collective migration of an epithelial monolayer in response to a model wound. *Proceedings of the National Academy of Sciences, 104*(41), 15988–15993. https://doi.org/10.1073/pnas.0705062104
27. Anon, E., Serra-Picamal, X., Hersen, P., et al. (2012). Cell crawling mediates collective cell migration to close undamaged epithelial gaps. *Proceedings of the National Academy of Sciences, 109*(27), 10891–10896.
28. Poujade, M., Grasland-Mongrain, E., Hertzog, A., et al. (2007). Collective migration of an epithelial monolayer in response to a model wound. *Proceedings of the National Academy of Sciences, 104*(41), 15988–15993.
29. Nikolic, D. L., Boettiger, A. N., Bar-Sagi, D., Carbeck, J. D., & Shvartsman, S. Y. (2006). Role of boundary conditions in an experimental model of epithelial wound healing. *American Journal of Physiology-Cell Physiology, 291*(1), C68–C75.
30. Vishwakarma, M., Di Russo, J., Probst, D., Schwarz, U. S., Das, T., & Spatz, J. P. (2018). Mechanical interactions among followers determine the emergence of leaders in migrating epithelial cell collectives. *Nature Communications, 9*(1), 1–12.
31. Gauquelin, E., Tlili, S., Gay, C., et al. (2019). Influence of proliferation on the motions of epithelial monolayers invading adherent strips. *Soft Matter, 15*(13), 2798–2810.
32. Murrell, M., Kamm, R., & Matsudaira, P. (2011). Tension, free space, and cell damage in a microfluidic wound healing assay. *PLoS ONE, 6*(9), e24283.
33. Block, E. R., Matela, A. R., SundarRaj, N., Iszkula, E. R., & Klarlund, J. K. (2004). Wounding induces motility in sheets of corneal epithelial cells through loss of spatial constraints role of heparin-binding epidermal growth factor-like growth factor signaling. *Journal of Biological Chemistry, 279*(23), 24307–24312.
34. Vitorino, P., & Meyer, T. (2008). Modular control of endothelial sheet migration. *Genes & Development, 22*(23), 3268–3281.
35. Binamé, F., Lassus, P., & Hibner, U. (2008). Transforming growth factor $β$ controls the directional migration of hepatocyte cohorts by modulating their adhesion to fibronectin. *Molecular Biology of the Cell, 19*(3), 945–956.
36. Haga, H., Irahara, C., Kobayashi, R., Nakagaki, T., & Kawabata, K. (2005). Collective movement of epithelial cells on a collagen gel substrate. *Biophysical Journal, 88*(3), 2250–2256.
37. Matsubayashi, Y., Ebisuya, M., Honjoh, S., & Nishida, E. (2004). ERK activation propagates in epithelial cell sheets and regulates their migration during wound healing. *Current Biology, 14*(8), 731–735.
38. Farooqui, R., & Fenteany, G. (2005). Multiple rows of cells behind an epithelial wound edge extend cryptic lamellipodia to collectively drive cell-sheet movement. *Journal of Cell Science, 118*(1), 51–63.
39. Klepeis, V. E., Cornell-Bell, A., & Trinkaus-Randall, V. (2001). Growth factors but not gap junctions play a role in injury-induced Ca2+ waves in epithelial cells. *Journal of Cell Science, 114*(23), 4185–4195.
40. Klepeis, V. E., Weinger, I., Kaczmarek, E., & Trinkaus-Randall, V. (2004). P2Y receptors play a critical role in epithelial cell communication and migration. *Journal of Cellular Biochemistry, 93*(6), 1115–1133.
41. Serra-Picamal, X., Conte, V., Vincent, R., et al. (2012). Mechanical waves during tissue expansion. *Nature Physics, 8*(8), 628.
42. Banerjee, S., Utuje, K. J. C., & Marchetti, M. C. (2015). Propagating stress waves during epithelial expansion. *Physical Review Letters, 114*(22), 228101.
43. Martin, P., & Parkhurst, S. M. (2004). Parallels between tissue repair and embryo morphogenesis. *Development, 131*(13), 3021–3034.

44. Solnica-Krezel, L. (2005). Conserved patterns of cell movements during vertebrate gastrulation. *Current Biology, 15*(6), R213–R228.
45. Young, P. E., Richman, A. M., Ketchum, A. S., & Kiehart, D. P. (1993). Morphogenesis in Drosophila requires nonmuscle myosin heavy chain function. *Genes & Development, 7*(1), 29–41.
46. Hutson, M. S., Tokutake, Y., Chang, M.-S., et al. (2003). Forces for morphogenesis investigated with laser microsurgery and quantitative modeling. *Science (80-), 300*(5616), 145–149.
47. Jankovics, F., & Brunner, D. (2006). Transiently reorganized microtubules are essential for zippering during dorsal closure in Drosophila melanogaster. *Developmental Cell, 11*(3), 375–385.
48. Kiehart, D. P., Galbraith, C. G., Edwards, K. A., Rickoll, W. L., & Montague, R. A. (2000). Multiple forces contribute to cell sheet morphogenesis for dorsal closure in Drosophila. *Journal of Cell Biology, 149*(2), 471–490.
49. Long, H. A., Boczonadi, V., McInroy, L., Goldberg, M., & Määttä, A. (2006). Periplakin-dependent re-organisation of keratin cytoskeleton and loss of collective migration in keratin-8-downregulated epithelial sheets. *Journal of Cell Science, 119*(24), 5147–5159.
50. Wong, P., & Coulombe, P. A. (2003). Loss of keratin 6 (K6) proteins reveals a function for intermediate filaments during wound repair. *Journal of Cell Biology, 163*(2), 327–337.
51. Christiansen, J. J., & Rajasekaran, A. K. (2006). Reassessing epithelial to mesenchymal transition as a prerequisite for carcinoma invasion and metastasis. *Cancer Research, 66*(17), 8319–8326.
52. Friedl, P., Hegerfeldt, Y., & Tusch, M. (2004). Collective cell migration in morphogenesis and cancer. *International Journal of Developmental Biology, 48*(5–6), 441–449.
53. Schmidt, M., Paes, K., De Mazière, A., et al. (2007). EGFL7 regulates the collective migration of endothelial cells by restricting their spatial distribution. *Development, 134*(16), 2913–2923.
54. Nabeshima, K., Inoue, T., Shimao, Y., et al. (2000). Front-cell-specific expression of membrane-type 1 matrix metalloproteinase and gelatinase A during cohort migration of colon carcinoma cells induced by hepatocyte growth factor/scatter factor. *Cancer Research, 60*(13), 3364–3369.
55. Soulie, P., Carrozzino, F., Pepper, M. S., Strongin, A. Y., Poupon, M.-F., & Montesano, R. (2005). Membrane-type-1 matrix metalloproteinase confers tumorigenicity on nonmalignant epithelial cells. *Oncogene, 24*(10), 1689–1697.
56. Wolf, K., Wu, Y. I., Liu, Y., et al. (2007). Multi-step pericellular proteolysis controls the transition from individual to collective cancer cell invasion. *Nature Cell Biology, 9*(8), 893–904.
57. Gaggioli, C., Hooper, S., Hidalgo-Carcedo, C., et al. (2007). Fibroblast-led collective invasion of carcinoma cells with differing roles for RhoGTPases in leading and following cells. *Nature Cell Biology, 9*(12), 1392–1400.
58. Condeelis, J., & Pollard, J. W. (2006). Macrophages: Obligate partners for tumor cell migration, invasion, and metastasis. *Cell, 124*(2), 263–266.
59. Macpherson, I. R., Hooper, S., Serrels, A., et al. (2007). P120-catenin is required for the collective invasion of squamous cell carcinoma cells via a phosphorylation-independent mechanism. *Oncogene, 26*(36), 5214–5228.
60. Wicki, A., Lehembre, F., Wick, N., Hantusch, B., Kerjaschki, D., & Christofori, G. (2006). Tumor invasion in the absence of epithelial-mesenchymal transition: Podoplanin-mediated remodeling of the actin cytoskeleton. *Cancer Cell, 9*(4), 261–272.
61. Vignjevic, D., Schoumacher, M., Gavert, N., et al. (2007). Fascin, a novel target of β-catenin-TCF signaling, is expressed at the invasive front of human colon cancer. *Cancer Research, 67*(14), 6844–6853.
62. Honda, H., & Eguchi, G. (1980). How much does the cell boundary contract in a monolayered cell sheet? *Journal of Theoretical Biology, 84*(3), 575–588.
63. Fletcher, A. G., Osterfield, M., Baker, R. E., & Shvartsman, S. Y. (2014). Vertex models of epithelial morphogenesis. *Biophysical Journal, 106*(11), 2291–2304.

64. Li, B., & Sun, S. X. (2014). Coherent motions in confluent cell monolayer sheets. *Biophysical Journal, 107*(7), 1532–1541.
65. Bi, D., Yang, X., Marchetti, M. C., & Manning, M. L. (2016). Motility-driven glass and jamming transitions in biological tissues. *Physical Review X, 6*(2), 21011.
66. Graner, F., & Glazier, J. A. (1992). Simulation of biological cell sorting using a two-dimensional extended Potts model. *Physical Review Letters, 69*(13), 2013.
67. Tarle, V., Gauquelin, E., Vedula, S. R. K., et al. (2017). Modeling collective cell migration in geometric confinement. *Physical Biology, 14*(3), 35001.
68. George, M., Bullo, F., & Campàs, O. (2017). Connecting individual to collective cell migration. *Science and Reports, 7*(1), 1–10.
69. Barton, D. L., Henkes, S., Weijer, C. J., & Sknepnek, R. (2017). Active vertex model for cell-resolution description of epithelial tissue mechanics. *PLoS Computational Biology, 13*(6), e1005569.
70. Noll, N., Mani, M., Heemskerk, I., Streichan, S. J., & Shraiman, B. I. (2017). Active tension network model suggests an exotic mechanical state realized in epithelial tissues. *Nature Physics, 13*(12), 1221.
71. Prost, J., Jülicher, F., & Joanny, J.-F. (2015). Active gel physics. *Nature Physics, 11*(2), 111.
72. Notbohm, J., Banerjee, S., Utuje, K. J. C., et al. (2016). Cellular contraction and polarization drive collective cellular motion. *Biophysical Journal, 110*(12), 2729–2738.
73. Arciero, J. C., Mi, Q., Branca, M. F., Hackam, D. J., & Swigon, D. (2011). Continuum model of collective cell migration in wound healing and colony expansion. *Biophysical Journal, 100*(3), 535–543.
74. Haeger, A., Wolf, K., Zegers, M. M., & Friedl, P. (2015). Collective cell migration: Guidance principles and hierarchies. *Trends in Cell Biology, 25*(9), 556–566.
75. Du Roure, O., Saez, A., Buguin, A., et al. (2005). Force mapping in epithelial cell migration. *Proceedings of the National Academy of Sciences, 102*(7), 2390–2395.
76. Gov, N. S. (2007). Collective cell migration patterns: Follow the leader. *Proceedings of the National Academy of Sciences, 104*(41), 15970–15971.
77. Trepat, X., Wasserman, M. R., Angelini, T. E., et al. (2009). Physical forces during collective cell migration. *Nature Physics, 5*(6), 426.
78. Petitjean, L., Reffay, M., Grasland-Mongrain, E., et al. (2010). Velocity fields in a collectively migrating epithelium. *Biophysical Journal, 98*(9), 1790–1800.
79. Evans, E., & Yeung, A. (1989). Apparent viscosity and cortical tension of blood granulocytes determined by micropipet aspiration. *Biophysical Journal, 56*(1), 151–160.
80. Theret, D. P., Levesque, M. J., Sato, M., Nerem, R. M., & Wheeler, L. T. (1988). The application of a homogeneous half-space model in the analysis of endothelial cell micropipette measurements. *Journal of Biomechanical Engineering, 110*(3), 190–199.
81. Müller, D. J., & Dufrêne, Y. F. (2011). Atomic force microscopy: A nanoscopic window on the cell surface. *Trends in Cell Biology, 21*(8), 461–469.
82. Pelling, A. E., & Horton, M. A. (2008). An historical perspective on cell mechanics. *Pflügers Archiv-European Journal of Physiology, 456*(1), 3–12.
83. Shin, D., & Athanasiou, K. (1999). Cytoindentation for obtaining cell biomechanical properties. *Journal of Orthopaedic Research, 17*(6), 880–890.
84. Daily, B., Elson, E. L., & Zahalak, G. I. (1984). Cell poking. Determination of the elastic area compressibility modulus of the erythrocyte membrane. *Biophysical Journal, 45*(4), 671–682.
85. Eekhoff, A., Bonakdar, N., Alonso, J. L., Hoffmann, B., & Goldmann, W. H. (2011). Glomerular podocytes: A study of mechanical properties and mechano-chemical signaling. *Biochemical and Biophysical Research Communications, 406*(2), 229–233.
86. Kollmannsberger, P., & Fabry, B. (2007). BaHigh-force magnetic tweezers with force feedback for biological applications. *Review of Scientific Instruments, 78*(11), 114301.
87. Van Vliet, K. J., Bao, G., & Suresh, S. (2003). The biomechanics toolbox: Experimental approaches for living cells and biomolecules. *Acta Materialia, 51*(19), 5881–5905.
88. Lincoln, B., Wottawah, F., Schinkinger, S., Ebert, S., & Guck, J. (2007). High-throughput rheological measurements with an optical stretcher. *Methods in Cell Biology, 83*, 397–423.

89. Bhat, S., Jun, D., Paul, B., & Dahms, T. E. S. (2012). Viscoelasticity in biological systems: a special focus on microbes. *Viscoelasticity—From theory to biological applications* (pp. 123–156). intech.
90. Yamada, S., Wirtz, D., & Kuo, S. C. (2000). Mechanics of living cells measured by laser tracking microrheology. *Biophysical Journal, 78*(4), 1736–1747.
91. Weitz, D., Wyss, H., & Larsen, R. (2007). Oscillatory rheology: Measuring the viscoelastic behaviour of soft materials. *GIT Laboratory Journal Europe, 11*(3–4), 68–70.
92. Lai, S. K., Wang, Y.-Y., Wirtz, D., & Hanes, J. (2009). Micro-and macrorheology of mucus. *Advanced Drug Delivery Reviews, 61*(2), 86–100.
93. Bulusu, K. V., Racan, S., & Plesniak, M. W. (2020). Macro-rheology characterization of gill raker mucus in the silver carp, hypophthalmichthys molitrix. *JoVE (Journal of Visualized Experiments), 161*, e61379.
94. Weigand, W. J., Messmore, A., Tu, J., et al. (2017). Active microrheology determines scale-dependent material properties of Chaetopterus mucus. *PLoS ONE, 12*(5), e0176732.
95. Patarin, J., Ghiringhelli, É., Darsy, G., et al. (2020). Rheological analysis of sputum from patients with chronic bronchial diseases. *Science and Reports, 10*(1), 1–10.
96. Gambini, C., Abou, B., Ponton, A., & Cornelissen, A. J. M. (2012). Micro-and macrorheology of jellyfish extracellular matrix. *Biophysical Journal, 102*(1), 1–9.
97. Madsen, C. D., & Cox, T. R. (2017). Relative stiffness measurements of tumour tissues by shear rheology. *Bio-Protocol, 7*(9), e2265–e2265.
98. Espinosa, G., López-Montero, I., Monroy, F., & Langevin, D. (2011). Shear rheology of lipid monolayers and insights on membrane fluidity. *Proceedings of the National Academy of Sciences, 108*(15), 6008–6013.
99. Fam, H., Bryant, J. T., & Kontopoulou, M. (2007). Rheological properties of synovial fluids. *Biorheology, 44*(2), 59–74.
100. Metzger, T. A., Shudick, J. M., Seekell, R., Zhu, Y., & Niebur, G. L. (2014). Rheological behavior of fresh bone marrow and the effects of storage. *Journal of the Mechanical Behavior of Biomedical Materials, 40*, 307–313.
101. Tran-Ba, K.-H., Lee, D. J., Zhu, J., Paeng, K., & Kaufman, L. J. (2017). Confocal rheology probes the structure and mechanics of collagen through the sol-gel transition. *Biophysical Journal, 113*(8), 1882–1892.
102. Burla, F., Dussi, S., Martinez-Torres, C., Tauber, J., van der Gucht, J., & Koenderink, G. H. (2020). Connectivity and plasticity determine collagen network fracture. *Proceedings of the National Academy of Sciences, 117*(15), 8326–8334.
103. Schmidt, F. G., Hinner, B., & Sackmann, E. (2000). Microrheometry underestimates the values of the viscoelastic moduli in measurements on F-actin solutions compared to macrorheometry. *Physical Review E, 61*(5), 5646.
104. Chen, D. T. N., Wen, Q., Janmey, P. A., Crocker, J. C., & Yodh, A. G. (2010). Rheology of soft materials. *Annual Review of Condensed Matter Physics, 1*(1), 301–322.
105. Fernández, P., Heymann, L., Ott, A., Aksel, N., & Pullarkat, P. A. (2007). Shear rheology of a cell monolayer. *New Journal of Physics, 9*(11), 419. https://doi.org/10.1088/1367-2630/9/11/419
106. Dakhil, H., Gilbert, D. F., Malhotra, D., et al. (2016). Measuring average rheological quantities of cell monolayers in the linear viscoelastic regime. *Rheologica Acta, 55*(7), 527–536.
107. Sander, M., Flesch, J., & Ott, A. (2015). Using cell monolayer rheology to probe average single cell mechanical properties. *Biorheology, 52*(4), 269–278.
108. Brückner, B. R., & Janshoff, A. (2018). Importance of integrity of cell-cell junctions for the mechanics of confluent MDCK II cells. *Science and Reports, 8*(1), 1–11.

Investigating Mechanical Anisotropy of Excised Porcine Skin Using Dog-Bone Specimen

Piyush Lakhani and Navin Kumar

Abstract The natural alignment of the collagen fibers and its failure mechanics for the skin plays important in plastic surgery, dermatology, and forensic science. Skin is an anisotropic composite material made of elastin and collagen fibers along with ground substance. Effect of collagen fibers orientation on the mechanical properties and its failure mechanics was investigated in the present article. Custom-built uniaxial testing setup was developed for the quantification of the mechanical properties. Imaging technique was used for the measurement of the local strain distribution over the specimen surface. Dogbone shaped specimens were prepared in four different orientation parallel, 30°, 60° and perpendicular to the collagen fibers direction. Stress strain relations obtained from experiments were used for the calculation of mechanical properties like ultimate tensile strength, elastic modulus in toe region and linear region, transition point and strain energy absorbed before failure. Elastic modulus in the linear region, ultimate tensile strength and strain energy were higher for the parallel orientation of collagen. Scanning electron microscopy (SEM) images explains the role of collagen as the load bearing material. In the direction of collagen alignment fibrils fractured plastically and display flat failure. While perpendicular to the alignment direction fibrils were fractured and display a fringes pattern. Finally, finite element modeling (FEM) along with the imaging technique with the experiment was used to explain the role of specimen shape on the measurement of mechanical properties. In short, the present study explains the failure mechanics and provides the essential data and information for the modeling of skin for finite element simulation using anisotropic constitutive model.

P. Lakhani (✉) · N. Kumar
Department of Mechanical Engineering, IIT Ropar, Rupnagar, India
e-mail: piyushlakhani1991@gmail.com

K. K. Dwivedi et al. (eds.), *Mechanics and Materials Science of Biological Materials*, Materials Horizons: From Nature to Nanomaterials,
https://doi.org/10.1007/978-981-96-5366-9_9

1 Introduction

Skin is three-layer structure consist of epidermis, dermis and hypodermis [1–3]. Collagen fibers are main structural part of skin which provides strength and contributed 75% dry weight of dermal tissue [4]. Various experimental techniques like uniaxial, biaxial, suction, inflation, indentation and imaging are used to study in vivo and in vitro mechanical properties of skin tissue from nano scale to macro scale properties [5–17]. Variation in mechanical properties of skin are subjected to humidity, location on body and position, age, nutrition, disease etc. [8, 18–26]. Stress strain relation for skin is nonlinear, direction and time dependent which represent the hyperelastic, anisotropic and viscoelastic properties, respectively [16, 17, 27].

Anisotropic properties of skin were recognized in 19th century. Langer was mapped the direction of natural tension lines (Langer lines) which represent the direction of collagen fibers [28]. Various reported study indicated the effect of Langer line on the properties of the skin. Further, these study reported the large variation in the tangent modulus, tensile strength and rupture strain along and perpendicular to the direction of collagen fibers [12, 15–17]. Microstructural changes at the collagen fibers and fibrils level can be monitored in-situ and ex-situ using the small angle X-ray scattering [29], Small angle light scattering [30, 31] and Scanning electron microscopy [14, 32–34]. Fibers were rotated and realign along the direction of loading, at higher strain fibrils and fibers were slide and delaminated [33]. These observations are reported for the cross section parallel to skin surface and out of plane deformation mechanics was rarely reported.

Uniaxial test is most common technique for mechanical testing because of its effortlessness of execution and standardized method. Even though the biaxial test can explain better anisotropy but has its own complications and limitations. Therefore, skin with the varying direction of Langer line is used for anisotropic quantifications. In vivo studies have limitation of loading where study of damage and failure is not possible. The accessibility of sample is also restricted for the in vivo studies therefore in vitro study are preferable. Uniaxial test of excised skin for anisotropic behavior of skin was introduced by Ridge et al. [35], where study was conducted along and perpendicular to the Langer lines of human cadaver skin. The mechanical properties at broad range of strains and strain rates were reported with the development of technology for uniaxial tests. These studies were used the rectangular [8, 29, 33, 34, 36–40] or dumbbell [12, 15, 17, 41–45] shape specimens. Further, these studies can be categorized based on load limits within linear zone [33, 36, 44] or up to the failure [8, 12, 15, 17, 34, 36–39, 41–43, 45] of specimens.

Study of the failure mechanics for skin and other soft tissue is unresolved problem. Reliable achievement of continuous and uniform clamping load distribution at the junction of specimen and clamp is challenging task. Difficulty arises due to low friction between the clamp and specimens. Friction force can be increased by higher clamping force which may leads to excessive deformation or premature failure during test for the rectangular specimens because of multiple reasons. First, sample is flattened and cross section area is reduced. Second, the tissue is dehydrated due to the squeezing of water. Various techniques were adopted to increase friction force such as improved surface geometry of clamp, adding the sandpaper and use of adhesives, etc. These all techniques for the rectangular or uniform cross-section specimens resulted in premature failure at the junction leads to wrong interpretation of mechanical properties. Limitations associated with the uniform shape specimens can be overcome with the use of dumbbell shape specimen.

Dumbbell shape specimen is used to apply uniform strain within the gauge length. Typical stress strain curve for uniaxial is J-shaped, it is used to extract the properties like tangent modulus, strain energy and anisotropy with preferred orientation of samples [12, 15, 17]. Additionally, imaging techniques like Digital Image Correlation (DIC) [46, 47] and Optical Coherence Tomography [48] are used to increase the accuracy of the measurement and understand the deformation mechanics.

In the present article, experiment is carried out on freshly excised pig skin, it has structural and histological similarity with the human skin [25, 49]. Experiment is done on custom built setup made from the syringe pump for uniaxial tensile test until the failure. Properties are direction dependent therefore specimen from various angle is tested to determine the effect of collagen orientation. Skin is soft material and properties are dependent on structural integrity, therefore strain distribution over the specimen is not likely to be uniform. 2D Digital Image Correlation (DIC) is used to calculate the local strain over the specimen. Microstructural changes near the failure site for different loading direction was observed using the SEM. Finally, Finite element modeling was implicating with hyperelastic material properties to observe the shape effect of specimen on the measured mechanical properties. This study article is aiming to provide the deeper understanding of the failure mechanics and quantitative values of engineering properties and anisotropy.

2 Material and Method

2.1 Tensile Specimen Preparation

Approximate 300 × 400 mm size skin sample was procured from the Loin body part of the pig in such a way that minimum fat tissue was attached without any damage, it was collected from the slaughterhouse within 1 h of the sacrifice. Sample preparation and experimental procedure is done as per approved protocol of institutional ethics committee. Sample preparation method is followed as described by the Tang and Bhushan [7]. Sample was kept in Phosphate Buffered Saline (PBS) and stored at 4 °C for 24 h before sample cutting. Average thickness of skin sample with epidermis and dermis layer was in between 3 and 4 mm through the whole specimen.

Dumbbell shape specimen was cut using the pneumatic sample cutter. Figure 1 shows the detail dimension of cutter and prepared skin sample for tensile test. Total 40 samples were prepared 10 from each four angles parallel, 30°, 60°, and perpendicular with respect to Langer line. The thickness of each sample was measured using the

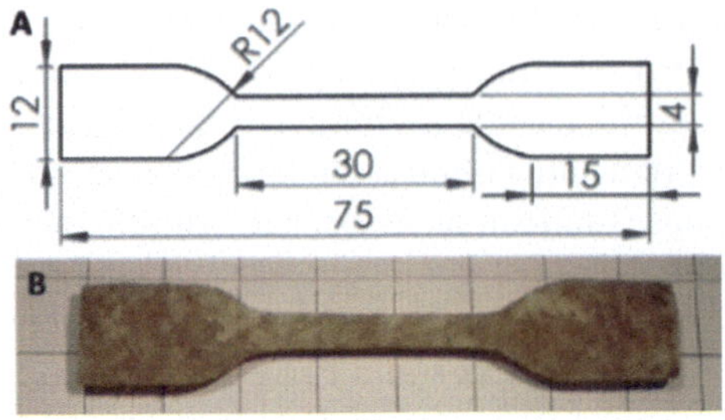

Fig. 1 **a** Detailed dimensions of the cutter dye in mm **b** dumbbell shape pig skin specimen after cutting on a pneumatic cutter

digital micrometer (Mitutoyo IP-65, 0.001 mm least count) at three different locations along gauge length and taken average. Sand paper of grit size P60 was glued at both end at the position of grip on both side using the low viscosity adhesive (Instant adhesive, Permabond) to make anti-slip surface to counter the leaning of slip from the grip. Experimental setup.

2.1.1 Tensile Test

As shown in Fig. 2, experimental setup consist of two syringe pumps (Harvard apparatus, pump 11 elite) were aligned on the aluminum base breadboard using the laser alignment tool to position the perfect orientation of two grips which were mounted on the custom built fixture. One of two grip was mounted on the 445 N (100 lb.) piezoelectric loadcell (Accuracy: 0.1%) to measure the load. Two syringe pump and load cell were controlled through the LabView (National Instrument, USA) and data were acquired using Data Acquisition (NI c-DAQ 9174, NI 9237) system.

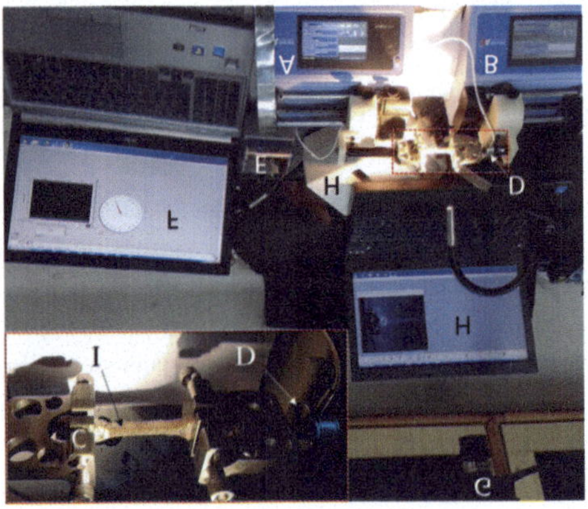

Fig. 2 Custom built experimental setup with detailed view of grip with skin specimen. **a** Syringe pump with grip fixture **b** Syringe pump with load cell and grip. **c** Tensile test grip. **d** Piezoelectric load cell. **e** Data acquisition system. **f** Computer system with LabView to acquire signal from load cell. **g** Camera for digital image capture. **h** Light source. **i** Dumbbell shape pig skin specimen

Displacement is measured by converting flow volume into the displacement per unit time. 50 ml syringe of diameter 29.2 mm chosen as syringe type, it leads to the displacement of 1.49329 mm per milliliter of fluid. Maximum and minimum traveling rate of slider on which grip was mounted having 2.65 and 2.5 nm/s respectively for each pump. Therefore, maximum, and minimum displacement rate of test setup was becoming 5.3, and 2.5 nm/s respectively and total stroke length is 149.33 mm. For quasi static tensile experiment, strain rate was 0.01/s. as gauge length is 30 mm displacement rate was 0.3 mm/s achieved by 0.2009 ml/s flow rate.

2.1.2 Digital Image Correlation

DIC is the non-contact type displacement measuring technique, it tracks the unique features of the gray scale visible surface image. It correlates the first image with other image or incremental images to measure the displacement and deformation. 5-megapixel Flir system Inc. camera was used to capture the images at 5 frames per second. Random spackle pattern of black India ink was generated by the air gun using tip size of 0.4 mm. Images were analyzed in Vic 2D (Correlated Solution, Inc) software to calculate the strain. Advantage of DIC to measure the strain over the measured from the traveling speed of syringe pump is that, DIC have capabilities to measure the full-field local strain distributed over the specimen.

2.2 *Hyperelastic Material Model*

Skin soft tissue shows the nonlinear hyperelastic relation between the stress and strain which was modeled by model the Neo-Hookean, Mooney-Rivlin, Yeoh, Ogden, etc. MATLAB (Math Works inc., USA) function 'lsqcurvefit' used for nonlinear least square procedure with 'Levenberg–Marquardt' algorithm to determine the best fitted constant with experiment stress strain curve. In present article Yeoh hyperelastic material model is used for incompressible, nonlinear elastic material. It is function of strain invariant [50–53]. Relation between stress and stretch for uniaxial tensile test in the form of stretch λ and constant C_{i0} is define by Eq. 1.

$$\sigma = 2 \cdot \left(\lambda - \lambda^{-2}\right) \cdot \sum_{i=1}^{3} \left(i \cdot C_{i0} \cdot \left(\lambda^2 + \frac{2}{\lambda} - 3\right)^{i-1} \right) \tag{1}$$

2.3 SEM

Strip of sample for the SEM was cut using the surgical blade. It was immersed in 2.5% glutaraldehyde for the 3-h duration to fix the structure. Ascending ethanol of 30, 50, 70 90, 95, 100 and 100 volume percentage for the duration of 30 min each was followed to dehydrate the sample [33].

2.4 Finite Element Modeling

Two different model were generated for the static analysis of dumbbell shaped specimen Abaqus. First, Computer Aided Design (CAD) tool (Solidworks, Dassault Systèmes, USA) was used to generate the 3D model of the dimension shown in Fig. 1 having the thickness of 2.5 mm and length between two grips 45 mm was kept for the analysis. Second model was generated using the micro computed tomography (CT, Nanotom S, Phonix General Electronics). Dumbbell shape specimen was dehydrated following similar procedure as described for SEM sample preparation. X ray tube was operated at 63 kV and 180 μA to generate the images. 3D model was generated from the 500 projection and exposure time was 500 ms. Images were captured at resolution of 30 μm. Gaussian smoothing filter of 2 pixel was applied while reconstruction of the 3D model from the images. Linear tetrahedral element C3D4H for the mesh. Yeoh material model was used for the analysis. One end of the model was encastred to represent the fix grip and at the other end predefined displacement was applied.

2.5 Statistical Analysis

Statistical analysis was performed to test the significant effect of Langer line on the properties like modulus and strain energy of the skin. One-way analysis of variance is used. Tukey–Kramer post hoc test is performed by 'multcompare' MATLAB function to compare the influence of collagen alignment direction on multiple group with each other. $p \leq 0.05$ is consider as the level of significance. To categorize the level of significance in figures, $p \leq 0.05$, $p \leq 0.01$, and $p \leq 0.001$ represented by *, **, and *** respectively.

3 Results and Discussion

Tensile test was performed on total 48 skin sample, out of that total 36 sample were used in result analysis. Sample with very low load capacity and failure near the grip are excluded from the analysis, unusual results occurred because of excessive load on the grip or damage to the tissue while sample preparation. Experiment gives the data of load vs deformation. Stress was calculated using force, and cross section area. Strain was calculated from the time, as traveling speed was constant. Figure 3 shows the typical stress strain curve obtained from the experiment having three distinct regions.

Mechanical testing technique was validated by simply comparing the result obtained through experiment with the previously reported modulus value for the skin. Modulus ratio and strength ratio of the linear region in the direction and perpendicular direction with respect to collagen alignment were fall within the comparable range as shown in Table 1. The mechanical properties calculated from the stress strain curve are summarized in the Table 2.

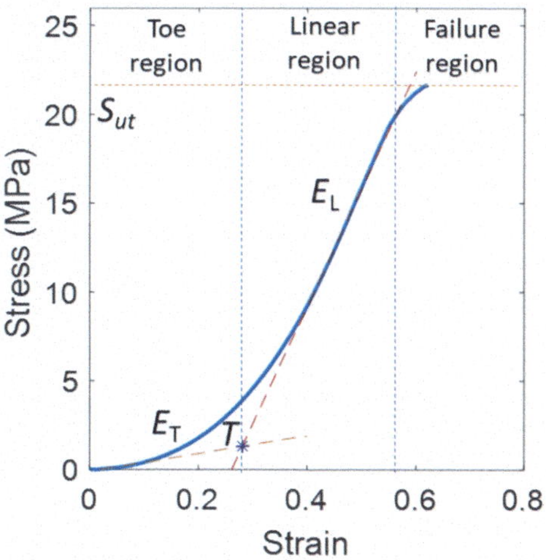

Fig. 3 Stress strain curve obtained after normalizing the load and displacement. Failure stress is define as tensile strength indicated by *S*, Slope of curve in Toe and Linear region is define as modulus E_T and E_L respectively, Intersection of linear fit in for E_T and E_L is define as the transition point indicated by *T*

Table 1 Comparison of the results obtained in the literature for the validation of the experiment Mean values are presented for the uniaxial tensile test

Linear elastic modulus (MPa)			Ultimate tensile strength (MPa)			Source
Parallel	Perpendicular	Ratio	Parallel	Perpendicular	Ratio	
80	50	1.6	20	15	1.3	Pissarenko [12]
108	37	2.91	32	11	2.90	Ottenio et al. [15]
73	37	1.97	17	10	1.70	Annaidh et al. [17]
38	29	1.31	9	6	1.5	Belkoff et al. [45]
68	35	1.94	22	12	1.83	Present study

Table 2 Shows mean and standard deviation properties for each test group

Angle with collagen fibers	Modulus (MPa)		Transition point		Strain energy (MJ/m^3)	Failure point	
	Toe region (E_T)	Linear region (E_L)	Strain	Stress (MPa)		Stress (MPa)	Strain
0°	3.40 ± 1.03	68.03 ± 8.68	0.22 ± 0.02	0.71 ± 0.21	4.52 ± 0.88	21.97 ± 1.61	0.56 ± 0.03
30°	3.08 ± 1.19	60.15 ± 3.31	0.23 ± 0.03	0.66 ± 0.19	3.49 ± 0.38	19.42 ± 1.31	0.56 ± 0.04
60°	3.01 ± 1.01	48.93 ± 3.21	0.22 ± 0.04	0.59 ± 0.14	3.63 ± 1.03	16.60 ± 1.53	0.58 ± 0.04
90°	2.95 ± 0.41	35.40 ± 3.57	0.23 ± 0.02	0.59 ± 0.09	2.47 ± 0.22	12.22 ± 0.87	0.59 ± 0.031

Nonlinear stress strain curve describe by the three regions[15], initially collagen fibers are very compliant and load is taken by the elastin fibers and collagen are rotate, became straight, and started to elongate, secondly in linear region loads are taken by the collagen fibers where fibers are stretched, slide and delaminate, and in third phase collagen fibers are fracture and stress strain curve became horizontal and start falling [33]. Range of modulus in linear region was found 35–108 MPa as detailed in Table 1. Elastic modulus of skin along and perpendicular to the Langer line was 68.03 ± 8.68 MPa and 35.40 ± 3.57, ultimate tensile strength was 21.97 ± 1.61 and 12.22 ± 0.87 respectively. Strain energy range between 2.5 and 4.5 MJ/ m^3. This large variation is observed due to anisotropic nature, test condition, testing method and biological variation. In vitro studies are subjected to the natural condition of skin while grafting and cleaning of surrounding fat tissue and hair. In vivo study are more realistic study but, having limitation as it does not allow to apply load up to the failure.

3.1 Failure Mechanism

SEM images and schematic of reduced numbers of collagen schematic for unstarched state of collagen is shown in Fig. 4a, b, f and g. Uniaxial load along the collagen direction were resulted in rotation and straightening of the fibers, it starts the resisting the load (Fig. 4c, and h) and exponential stiffening of skin took place (Fig. 4k). Upon application of the excessive load fibrils were fractured and recovered back up to some extent. Orientation of collagen fibers with the loading direction shows different behavior near fracture site. Loading in the direction of collagen fibers (Fig. 4b–d), nearly all fibers were aligned in the direction, and fibrils were fails plastically and does not recover marginally. At the fracture site fibrils were aligned along the direction of collagen (Fig. 4d and e). Loading perpendicular to the collagen direction (Fig. 4g–i), several fibers were fails plastically, remaining perpendicular orientation collagen does not took part in failure mechanism and shows rugged fringe pattern distribution near the fracture site (Fig. 4i and j) represent the orientation of collagen perpendicular to the loading direction in natural state.

DIC provides the full-field strain distribution all over the specimen that used to validate the theoretically measured strain, and strain concentration near the failure site. It was used in several in vivo and investor studies to validate the computational results for soft tissue [54–56]. Strain measured from the digital image correlation was not consistent with the theoretically measured strain. Theoretical calculation considers the gauge length and the displacement of the end grips, while DIC gives the value of local strain. Figure 5 shows the full-field strain distribution just before the rupture, deeper inspection near the failure site indicates considerably higher local strain near the failure point. It was observed 20–50% higher as compared to nominal strain calculated theoretically.

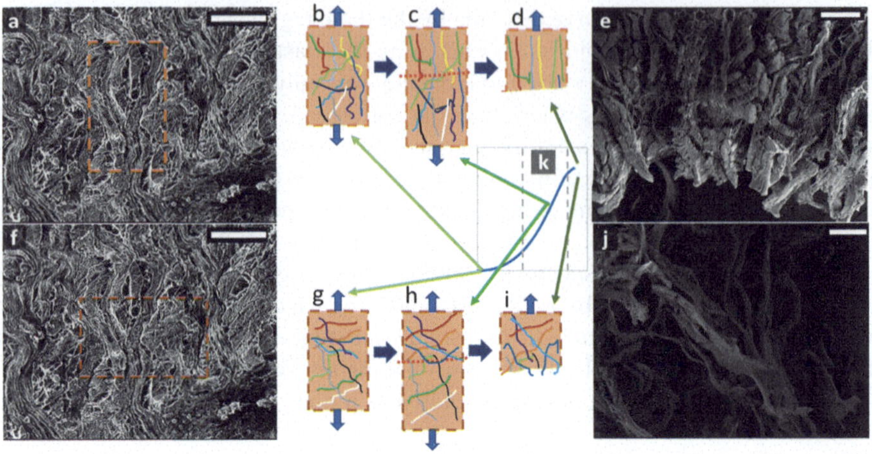

Fig. 4 Failure mechanism of the skin near the fracture site. SEM images (**a**, **e**, **f**, **j**) and schematic diagram with reduced number of collagen (**b**–**d**, **g**–**i**), and typical stress strain curve (**k**). Red dotted line **c**, **h** represent the fracture location. Collagen were rotated and makes aligned large number of fibers for uniaxial load along the fibers direction (c), fractured state shows the alignment in single direction (**d**, **e**). For the loading perpendicular to collagen direction several fibers (not all) were rotated and start resisting the loads (**h**), fractured state (**i**, **j**) shows rugged fringes pattern, it signify the collagen orientation perpendicular to the loading direction. Scale bar for SEM images were 100 μm

Fig. 5 Full-field distribution of the Lagrange strain along the load direction just before the rupture Localized higher strain was occurred near the failure zone

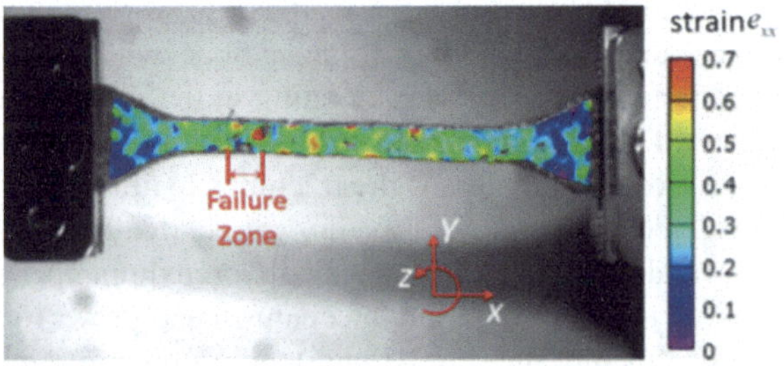

3.2 *Inter Group and Intra Group Variation*

Effect of collagen fiber direction and Langer lines on the mechanical properties are well defined qualitative phenomenon, present study shows the quantitative variation in the modulus, strength, energy, transition, and failure stress, and strain at four different angles with statistical analysis to signify the effect level. Variation in biological samples are affected by the several parameters. Intra group variation is due to anisotropy, body site, natural tension state, humidity level, etc. Large variation in the intergroup specimen was primarily due to effect of collagen fiber orientation and anisotropy of skin. Figure 6 shows the intragroup and intragroup variation in

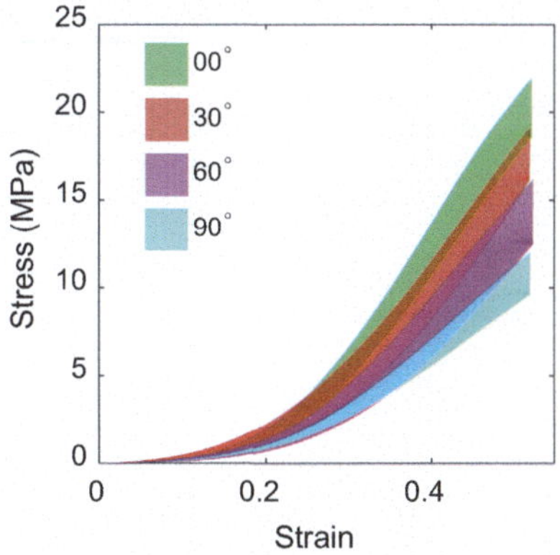

Fig. 6 Stress–strain relation for the all tested specimen (N = 36). Color represent the different direction of the specimen with respect to collagen fibers direction. Wide span of same color represents the intra group variation and wide span of different color represent the intergroup variation

the form of stress strain curve. Width of color represent the intragroup variation and difference between two color spans represent the intergroup variation.

3.3 *Influence of Orientation on Mechanical Properties*

Four group of samples parallel, 30°, 60°, and perpendicular to the collagen direction were studied. Figure 7 shows the influence of collagen distribution direction on the engineering properties of the skin. Collagen fiber alignment does not have significant effect on the modulus in toe region, transition stress and strain, and failure strain. Modulus in the linear region varies considerably across the parallel orientation to 60° ($p \leq 0.01$) and 90° ($p \leq 0.001$), 30°–60° ($p \leq 0.01$) and 90° ($p \leq 0.001$), and 60°–90° ($p \leq 0.01$). Strain energy have significant effect of the orientation between 0°–90° ($p \leq 0.001$) and 30°–90° ($p \leq 0.01$). Failure stress varies slightly across the parallel orientation to 30° ($p \leq 0.05$) and considerably across parallel to 60° ($p \leq 0.01$) and 90° ($p \leq 0.001$). Similarly, significant variation found across 30°–60° ($p \leq 0.01$) and 90° ($p \leq 0.001$), and 60°–90° ($p \leq 0.01$).

Statistical data shows the higher failure stress parallel to the collagen direction as compare to other direction shows the contribution of the collagen fiber strength on the ultimate tensile strength. Modulus in the linear region having the significant effect of collagen fiber orientation as linear region is collagen dominated region. Toe region modulus is elastin dominated; hence value does not show significant effect of collagen orientation. Transition point is tentative representation of the intermediate point between the elastin dominated elastin region and collagen dominated linear region. Statistical data shows no significant differences for the transition strain and transition stress.

In present article, Langer lines direction is considered perpendicular to the body axis at the spare ribs from where sample was excised. Noninvasive optical method accurately determine the collagen direction makes this kind of study more practicable

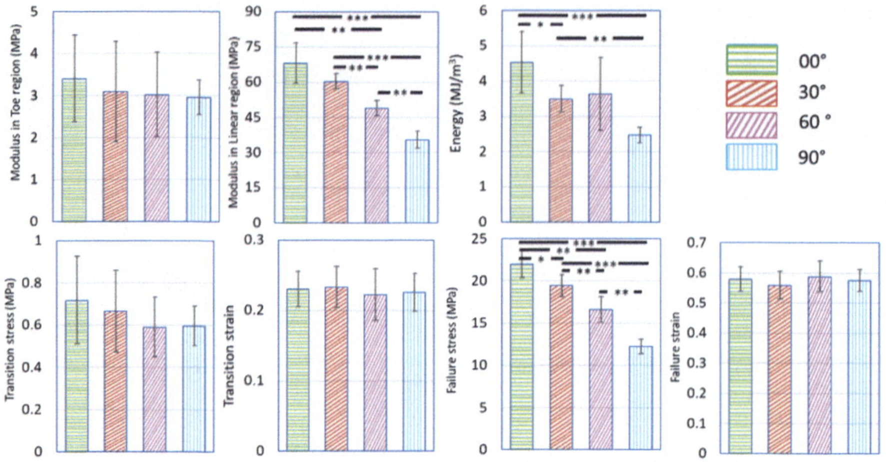

Fig. 7 Influence of collagen direction on the mean of modulus in toe and linear region, strain energy, transition stress and strain, failure stress and strain. Error bar represent the standard deviation

[30, 57]. Quantitative data on effect of the collagen on the material strength and skin mechanical properties were simplified the computational anisotropic modeling of the skin. Stress strain relation is beneficial in the structure base modeling of skin.

3.4 *Hyperelastic Properties*

The constant parameter was calculated from the data obtained through experiments, an algorithm was optimized to get the goodness of fit (R^2) greater than 0.9970 by defining the initial values, lower and upper boundary range for the constants [44, 58]. Table 3 shows the material constant with the mean and standard deviation for the all four different orientation of the collagen fibers. These properties are useful in the computational modeling of skin [44, 58–66]. Combination of four directions large deformation mechanical properties can be used for structure based computational anisotropic model [58].

3.5 *Effect of Specimen Shape*

Model developed using the CT has end to end length between the grip was 37.31 mm, reduced due to shrinkage while dehydration process. 15 mm (nearly 60% of gauge length) displacement was applied. Similarly, for CAD model have a total length 45 mm and displacement of 18 mm (60% of original gauge length) was applied. Material properties are taken from Table 3 for incompressible hyperelastic Yeoh

Table 3 Optimized value of material constant with mean and standard deviation for four different collagen alignment specimens

Orientation	Material constants			R-square
	C10 (MPa)	C20 (MPa)	C30 (MPa)	
00°	−0.6431 ± 0.4251	12.4799 ± 2.5485	−4.8966 ± 1.8142	0.9989
30°	−0.3325 ± 03,870	9.4214 ± 2.1764	−2.9652 ± 1.6026	0.9991
60°	−0.0499 ± 0.6541	8.6726 ± 1.6869	−3.4325 ± 1.3228	0.9986
90°	−0.1510 ± 0.1836	6.6371 ± 1.3345	−2.6403 ± 0.9550	0.9995

model. Strain pattern measured from the DIC and computational having the similar pattern as described in Fig. 8a–c. Engineering strain calculated through the experiment was not consistence with the average strain calculated in the gauge length from the DIC and FEM. Measurement of gauge length was with the ideally zero strain assumption outside the selected length.

Shape of the specimen is affected the stress and strain results as shown in Fig. 8d. Stress was taken from the experiment and strain was taken as the mean of strain for at least 200 points or node within the gauge length. Results for the DIC, CT model and CAD model shows the good consistency with each other. Experimental stress strain curve was followed different path because of assumption of gauge length. To overcome inaccuracy in the measurement experimental method should be followed by the computation or full-field imaging technique for the measurements.

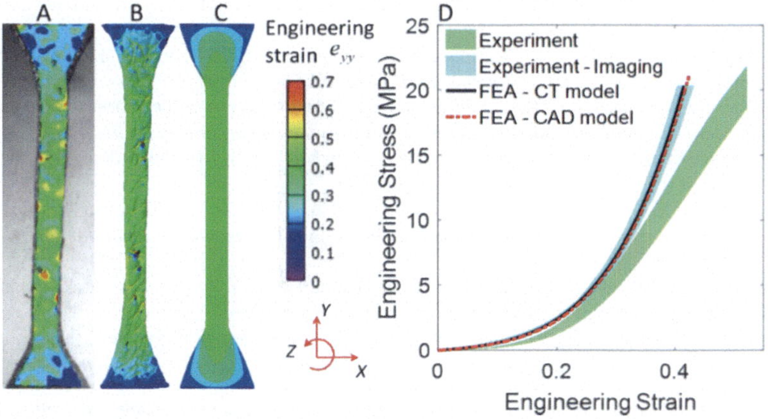

Fig. 8 Computed engineering strain from **a** imaging technique **b** FE analysis of CT model **c** FE analysis of CAD model. **d** Custom-built between the engineering stress calculated from the experiment, and average of engineering strain in gauge length obtained from **a**, **b**, **c**

4 Conclusion

Present article concluded about role of the collagen fibers and its alignment on the mechanical properties, and failure mechanics of the material. Initially distribution of collagen fibers in the skin found to be highly disorder and curly. Load makes them aligned in the particular direction. Fracture along the collagen direction makes fibers highly aligned, and fracture perpendicular to the collagen direction makes rugged fringes pattern which does not took part in resistance to the load.

Data Availability Not applicable.

Declaration of Competing Interest The authors have no conflicts to disclose.

References

1. Elsner, P., Berardesca, E., & Maibach, H. I. (Eds.). (1994). *Bioengineering of the skin: Water and the Stratum Corneum.*
2. Wilhelm, K.-P., Berardesca, E., Elsner, P., & Maibach, H. I. (1997). *Biengineering of the skin: Skin imaging and analysis.* Boca Raton: CRC Press.
3. Foster, T. (2002). *Cosmetic lipids and the skin barrier.* Marcel Dekker.
4. Wilkes, G., Brown, I., & Wildnauer, R. (1973). The biomechanical properties of skin. *CRC Critical Reviews in Bioengineering, 1*(4), 453–495.
5. Piérard, G. E., Piérard, S., Delvenne, P., & Piérard-franchimont, C. (2013). In vivo evaluation of the skin tensile strength by the suction method: pilot study coping with hysteresis and creep extension. *ISRN Dermatology, 2013*, 1–7.
6. Del Prete, Z., Antoniucci, S., Hoffman, A. H., & Grigg, P. (2004). Viscoelastic properties of skin in Mov-13 and Tsk mice. *Journal of Biomechanics, 37*, 1491–1497.
7. Tang, W., & Bhushan, B. (2010). Adhesion, friction and wear characterization of skin and skin cream using atomic force microscope. *Colloids Surfaces B Biointerfaces, 76*, 1–15.
8. Pawlaczyk, M., Lelonkiewicz, M., & Wieczorowski, M. (2013). Age-dependent biomechanical properties of the skin. *Postępy Dermatologii, 30*(5), 302–306.
9. Regoire, J. M. G., Aillant, L. V., All, Y. G., & Atat, F. P. (1998). An in vivo method for measuring the mechanical properties of the skin using ultrasound. *Ultrasound in Medicine and Biology, 24*(2), 215–224.
10. Surgery, P., & Hospital, A. R. (1999). Tensile properties of relaxed exicised skin exhibiting striae distensae. *Journal of Medical Engineering & Technology, 23*(2), 69–72.
11. Dombi, G. W., Haut, R. C., & Sullivan, W. G. (1993). Correlation of high- speed tensile strength with collagen content in control and lathyritic rat skin.pdf. *Journal of Surgical Research, 54*, 21–28.
12. Pissarenko, A., Yang, W., Quan, H., Brown, K. A., & Williams, A. (2019). Tensile behavior and structural characterization of pig dermis. *Acta Biomaterialia, 86*, 77–95.
13. Gahagnon, S., Mofid, Y., Josse, G., & Ossant, F. (2012). Skin anisotropy in vivo and initial natural stress effect: A quantitative study using high-frequency static elastography. *Journal of Biomechanics, 45*(16), 2860–2865.
14. Kalra, A., & Lowe, A. (2016). An overview of factors affecting the skins Youngs modulus. *Journal of Aging Science, 4*(2), 1000156.
15. Ottenio, M., Tran, D., Ní Annaidh, A., Gilchrist, M. D., & Bruyère, K. (2015). Strain rate and anisotropy effects on the tensile failure characteristics of human skin. *Journal of the mechanical behavior of biomedical materials, 41*, 241–250.

16. Tonge, T. K., Atlan, L. S., Voo, L. M., & Nguyen, T. D. (2013). Full-field bulge test for planar anisotropic tissues: Part I—Experimental methods applied to human skin tissue. *Acta Biomaterialia, 9*(4), 5913–5925.
17. Ní, A., Bruyère, K., Destrade, M., & Gilchrist, M. D. (2012). Characterization of the anisotropic mechanical properties of excised human skin. *Journal of the Mechanical Behavior of Biomedical Materials, 5*(1), 139–148.
18. Tilleman, T. R., Tilleman, M. M., & Neumann, M. H. A. (2004). The elastic properties of cancerous skin: Poisson's ratio and Young's modulus. *The Israel Medical Association Journal, 6*, 753–755.
19. Miller, K. W., Farage, M. A., Elsner, P., & Maibach, H. I. (2013). Characteristics of the aging skin. *Advances in Wound Care, 2*(1), 5–10.
20. Xu, F., Seffen, K. A., & Lu, T. J. (2008). Temperature-dependent mechanical behaviors of skin tissue. *IAENG International Journal of Computer Science, 35*(1).
21. Kvistedal, Y. A., & Nielsen, P. M. F. (2009). Estimating material parameters of human skin in vivo. *Biomechanics and Modeling in Mechanobiology, 8*, 1–8.
22. Branski, L. K., et al. (2009). A porcine model of full-thickness burn, excision and skin autografting. *Burns, 34*(8), 1119–1127.
23. Mirrashed, F., & Sharp, J. C. (2004). In vivo quantitative analysis of the effect of hydration (immersion and Vaseline treatment) in skin layers using high-resolution MRI and magnetisation transfer contrast. *Skin Research and Technology, 10*, 14–22.
24. Daly, C. H., & Odland, G. F. (1979). Age-related changes in the mechanical properties of human skin. *The Journal of Investigative Dermatology, 73*(1), 84–87.
25. Avon, S. L., & Wood, R. E. (2005). Porcine skin as an in-vivo model for ageing of human bite marks. *The Journal of Forensic Odonto-Stomatology, 23*(2), 30–39.
26. Ryu, H. S., Joo, Y. H., Kim, S. O., Park, K. C., & Youn, S. W. (2008). Influence of age and regional differences on skin elasticity as measured by the Cutometer. *Skin Research and Technology, 14*, 354–358.
27. Barbenel, J. C., & Evans, J. H. (1977). The time-dependent mechanical properties of skin. *The Journal of Investigative Dermatology, 69*(3), 318–320.
28. Langer, K. (1861). On the anatomy and physiology of the skin. *British Journal of Plastic Surgery, 31*(1), 3–8.
29. Sherman, V. R., Tang, Y., Zhao, S., Yang, W., & Meyers, M. A. (2017). Structural characterization and viscoelastic constitutive modeling of skin. *Acta Biomaterialia, 53*, 460–469.
30. Girard, M. J. A., et al. (2011). Quantitative mapping of scleral fiber orientation in normal rat eyes. *Investigative Ophthalmology and Visual Science, 52*(13), 9684–9693.
31. Pant, A. D., Thomas, V. S., Black, A. L., Verba, T., John, G., & Amini, R. (2017). Pressure-induced microstructural changes in porcine tricuspid valve leaflets. *Acta Biomaterialia, 67*, 248–258.
32. Sherman, V. R., Yang, W., & Meyers, M. A. (2015). The materials science of collagen. *Journal of the Mechanical Behavior of Biomedical Materials, 52*, 22–50.
33. Yang, W., et al. (2015). On the tear resistance of skin. *Nature Communications, 6*, 1–10.
34. Jansen, L. H., & Rottier, P. B. (1958). Some mechanical properties of human abdominal skin measured on excised strips. *Dermatologica, 117*, 65–83.
35. Ridge, M. D., & Wright, V. (1966). The directional effects of skin: a bio-engineering study of skin with particular reference to lancer's lines. *The Journal of Investigative Dermatology, 46*(4), 341–346.
36. Unit, B. (1977). Directional variations in the extensibility of human skin. *British Journal of Plastic Surgery, 30*, 105–114.
37. Zhou, A. B., et al. (2019). Strain rate sensitivity of skin tissue under thermomechanical loading. *Philosophical Transactions of the Royal Society A: Mathematical, Physical and Engineering Sciences, 368*(1912), 679–690.
38. Dunn, M. G., & Silver, F. H. (1983). Viscoelastic behavior of human connective tissues: relative contribution of viscous and elastic components. *Connective Tissue Research, 12*(1), 59–70.
39. Lim, J., & Chen, W. W. (2011). Mechanical response of pig skin under dynamic tensile loading.

40. Wong, V. W., Levi, K., Akaishi, S., Dauskardt, R. H., Schultz, G., & Dauskardt, R. H. (2012). Scar zones: Region-specific differences in skin tension may determine incisional scar formation. *American Society of Plastic and Reconstructive Surgery, 129*(6), 1272–1276.
41. Haut, R. C. (2016). The effects of orientation and location on the strength of dorsal rat skin in high and low speed tensile failure experiments. *Journal of Biomechanical Engineering, 111*(1989), 136–140.
42. Ankersen, J., Birkbeck, A. E., Thomson, R. D., & Vanezis, P. (1999). Puncture resistance and tensile strength of skin simulants. *Proceedings of the Institution of Mechanical Engineers, Part H: Journal of Engineering in Medicine, 213*(6), 493–501.
43. Jacquemoud, C., Bruyere-garnier, K., & Coret, M. (2007). Methodology to determine failure characteristics of planar soft tissues using a dynamic tensile test. *Journal of Biomechanics, 40*, 468–475.
44. Remache, D., Caliez, M., Gratton, M., & Dos Santos, S. (2018). The effects of cyclic tensile and stress-relaxation tests on porcine skin. *Journal of the Mechanical Behavior of Biomedical Materials, 77*, 242–249.
45. Belkoff, S. M., Naylor, E. C., Walshaw, R., Lanigan, E., Colony, L., & Haut, R. C. (1995). Effects of subcutaneous expansion on the mechanical properties of porcine skin. *Journal of Surgical Research, 58*(2), 117–123.
46. Boyer, G., Molimard, J., Ben Tkaya, M., Zahouani, H., Pericoi, M., & Avril, S. (2013). Assessment of the in-plane biomechanical properties of human skin using a finite element model updating approach combined with an optical full-field measurement on a new tensile device. *Journal of the Mechanical Behavior of Biomedical Materials, 27*, 273–282.
47. Laiacona, D., et al. (2019). Non-invasive in vivo quantification of human skin tension lines. *Acta Biomaterialia, 88*, 141–148.
48. Hendriks, F. M., Brokken, D., Oomens, C. W. J., Bader, D. L., & Baaijens, F. P. T. (2006). The relative contributions of different skin layers to the mechanical behavior of human skin in vivo using suction experiments. *Medical Engineering & Physics, 28*(3), 259–266.
49. Ranamukhaarachchi, S., Lehnert, S. A., et al. (2016). A micromechanical comparison of human and porcine skin before and after preservation by freezing for medical device development. *Scientific Reports, 6*(1), 32074.
50. Yeoh, O. H. (1993). Some forms of the strain energy function for rubber. *Rubber Chemistry and Technology, 66*, 754–771.
51. Shahzad, M., Kamran, A., Siddiqui, M. Z., Farhan, M., Materials, A., & Highway, I. (2015). Mechanical characterization and FE modelling of a hyperelastic material. *Materials Research, 18*(5), 918–924.
52. Sasso, M., Palmieri, G., Chiappini, G., & Amodio, D. (2008). Characterization of hyperelastic rubber-like materials by biaxial and uniaxial stretching tests based on optical methods. *Polymer Testing, 27*(8), 995–1004.
53. Lapeer, R. J., Gasson, P. D., & Karri, V. (2011). A hyperelastic finite-element model of human skin for interactive real-time surgical simulation. *IEEE Transactions on Biomedical Engineering, 58*(4), 1013–1022.
54. Moerman, K. M., Holt, C. A., Evans, S. L., & Simms, C. K. (2009). Digital image correlation and finite element modelling as a method to determine mechanical properties of human soft tissue in vivo. *Journal of Biomechanics, 42*(8), 1150–1153.
55. Maiti, R., et al. (2016). In vivo measurement of skin surface strain and sub-surface layer deformation induced by natural tissue stretching. *Journal of the Mechanical Behavior of Biomedical Materials, 62*, 556–569.
56. Tonge, T. K., Voo, L. M., & Nguyen, T. D. (2013). Full-field bulge test for planar anisotropic tissues: Part II—A thin shell method for determining material parameters and comparison of two distributed fiber modeling approaches. *Acta Biomaterialia, 9*(4), 5926–5942.
57. Murdock, K., Martin, C., & Sun, W. (2018). Characterization of mechanical properties of pericardium tissue using planar biaxial tension and flexural deformation. *Journal of the Mechanical Behavior of Biomedical Materials, 77*, 148–156.

58. Groves, R. B., Coulman, S. A., Birchall, J. C., & Evans, S. L. (2013). An anisotropic, hyperelastic model for skin: Experimental measurements, finite element modelling and identification of parameters for human and murine skin. *Journal of the Mechanical Behavior of Biomedical Materials, 18*, 167–180.
59. Dwivedi, K. K., Lakhani, P., Sihota, P., Kumar, S., & Kumar, N. (2023). The multiscale characterization and constitutive modeling of healthy and type 2 diabetes mellitus (t2dm) sprague dawley. *Acta Biomaterilia, 158*, 324–346. https://doi.org/10.1016/j.actbio.2022.12.037
60. Dwivedi, K. K., Lakhani, P., Yadav, A., Mehta, D., Kumar, S., & Kumar, N. (2023). Location specific multi-scale characterization and constitutive modeling of pig aorta. *Journal of the Mechanical Behavior of Biomedical Materials, 142*, 105809. https://doi.org/10.1016/j.jmbbm.2023.105809
61. Dwivedi, K. K., Lakhani, P., Kumar, S., & Kumar, N. (2022). A hyperelastic model to capture the mechanical behaviour and histological aspects of the soft tissues. *Journal of the Mechanical Behavior of Biomedical Materials, 126*, 105013. https://doi.org/10.1016/j.jmbbm.2021.105013
62. Dwivedi, K. K., Lakhani, P., Parashar, A., & Kumar, N. (2022). Effect of collagen fiber orientation on the Poisson's ratio and stress relaxation of skin: An ex-vivo and in-vivo study. *Journal of Royal Society Open Science, 9*, 211301. https://doi.org/10.1098/rsos.211301
63. Lakhani, P., Dwivedi, K. K., Parashar, A., & Kumar, N. (2021). Non-invasive in vivo quantification of directional dependent variation in mechanical properties for human skin. *Frontiers in Bioengineering and Biotechnology, 979*(* Contributed equally). https://doi.org/10.3389/fbioe.2021.749492
64. Dwivedi, K. K., Lakhani, P., Kumar, S., & Kumar, N. (2020). Effect of strain rate on the stress relaxation of the pig dermis: A hyper-viscoelastic approach. *Journal of Biomechanical Engineering, 142*(9), 091006. https://doi.org/10.1115/1.4046205
65. Dwivedi, K. K., Lakhani, P., Kumar, S., & Kumar, N. (2020). Frequency dependent inelastic response of collagen architecture of pig dermis under cyclic tensile loading: An experimental study. *Journal of the Mechanical Behavior of Biomedical Materials, 112*, 104030. https://doi.org/10.1016/j.jmbbm.2020.104030
66. Lakhani, P., Dwivedi, K. K., & Kumar, N. (2020). Directional dependent variation in mechanical properties of planar anisotropic porcine skin tissue. *Journal of the mechanical behavior of biomedical materials, 104*, 103693. https://doi.org/10.1016/j.jmbbm.2020.103693

Bioinspired Materials for Sustainable Technological Development

Deepa Singh, Sonika Chauhan, Neeraj Bisht, Nisha Rani, Rakesh Chandra, and Shailesh Yadav

Abstract Biomimicry technique of creating technical advancements by mimicking natural phenomena is known as this method focuses at biological systems in order to create solutions for issues or challenges that affect humans. Notable examples include the Japanese bullet train, which was designed to reduce noise and enhance speed by modeling its design after the beak of a kingfisher, and airplanes that replicate the hook-and-loop mechanism of burrs. By emulating biological designs, scientists and engineers create sustainable and efficient technologies. The intersection of natural and synthetic materials holds potential for innovative developments in various fields, leveraging unique properties such as strength, toughness, and wear resistance found in natural materials like shells, bones and enamel. The developments of materials with improved properties are influenced by the study of microstructures that are found in nature, such as composites inspired by nacre. Understanding and reproducing these features it becomes easier by different methods such as Extended Finite Element Method (XFEM), simulations and Molecular Dynamics (MD). Developments in bioinspired materials, especially for the applications that required resistance to blasts, shows how biomimicry can be used to develop robust defenses. The performance and design of composite structures are further improved via optimization techniques which are essential in modern engineering. This combination of artificial innovation and natural inspiration is about to transform material science and engineering, producing cutting-edge technologies with never-before-seen performance and adaptability. This study examines optimization and simulation methods that are critical to the development of bioinspired materials.

Keywords Biomimicry · Simulation · Optimization · Bioinspired materials

D. Singh · S. Chauhan (✉) · N. Bisht · N. Rani · R. Chandra · S. Yadav
Mechanical Engineering Department, College of Technology, G. B. Pant University of Agriculture and Technology, Pantnagar, India
e-mail: sonika.chauhan.me@gmail.com

K. K. Dwivedi et al. (eds.), *Mechanics and Materials Science of Biological Materials*, Materials Horizons: From Nature to Nanomaterials,
https://doi.org/10.1007/978-981-96-5366-9_10

1 Introduction

Nature has continuously served as an inspiration for human technical innovation. It is possible to find solutions of the issues that people face by studying biological structures and processes; this approach is known as biomimicry. Aeroplane design, which was influenced by eagle wings is a classic example of biomimicry. Analysis in the composition and operation of eagle wings engineers were able to maximize an aircraft's aerodynamics, allowing for controlled flying. The beak of the kingfisher serves as an inspiration for the Japanese bullet train, which is well-known for its efficiency and speed. Engineers noticed that kingfisher makes very less splashing or noise when it dives into the water. Engineers noticed that kingfisher's beak makes very less splashing or noise when it dives into the water. Engineers were able to improve the performance of high-speed trains by reducing noise and increasing speed by implementing this theory into design of trains. Velcro, the ubiquitous bonding material is utilized in a variety of applications. The burrs that adhere to garments and fur served as the inspiration for Velcro [1]. George de Mestral, a Swiss engineer, noticed that burrs had a hook-like structure and created a synthetic substitute made of hooks and loops that interlock when squeezed together to provide a robust but reversible fastening mechanism. These illustrations show how important it looks for the nature to provide innovative solutions to the challenging problems. Scientists and engineers can create environmentally friendly and efficient sustainable technology by studying and mimicking biological designs and processes. In addition to the advancing technology, biomimicry increases awareness of the complex systems and designs present in the natural world. Technological innovation seems likely to occur when designs from natural materials are combined with artificial structures to create new materials and microstructures that are specifically made for engineering uses. The intrinsic qualities and functions of natural materials are highly inspiring since they frequently beat synthetic counterparts. The development of novel materials with improved performance and qualities exists at the interface of synthetic and natural materials [2]. The capability to display different combinations of mechanical properties that are often difficult to duplicate in synthetic materials is one of their main advantages. Materials like bone and shell are strong and robust that they frequently challenge to accomplish with synthetic substitutes. Furthermore, antlers and horns show remarkable resilience to injury, offering valuable information about improving the resilience and durability of materials [3]. Through the utilization of natural materials' intrinsic design principles, researchers and engineers can open opportunities for the development of sophisticated artificial structures and microstructures. Researchers can use biomimicry and bioinspired engineering to develop synthetic materials having specific traits and capabilities by studying their composition, hierarchical structures, and mechanical behavior. Incorporating the design of natural material into synthetic structures presents enormous opportunities in a variety of industries, such as biomedical engineering, automotive, aerospace, and more. Future advancements in engineering and material science will completely transform the field by utilizing the advantages of both synthetic and natural materials. This will develop innovative technologies with previously unheard-of levels of performance and adaptability. A well-studied method inspired by nature involves replicating structural elements found in natural systems. Scientists have demonstrated that mimicking the brick-and-mortar composite microstructures inspired by nacre can enhance elastic modulus of material and improve its toughness. By inducing crack deflection through the composite structure's geometric design, this enhancement is made possible [4]. Stomatopod-inspired structures with helicoidal composite microstructures possess higher energy absorption capacities. Their geometries are responsible for promoting longer cracking paths. Similarly, composite microstructures modeled after conch shells, with cross-lamellar arrangements had enhanced the stiffness and impact resistance. Additionally, tubular composite microstructures inspired by osteons have shown strength and toughness of materials through various toughening mechanisms [5]. However, a key drawback of this method is that many of these bio-inspired geometries was constructed using a rigid building blocks that were repetitively used within a structure. These often contain scaled-down representations of the complex microstructure architectures found in nature. Simplified geometries can enhance a material's characteristics, but one major disadvantage is that they are less flexible for use of certain engineering. In simple terms, these structures are incapable of being precisely adjusted or tailored to meet the needs of various engineering scenarios. Although this method is successful in simulating basic geometries present in natural composites. It is not successful in reproducing the essential adaptive character and geometrical complexities found in the natural design processes.

This review aims to present adaptive methodologies for designing bio-inspired structures by employing a pattern generation algorithm inspired by nature. Instead of creating composite microstructure geometries through a laborious element-by-element design process, the proposed approach utilizes bio-inspired algorithm for generation of patterns [6].

2 Simulation Techniques

2.1 Machine Learning, Genetic Algorithm, Convolutional Neural Network

Machine learning (ML) is a branch of an artificial intelligence (AI) focused on identifying patterns within intricate datasets, has demonstrated its utility across numerous research domains over the past few decades [7]. Their method employs an algorithm of self-learning to identify high-performance of materials while phasing out inferior designs. Machine learning algorithms have proven useful in various areas such as spam detection, drug discovery and image recognition [8, 9]. Gu et al. [9] proposed a novel method or technique, machine learning to design hierarchical materials, leveraging a database of thousands of structures from FEA. On the other hand, Convolutional neural networks (CNN) were utilized to forecast material properties of composite, demonstrating the machine learning (ML) model produces superior designs with significantly improved properties of material as compared to trained data. CNN enables an examination of the hierarchical systems even without full characterization of their microstructure, thereby expediting the prediction of material properties. Blum and Langley [10] survey explores how machine learning addresses datasets with large amounts of irrelevant information, focusing on selecting relevant features and examples. It also outlines future challenges in this field. Recently, different properties of material like fatigue life, elasticity and plasticity have been explored [11]. Previous studies mainly focused upon predicting material properties rather than developing new materials with the desired characteristics. There is limited validation of these ML approaches through experimental analysis [12]. Chiu et al. [13] used a combination of Genetic Algorithm (GA) and Conditional Variational Autoencoder (CVAE) to investigate a two-dimensional (2D) binary composite system with an edge crack and grid-like structure. This approach enables the design of composites with desired properties as toughness and stiffness. Comparing predicted toughness and stiffness to the desired values determines the fitness of each design, utilizing the negative mean square error. FE simulations are employed for generation of machine-learning datasets while tensile tests are employed on 3D-printed specimens to validate the results. Incorporating soft material behind the crack tip significantly enhances the overall toughness of a composite. While both GA and CVAE produces composite designs with high accuracies. CVAE achieves this outcome in considerably less time (~1/7500) making it more efficient approach. In Genetic Algorithms (GAs), superior designs ("offspring") evolve over generations through iterative "crossover, selection and mutation", repeating undirected process of natural selection. Whereas, in a systems with wide numbers of design parameter, these conventional techniques can be excessively takes time, resembling brute-force methods. For example, in the game of Go, which features a 19×19 grid size with three potential states at each intersection point, there are approximately 2.08×10^{170} legal positions [14]. For extremely larger systems, the machine learning emerges as viable solution that had been suggested in academic literature. A notable example is Alpha Go, developed by Google Deep Mind, a subsidiary of Alphabet Inc. [15]. Machine learning (ML) can establish intricate numerical relationships between output and input data from complex and large datasets by learning from previous experiences. The utilization of fingerprints derived from either chemo-structural data (configurational and compositional information) or from distribution of electronic charge density has been demonstrated to enable accurate and ultra-fast predictions of the property. Making use of these learning paradigms improves current attempts to systematically explore and draw insights from large chemical domains. Consequently, such approaches hold the potential to substantially expedite the discovery of new materials tailored for specific applications. As a result, numerous researchers opt to substitute traditional simulation methods with ML techniques to expedite the development process across various tasks, including predicting the properties of chemical molecules [16, 17] and composites [18, 19]. The ability to customize materials is a key advantage of composites. By integrating novel bio-reinforcements into composites, we can utilize agricultural waste and develop tougher and more dependable materials. However, the vast array of potential combinations of natural materials presents challenges in identifying optimal composite designs. In this context, machine learning was utilized to accurately predict the fracture toughness properties of multiscale bio-nano-composites [20]. Furthermore, recent studies have highlighted machine learning's remarkable ability to tackle inverse design problems. Specifically, ML models like Generative Adversarial Network (GAN) have shown promise in this domain. In order to create microstructural materials, a deep learning model that combines a convolutional neural network (CNN) and a deep convolutional generative adversarial network (DCGAN) has been introduced. While CNN acts as a substitute model to determine the connection between the microstructure and its properties, DCGAN produces design options that adhere to geometrical constraints [21]. With the Conditional Variational Autoencoder (CVAE), structures of chemical molecules or composite structures acquired by inputting the desired requirements into ML model [22]. However, machine learning (ML) has demonstrated remarkable accuracy and precision while solving the problems of inverse design, yet interpreting model predictions accurately and extracting design guidelines is a challenge. Determining which design elements are more important and contribute to the desired properties is still a major challenge. To address this issue, various methods, often referred to as Explainable Artificial Intelligence (XAI), have been proposed. These methods aim to provide insights into how ML models arrive at their predictions and help interpret their decision-making processes [23]. In order to capture the nonlinear mapping between a material's stiffness, macroscale (effective) and its three-dimensional microstructure, Yang et al. [24] established deep learning models. It was believed that output of micromechanical finite element models may approximate the ground truth in the lack of an appropriate experimental dataset. A convolutional neural network (ML/deep model) and a multilayer perceptron neural network (ML model) were used in this method. According to the results, the convolutional neural network that was suggested performed up to 54% better in terms of the mean absolute stiffness inaccuracy during testing than advanced physics-based conventional techniques. However, for some mid-range effective stiffness values, the deep learning approach showed limits.

2.2 Extended Finite Element Method (XFEM)

Finite element method has brought about a revolution in material designing by significantly reducing the design costs and time through structural and prefabrication optimization. It enables evaluation of mechanical properties critical to engineering endeavors by determining maximum load capacity of a bridge, to identify potential failure points along with assessing the fatigue life of components like automobile suspensions. However, despite its utility, standard FEM falls short in fully capturing fracture behavior, including phenomena such as crack nucleation and propagation. Over the recent decades, several methodologies have emerged to address this gap. Notably, following the observations of distribution of stresses around the cracks, contour integral method (CIM) and variations had developed to compute stress intensity factor (SIF) in cracked materials. These approaches embedded in commercial softwares including CT-integral, J-integral and T-stress which are closely related to CIM. These techniques offer valuable insights into fracture mechanics by enhancing our understanding on behavior of materials under fracture and providing essential tools for structural analysis and design optimization.

On the hand, Belytschko and Black [25] and Moeset al. [26] have first developed and proposed XFEM to obtain the solution of the problems related to singularity and discontinuity in the field of material. To account for the occurrence of cracks, discontinuous enrichment functions are incorporated into finite element approximation. This technique enables cracks to be arbitrarily positioned within the mesh. In highly curved cracks remeshing might be necessary to make the process simpler [25]. Many researchers used XFEM, a modern numerical method, to model crack propagation in nacre, a shell material. XFEM incorporates a traction–separation law and additional features like jump function to accurately simulate crack behavior. Strength and toughness are used to assess nacre's performance. XFEM enhances traditional FEM by reflecting displacement discontinuity near crack tips [27]. Fries and Belytschko [28] used the XFEM to investigate the impact of structural parameters on toughness and strength of materials. Toughening mechanism associated with various morphologies of the fundamental staggered unit was discussed. The simulation findings indicate maximum failure stress of 3% higher compared with experimental test, while 1.17% failure strain was noticed that was lower than the experimental result. The specimen's ability to absorb energy is evaluated by three-point bending tests, which show an 11% improvement over a single plastic specimen. Three-point bending tests are replicated using finite element technique simulations. Nacre-like structures are modeled experimentally to determine material properties [29]. Hexahedral 8-noded solid elements are utilized to discretize the numerical model. FE model replicates the conditions of tensile test in order to incorporate the measured bonding force into numerical simulation. The commercial FEM program LS-Dyna was used for the numerical analysis. Only the ungripped portions are modeled, and the top and bottom surface displacements are controlled, to improve numerical efficiency [30]. Three-point bending tests are used to assess the energy absorption capacity of the specimen, which is determined to be 11% higher compared to that of a single plastic specimen. Nacre-like structures are modeled to reflect experimentally determined material properties, and finite element method simulations are well employed to replicate three-point bending tests. The two supports are fixed in place, and the indenter applies a load at a velocity of 2 mm/min in the vertical, downward direction [29]. The design of bone-like materials, inspired by its osteonal feature [31] or lamellar characteristic [32], has led to improvements in material toughness. With this concept, a composite model inspired by the ultrastructure of bone to further enhance material properties was proposed by the researchers. Additionally, an analysis was conducted on the proposed composite's finite element (FE) model under tensile stress conditions. The bioinspired composite was then compared to a traditional composite that had the same components and volume percentages of soft and hard phases. The geometry and finite element (FE) model of the composite were created using the ABAQUS (version

2016) FE program. The authors used the cohesive element (designated as COH2D4) from the ABAQUS library to model the thin sticky phase. Cohesive elements have the ability to capture the beginning and progression of damage within the adhesive substance. By establishing a connection between cohesive traction and element separation, the behavior of these elements is controlled [33]. The dataset is generated using ABAQUS and MATLAB. Properties such as mechanical strengths can be obtained from Finite Element Analysis (FEA). The primary failure modes of a brick-and-mortar composite include failure in the vertical, horizontal, or combined orientations of the bricks [34]. XFEM has been widely used for simulating behavior of fractures across various materials, including rocks [35, 36], steel [37–39], cast iron [40, 41], aluminum [42], silicon oxycarbide [43], and composites [44] XFEM has improved our knowledge on fatigue crack propagation on bridges [37], evaluated the impact characteristics of high-pressure gas transportation pipes [38], and examined the flexural response of coated steel used in yacht construction [39] in the steel industry [39]. Chang et al. [40] demonstrated that XFEM solutions align well with experimental uniaxial tests in the context of cast iron, showing accurate representation elastoplastic behavior of the materials. Under a more intricate loading scenario involving an eccentric impact stress on a cast iron structure using three-point bending, Tsuda et al. [41] emphasized adaptability of XFEM in addressing unconventional cases. When applied in the sample of an aluminum alloy, Gairola and Jayaganthan [42] successfully predicted the fracture toughness of a material suggesting a slight superiority of 3D XFEM simulations over 2D counterparts. By simulating the fracture resistance curves of compact tension specimens, Kim et al. [45] investigated on an aluminum alloy the effect of hydrogen gas on the fracture behavior.. The XFEM load and displacement curves nearly matched experimental circumstances within a 5% margin. In a recent study, Xie et al. [43] effectively obtained the fracture behavior and mechanical properties of silicon carbide film, demonstrating the efficiency of XFEM in evaluating interfacial strength.

Amazing qualities found in natural materials include tissue regeneration [46], color shifting [44], self-cleaning [47], and fracture resistance [48]. Recently, discoveries on the behavior of biological materials during fracture have also gained attention. The mantis shrimp, for example, attacks its victim by unleashing a tremendous amount of energy from its saddle [49]. Direct contact with its prey is accomplished by a dactyl club, an appendage resembling a club. In order to scatter energy and minimize the possibility of damage of self-infliction it is highly mineralized and possesses a specific microstructural architecture [50, 51]. A Spiders possess a dual attachment system that allows them to adhere to various surfaces. Van der Waals forces allow tiny and small hairs on the tip of their tarsus to adhere to smooth surfaces. However, when encountering rough surfaces these forces are insufficient, spiders employ their claws to enhance friction with the surface. These claws are composed of hierarchical material containing inclusion of metal ion resulting in a structure with higher wear resistance [52]. Bones of Mammal, particularly cortical bone tissue serve as a structure for load-bearing and composed mostly of osteons at the microscale. Osteons are cylindrical structures with cement lines, or weaker outer layers, around a concentric inner channel. The total fracture toughness of bone tissue is greatly increased by this architectural design, which permits fracture toughness mechanisms such crack deflection and arrest [53, 54]. It is clear that biological tissues microstructure affects the characteristics of macroscale materials. It might be difficult to comprehend the mechanisms underlying the improvement in mechanical characteristics, and a mix of computational and experimental investigations is frequently needed. Thorough XFEM investigation examined six different biological structures under bending loads [55]. These structures are frequently seen in natural materials, such as layered materials (such as nacre shells insect exoskeletons), columnar materials (such as tendons and spider silk), tubular materials (such as dentin and horns), helical materials (such as mantis shrimp dactyl clubs), sutured materials (such as turtle shells), and sandwich materials (such as toucan beaks). The analysis correctly anticipated each design's fracture behavior even after taking into account additional necessary simplifications and simplified designs with only two components—rigid and soft building blocks. It also emphasized the impact of biological materials soft-stiff component arrangement on mechanical characteristics. Chen et al.'s study [55] demonstrates that the helical column structure deflects cracks in the soft matrix of the material. This work not only demonstrated how well XFEM can represent complicated composites, but it also provided inspiration for the development of bioinspired materials.

Modeling inner structure of a composite is crucial for accurately simulating its fracture behavior. However, many studies simplified the bone microstructure for the analysis of computational. For instance, they often model perfectly one circular osteon. Gustafsson et al. [56] took a different approach by employing X-ray microtomography for extraction of local tissue of bone, which served as a input for analysis of XFEM. This method yielded highly real crack paths as compared to experimental results [57]. Similarly, Yadav et al. [58], utilized data from microtomography of cortical bone by demonstrating the crack behavior variation with the bone age. To extract the characteristics of osteon, interstitial matrix and cement line uses measurements of nanoindentation. While the mechanical characteristics of young and old bone changed only small, the structural comparison showed notable differences: aged bone had a 15% lower bone volume percentage and an 80% smaller osteon wall thickness. Bioinspired elements including aerodynamic train designs modeled after bird beaks [59], most efficient ventilation systems for building inspired by termite mounds [60], and movements of robot that mimics humans [61] have evolved in a recent time. Understanding a material's fracture behavior it is essential to understand its deformation and yielding mechanics in the field of structural materials where novel bioinspired materials are constantly being produced [62]. Extended Finite Element Method (XFEM) serves is an important tool of computational for predicting various mechanical properties of the structures of bioinspired materials. Libonati et al. [63] demonstrated variations in local collagen fiber by osteon morphotypes and significantly influence an overall performance of a material. Finite Element Method and 3D printing highlighted that vertical morphotypes (VOM) exhibits superior performance under tensile loading, while twisted morphotypes (TOM) demonstrate greater resilience under compression. Libonati et al. [63] and Vellwock et al. [64] delved into design of composite using Extended Finite Element Method. The first computational model of composite designs showed a strong degree of agreement with experimental findings in a three-point bending and tensile tests. When the initial crack comes in contact with cement-like structure during tensile test simulation, it was deflected and arrested. The crack progressed between two osteonic zones in a three-point bending test by mimicking the experimental test results. XFEM analysis highlighted the significance of osteon-like area in distributing stresses. In the second investigation, fracture behavior of composites under external bending involves additional interlayered glass fibers. In this case, lower surface of the composite is where the break started. It then proceeded higher partially redirecting by an area inspired by an osteon and was finally stopped and eliminated entirely. It then progressed upward partially deflected by an area inspired by an osteon and was finally stopped and totally deflected by an additional osteonic structure. Hence, several studies showed that XFEM analysis not only replicates experimental conditions [65] but also highlights mechanism of fracture toughness applicable to other materials designs.

2.3 Gray Scott Model

A mathematical model called "Gray-Scott" is used to mimic the processes of reaction–diffusion that is seen in a range of chemical reactions. It is composed of primarily two variables, usually called "activator" and "inhibitor," which indicate the concentrations of reacting chemicals. The model explains the time-dependent reactions and diffusion of these chemicals in a spatial domain [66]. The Gray-Scott model was chosen to generate bio-like composite microstructure geometries owing to its remarkable capability to replicate a diverse range of patterns observed in the nature through mechanism of biological pattern generation. The model was originally developed by Scott and Gray to investigate autocatalytic chemical reactions within a well-stirred, continuously fed, open-tank reactor, this model focused on isothermal systems [67–69]. The observed phenomena encompass a spectrum of intriguing behaviors, including hysteresis, multistability, critical extinctions, critical ignitions and anomalous relaxation times though infinite values. Stationary state patterns influenced by time of residence can manifest as mushrooms and isolas. These phenomena lend themselves straightforward algebraic analysis, facilitating a deeper understanding of the underlying dynamics [67]. However, Pearson expanded this model to two-dimensional demonstrating a multitude of turning and chaotic patterns emerge over space and time in reaction to perturbations in the steady-state system [70]. In a set of governing equations, kill rate and feed rate determines the type of pattern generated in a system while a diffusion coefficient regulates scale length of these patterns. The concept of pattern formation through kinetics of reaction–diffusion was initially explored in the seminar work of Turing on morphogenesis. Turing's research introduced patterns such as stripes, spots, and spirals and hexagons which came to be known as "Turing patterns" [71]. The primary focus of the investigation lies in understanding the onset of instability within the system. It has been observed that there are six distinct forms in which this instability may manifest. The emergence of stationary waves on the ring structure is of particular interest. It was proposed that this phenomenon could potentially explain patterns such as Hydra's tentacle patterns and the formation of whorled leaves. Consequently, previous investigations utilizes Gray-Scott model to primarily concentrate on characterizing and understanding development of these patterns [72]. In the context of bioinspired composites, Gray-Scott model was applied to simulate formation and evolution of complex microstructures. By incorporating the patterns generation algorithms for bio-inspired materials into model researchers can simulate development of composite microstructures with the required properties. This approach allows an exploration of novel designs and optimization of material properties based upon the principles observed in natural systems. Overall, the Gray-Scott model provides a powerful tool to study the behavior of bioinspired composites and guiding the design process toward achieving enhanced structural and mechanical characteristics.

2.4 Molecular Dynamic Simulation (Integration with Higher-Scale Simulation and Atomic-Scale)

The study of multifunctional composites has faced several obstacles as well as opportunities due to the introduction of MXenes and other 2D materials. In order to explore the possibility of bioinspired design in engineered 2D composites this research focuses on the highly mineralized structure of 2D natural composites. Beyond its common brittle mineral phase, the study emphasizes the unique characteristics of the natural system, focusing in particular on the structural organization and the existence of intermittent soft organic phases [73]. In general, brick–mortar structure that are found in nacre and wood-based composites presents one of the most accessible systems to replicate and compare to more intricate hierarchical arrangements found in bones as well as in plants. Nonetheless, while replicating bioinspired design features in engineered systems holds tremendous promise, it also poses significant challenges that have yet to be fully addressed [74]. The vast array of hetro and homo nuclear 2D materials along with diverse layers of surface termination and innovative synthesis methods presents immense opportunities for the atomic-level systematic integration along with experimental techniques. This integration can guide material and process selection and establish a comprehensive testing, development and validation platform. While experimental successes in realizing various 2D materials are notable, atomic-level simulations for most 2D materials are currently limited but demonstrate promising predictive capabilities. MXenes, recent molecular dynamics (MD) simulations focused on titanium-carbide based MXene that have provided insights into its mechanical properties. These simulations predicted a tensile modulus of elasticity (500–600 GPa), nonlinear strain transition below 1%, high strain to failure ratios (6–9%), as well as strain-rate sensitivity and temperature [75, 76] and binding energy (0.8–2 × 10–18) [77]. Experiments on indentation conducted in Ti3C2Tx MXene obtained lower elastic modulus (300 GPa) compared to predictions from atomistic simulations [75, 76]. To address these discrepancies and provide a comprehensive understanding of MXene properties, atomistic simulation studies must be expanded. This expansion should lead to the creation of a database correlating MXene compositions, electromechanical/electrochemical properties and surface terminations bridging the gap between in-situ experimental values and atomistic predictions. This database would enhance modeling accuracy, validate simulation approaches, and mitigate experimental uncertainties such as surface roughness and interfacial chemistry variations. Furthermore, higher-order analysis techniques are crucial in integrating atomic models to model realistic processing scenarios and geometries. Molecular dynamics analysis combination with multiscale continuum mechanics-based finite element analysis (FEA) addresses size limitations of atomistic simulations by providing insights into real material behavior. For instance, analysis of higher-order continuum requires stress–strain responses of two-dimensional sheets to measure strength of interface (e.g., frictional/shear response, surface interaction energies) to model interactions between polymer matrices and MXene flakes. These properties are difficult to measure experimentally, can be effectively captured through integrated MD/FEA simulations. Without detailed and specific information of material, FEA applications for composite design in 2D would be limited to rigid contact interactions and elastic response. To integrate models at atomic-level with analysis of higher-order it is crucial for accurately model realistic geometries, larger and processing conditions. Molecular dynamics analysis is coupled with multiscale continuum mechanics-based finite element analysis (FEA) to overcome size limitations inherent in atomistic simulations and provide realistic properties of a material. Particularly, analysis of higher-order continuum becomes necessary for stress–strain response of two-dimensional sheets and also for characterization of strength of interfaces, including frictional/shear responses and surface interaction energies. These parameters are essential for modeling phenomena such as adhesion and slippage between flakes of MXene and polymer surfaces. Given the complexity and time-consuming nature of measuring these properties experimentally, integrated MD/FEA simulations offer a dual benefit. They enable the use of real interactions and properties of material while also bridging outcomes of experimental. Without detailed information of specific material, FEA applications for 2D composite design would be restricted to rigid contact interactions and elastic responses. Therefore, integration of MD and FEA methodologies is essential for accurate modeling of 2D composite materials [78–80].

2.5 *Automated Simulation Techniques Under Blast Loading*

Structural engineers have many challenges as a result of blast loading, especially when it comes to safety of peoples and infrastructures. Researchers are looking to develop bioinspired composites as a possible answer to this issue. These composites are inspired by naturally occurring materials and structures through millions of years endure dynamic and impact loads. The mechanical properties of natural materials like nacre, bone, and plant fibers are utilized to develop bioinspired composites under blast loading. These materials are excellent models for creating blast-resistant composites because of their exceptional toughness, strength, and energy absorption capacities.

Composites inspired by nacre imitate the alternating layers of soft organic layers and hard mineral plates that develop nacre's brick-and-mortar structure. The composite has extraordinary toughness because of its hierarchical structures as it distributes energy and redirects cracks. In the same way composites inspired by bone imitate the collagen-mineral composition of real bone. These composites reduce possibility of fracture and catastrophic failure under blast loading circumstances, allowing them to efficiently absorb and transmit energy. Plant-inspired fibers, such those in palm or bamboo trees, provide an additional route for the creation of bioinspired composites. Due of their strength and flexibility hierarchical organization of fibers makes them ideal for blast-resistant applications.

The catastrophic effects of blast loading on buildings and human lives highlight the urgent need for innovative and resilient protective solutions [81, 82]. Protective structural engineering has seen the growth of biomimicry as a promising field with great potential to meet this essential need. Current efforts to mimic nature for high loading scenarios take several forms:

1. Nacre-like composite panels was developed to withstand blast [83–85] and ballistic impact [86].
2. 3D Composite printed prototypes, resembling conch shells, have been tested for resilience against low-velocity impact [87].
3. The usefulness of fish scale-inspired panels is to survive against ballistic impact and puncture has been investigated [88–91].
4. The response of crustacean-like fiber-reinforced composites to impacts at low velocities has been investigated [92, 93].
5. To reduce ballistic shock, functionally graded foams that resemble bones have been developed [94].
6. The behavior of lightweight cellular solids that resemble porcupine quills under blast loading has been studied [95].

These investigations show great progress toward bioinspired protection systems, their application is still limited. To fully realize the possibility of biomimicry in developing robust protection systems that may reduce the damaging impacts of blast loading, more investigation and development are required.

A Bioinspired cellular solid, which are lightweight and have excellent energy absorption capacities, shows potential benefits as blast-resilient protective systems. Comprehensive research to completely comprehend their capacity under tremendous stresses is still lacking. On the other hand, a great deal of research has been done on manufactured cellular structures with regular geometries for blast applications [96, 97]. However, these structures frequently have restrictions in their buckling and collapse mechanisms which can limit their ability to absorb energy. Applications related to cellular structures design includes shielding military vehicle base plates against improvised explosive devices and concrete slabs from localized blast detonations [98]. Another common application that makes use of cellular structures' capacity for energy absorption is crashworthiness [99, 100].

Studies evaluating the performance of various topologies of cellular structures under impact loading and blast conditions such as re-entrant auxetic and hexagonal structures [98, 101, 102], have been conducted. These studies have revealed that capacity of energy absorption in cellular structures is facilitated by unique mechanism of collapsing and buckling features of various geometrical arrangements. However, a significant drawback lies in the fact that these structures are typically constructed using a limited parameters that restricts the space design considerably. Ghazlan et al. [103] studied that structural characteristics of trabecular bones are replicated to investigate dissipation mechanisms of an effective energy under blast loading conditions. To achieve this goal, researchers identified various crucial design parameters for developing unit cell as bone. They then built a computer script that could automatically produce FE models that resembled bones and had different combinations of design parameters. We do simulations using blast loading conditions applied to these models. By using this method, the researchers hope to find the ideal structure that performs better at dissipating energy than conventionally designed cellular structures. These conventional constructions which are frequently employed in blast-resistant applications comprise re-entrant, hexagonal, and square cellular shapes. Research on bioinspired composites for blast loading is still in its early phase. Though encouraging progress has not yet been made. Engineers want to create materials that are resilient, strong, and light enough to withstand blast forces by studying and imitating the structural principles found in nature. These bioinspired materials have the power to completely transform the infrastructure protection systems and blast-prone protective systems settings for military use, infrastructure, and civilian safety.

3 Optimization Techniques

Ability to combine unique mechanical properties and adaptable functions, composite structures has become essential parts of many engineering applications. The complex manipulation of composition and arrangement involves the harmonious blending of two or more base materials. Through meticulous manipulation of the properties and arrangement of these component parts engineers can adjust the properties of composites to meet specific requirements in a range of sectors (such as renewable energy, automotive, aerospace, and construction). Because of its capability to produce customized performance qualities, composite structures are now considered as essential components of contemporary engineering designs, enabling improvements in sustainability, durability, and efficiency for a variety of uses [104]. Canale et al. [105] developed a method for optimizing the stacking sequence of blades of composite fan. The proposed optimization procedure involves two levels. In the first level, sequence of stacking is optimized to maximize stiffness, thereby minimizing deformation. Constraints in this level are less strict. In the second level, tapering or blending of each ply within sequence of stacking is examined. A method was developed by Herencia et al. [106] to optimize a composite wing using two step level methods. Mathematical programming (MP) is used at the initial level of composite optimization. Anisotropy and transverse shear stiffness of the skin and stiffeners are taken into account while modeling them using lamination parameters. With ply angles of 0^0, 90^0, 45^0, and -45^0, it is expected that skin and stiffener laminates are both symmetric and mid-plane symmetric. This method does complicated structural analysis by utilizing Mathematical Programming (MP) at the first level. At second level, Genetic Algorithms (GA) is employed to address discrete lay-up combinational problem [106]. However, for manufacturing composites conventional methods have been typically restricted to layup operations, where laminate composites are produced by stacking layers of resins and plies [107]. The significant progress in 3D printing technology has enabled the manufacturing of composites along with more intricate structures and increased design flexibility [108]. For example, various researchers have developed composites of bioinspired materials with intricate structure resembling diatom [109], nacre [110], bone [111] and other natural structures. The wide-ranging design possibilities facilitated by 3D printing introduce a new challenge: efficiently and effectively designing structures and selecting materials to meet mechanical requirements. Traditional optimization techniques like greedy optimization [112], genetic algorithms (GAs) [18] and others have been employed for topological optimization. Genetic algorithm is inspired by Darwin's principles of naturally selection search-based algorithms that have proved successful in biological evolution [113]. Also, it is utilized in the optimization of parameter across various engineering and scientific domains [114, 115]. Improved designs, referred to as "offspring," in genetic algorithms undergo generational evolution by means of crossover, mutation, and repetition of selection, simulating the mechanism of undirected natural selection.

Artificial intelligence (AI) techniques with EFEM hold considerable promising in future. Using neural networks for optimization tasks with XFEM simulation results to improve structure designs, speed up the design process, and broaden the domain design is one prominent example. As underscored in the current review, even minor fluctuations in the properties of composite material can exert substantial influence on failure characteristics. However, these properties are often plagued by uncertainties, stemming from simplifications and challenges in precise measurement. In addressing these uncertainties, Bayesian networks had provided a useful method for managing these types of uncertainties by integrating probability distributions for particular variables like yield strength, defect orientation, porosity distribution and material elastic modulus. Also, it offers a valuable approach by incorporating probability distributions for specific variables such as material yield strength, porosity distribution, elastic modulus and flaw orientation. This integration encourages a material's mechanical behavior that is more thorough. Although XFEM models using Bayesian networks have been proposed recently, it is important to remember that these models may have a tendency to provide simplified structural representations.

One of the primary challenges in utilizing XFEM and AI together to improve engineering processes is striking a balance between the computing efficiency and the subtleties of material behavior [116].

3.1 Q-Learning

Biomaterials derive their exceptional properties from their intricate microstructural composition. Experimental evidence often reveals that greater complexity in material architecture necessitates higher resolutions. This heightened resolution results in smoother strain and stress fields within materials, thereby exerting a more significant effect on its stress behavior. The number of possible material combinations grows unmanageable as the system gets larger. As a result, time needed for a model to find the best designs also increases. In addition, increased environmental complexity deteriorates training convergence rate by developing optimization problems in the large design space making it difficult to solve [117].

Q-learning and the algorithms associated with it plays a prominent role in realm of reinforcement learning. [118, 119]. It utilizes a Q-value function for computation of anticipated value of rewards to guide subsequent optimization strategies. In this context, the Q-value signifies the reward achievable after specific action in a given state. Through Q-value function both current and past values are updated iteratively for Q-values. Several algorithms for reinforcement learning have origins in Q-learning. Notable uses occur in a variety of fields:

1. Game Theory: Google Deep Mind has made significant contributions to this field of study [120], and Q-learning has applications in game theory [121].
2. Robotics: Q-learning provides a structured framework for creating intricate robotic actions. [122].
3. Material Design: Researches uses geometric or parametric control techniques along with Q-learning-related algorithms [123].
4. Industrial Processes: Industrial processes are optimized and controlled by application of Q-learning [124].
5. Neural Network Designs: Q-learning is applied to neural network training and optimization [125].
6. General Control Theories: Q-learning is useful in other control theory contexts and forms the basis for many improved algorithms [126].

According to Yu et al. [117], they introduced a hierarchical design protocol aimed to mitigating the issues related to increased time cost. This protocol draws inspiration from the progressive growing training approach employed in Generative Adversarial Networks (GANs) [127]. The primary objective of the Q-learning model is to facilitate the design of materials characterized by high fracture toughness while minimizing stress concentration at the tip of preexisting cracks. This objective directly handles the challenges associated with defects of materials. Such flaws can leads to stress concentration by ultimately making the material more susceptible to damage

[128]. The framework outlined by Yu et al. [117] presents a potential alternative to traditional composite optimization methods, which frequently encounter challenges associated with high dimensionality. Moreover, conventional techniques often struggle to accurately predict high-resolution designs due to algorithmic limitations in escaping low-resolution local minima. However, algorithmic constraints in avoiding low-resolution local minima frequently make conventional methodologies impossible to anticipate high-resolution designs with any degree of accuracy. Furthermore, Q-learning is the foundation of several enhanced algorithms showing its effectiveness in resolving issues with simple situations. Its simplicity makes it generalized to other algorithms easier [129]. This demonstrates the adaptability and durability of Q-learning in a range of applications.

4 Conclusions

Biomimicry, the study of nature's designs and processes, offers innovative solutions for engineering challenges. By mimicking biological structures, such as the wings of eagles or the beak of a kingfisher, engineers have developed more efficient technologies like aircraft and high-speed trains. The integration of natural and synthetic materials promises the creation of advanced, high-performance composites. Techniques such as machine learning, finite element analysis, and molecular dynamics simulations are enhancing our ability to design bio-inspired materials with superior properties. However, challenges remain in replicating the complex adaptability and intricacies of natural designs. Future advancements in biomimicry and optimization techniques will likely lead to transformative developments across various industries, enabling the creation of more sustainable and efficient technologies.

Data Availability Not applicable.

Declaration of Competing Interest The authors have no conflicts to disclose.

References

1. Hwang, J., Jeong, Y., Park, J. M., Lee, K. H., Hong, J. W., & Choi, J. (2015). Biomimetics: Forecasting the future of science, engineering, and medicine. *International Journal of Nanomedicine, 10*, 5701–5713. Dove Medical Press Ltd.
2. Xia, Z. (2016). *Biomimetic principles and design of advanced engineering materials*. John Wiley & Sons Inc.
3. Bhushan, B. (2009). Biomimetics: Lessons from nature—An overview. *Philosophical Transactions of the Royal Society, 367*, 1445–1486.
4. Jia, Z., Yu, Y., & Wang, L. (2019). Learning from nature: Use material architecture to break the performance tradeoffs. *Materials & Design, 168*, 107650.

5. Libonati, F., Gu, G. X., Qin, Z., Vergani, L., & Buehler, M. J. (2016). Bone-inspired materials by design: Toughness amplification observed using 3D printing and testing. *Advanced Engineering Materials, 18*(8), 1354–1363.
6. Gu, G. X., Dimas, L., Qin, Z., & Buehler, M. J. (2016). Optimization of composite fracture properties: Method, validation, and applications. *Journal of Applied Mechanics, 83*(7), 071006.
7. Michalski, R. S., Carbonell, J. G., & Mitchell, T. M. (2013). *Machine learning: An artificial intelligenceapproach*, Springer Science & Business Media.
8. Witten, I. H., Frank, E., Hall, M. A., & Pal, C. J. (2016). *Data mining: Practical machine learning tools and techniques*, Morgan Kaufmann.
9. Gu, G. X., Chen, C. T., Richmond, D. J., & Buehler, M. J. (2018). Bioinspired hierarchical composite design using machine learning: Simulation, additive manufacturing, and experiment. *Materials Horizons, 5*(5), 939–945.
10. Blum, A. L., & Langley, P. (1997). *Artificial Intelligence, 97*, 245–271.
11. Agrawal, A., & Choudhary, A. (2016). *APL Materials, 4*, 053208.
12. Goh, G. B., Hodas, N. O., & Vishnu, A. (2017). *Journal of Computational Chemistry*.
13. Chiu, Y. H., Liao, Y. H., & Juang, J. Y. (2023). Designing bioinspired composite structures via genetic algorithm and conditional variational autoencoder. *Polymers, 15*(2), 281.
14. Tromp, J. (2016). *The number of legal go positions. in proceedings of international conference on computers and games* (pp. 183–190). Cham, Switzerland: Springer.
15. Silver, D., Huang, A., Maddison, C. J., Guez, A., Sifre, L., Van Den Driessche, G., Schrittwieser, J., Antonoglou, I., Panneershelvam, V., Lanctot, M., & Dieleman, S. (2016). Mastering the game of Go with deep neural networks and tree search. *Nature, 529*(7587), 484–489.
16. Pilania, G., Wang, C., Jiang, X., Rajasekaran, S., & Ramprasad, R. (2013). Accelerating materials property predictions using machine learning. *Science and Reports, 3*, 2810.
17. Ahneman, D. T., Estrada, J. G., Lin, S., Dreher, S. D., & Doyle, A. G. (2018). Predicting reaction performance in C-N cross-coupling using machine learning. *Science, 360*, 186–190.
18. Yu, C.-H., Qin, Z., & Buehler, M. J. (2019). Artificial intelligence design algorithm for nanocomposites optimized for shear crack resistance. *Nano Futures, 3*, 035001.
19. Chen, C.-T., & Gu, G. X. (2019). Machine learning for composite materials. *MRS Communications, 9*, 556–566.
20. Daghigh, V., Lacy, T. E., Jr., Daghigh, H., Gu, G., Baghaei, K. T., Horstemeyer, M. F., & Pittman, C. U., Jr. (2020). Machine learning predictions on fracture toughness of multiscale bio-nano-composites. *Journal of Reinforced Plastics and Composites, 39*(15–16), 587–598.
21. Tan, R. K., Zhang, N. L., & Ye, W. (2020). A deep learning-based method for the design of microstructural materials. *Structural and Multidisciplinary Optimization, 61*, 1417–1438.
22. Skalic, M., Jiménez, J., Sabbadin, D., De Fabritiis, G. (2019). Shape-based generative modeling for De Novo drug design. *Journal of Chemical Information and Modeling, 59*, 1205–1214.
23. Ribeiro, M. T., Singh, S., & Guestrin, C. (2016). "Why should I trust you?" Explaining the predictions of any classifier. In *Proceedings of the 22nd ACM SIGKDD international conference on knowledge discovery and data mining* (pp. 1135–1144), San Francisco, CA, USA, 13–17 August 2016.
24. Yang, Z., et al. (2018). Deep learning approaches for mining structure-property linkages in high contrast composites from simulation datasets. *Computational Materials Science, 151*, 278–287.
25. Belytschko, T., & Black, T. (1999). Elastic crack growth in finite elements with minimal remeshing. *International Journal for Numerical Methods in Engineering, 45*(5), 601–620.
26. Moës, N., Dolbow, J., & Belytschko, T. (1999). A finite element method for crack growth without remeshing. *International Journal for Numerical Methods in Engineering, 46*(1), 131–150.
27. Tang, B., Niu, S., Yang, J., Shao, C., Wang, M., Ni, J., Zhang, X., & Yang, X. (2022). Investigation of bioinspired nacreous structure on strength and toughness. *Biomimetics, 7*(3), 120.

28. Fries, T. P., & Belytschko, T. (2010). The extended/generalized finite element method: An overview of the method and its applications. *International Journal for Numerical Methods in Engineering, 84*(3), 253–304.
29. Ko, K., Jin, S., Lee, S. E., Lee, I., & Hong, J. W. (2019). Bio-inspired bimaterial composites patterned using three-dimensional printing. *Composites Part B: Engineering, 165*, 594–603.
30. Hallquist, J. O. (2012). *LS-DYNA keyword user's manual* (Vol. 1). Livermore Software Technology Corporation.
31. Libonati, F., Gu, G. X., Qin, Z., Vergani, L., & Buehler, M. J. (2016). Bone-inspired materials by design: Toughness amplification observed using 3D printing and testing. *Advanced Engineering Materials, 18*, 1354–1363. https://doi.org/10.1002/adem.v18.8
32. Dimas, L. S., Bratzel, G. H., Eylon, I., & Buehler, M. J. (2013). Tough composites inspired by mineralized natural materials: Computation, 3D printing, and testing. *Advanced Functional Materials, 23*, 4629–4638. https://doi.org/10.1002/adfm.v23.36
33. Maghsoudi-Ganjeh, M., Lin, L., Wang, X., & Zeng, X. (2019). Bioinspired design of hybrid composite materials. *International Journal of Smart and Nano Materials, 10*(1), 90–105.
34. Morsali, S., Qian, D., & Minary-Jolandan, M. (2020). Designing bioinspired brick-and-mortar composites using machine learning and statistical learning. *Communications Materials, 1*(1), 12.
35. Sivakumar, G., & Maji, V. B. (2016). *Simulation of crack propagation in rocks by XFEM* (pp. 291–296). Atlantis Press: Amsterdam, The Netherlands.
36. Wang, T., Liu, Z., Zeng, Q., Gao, Y., & Zhuang, Z. (2017). XFEM modeling of hydraulic fracture in porous rocks with natural fractures. *Science China Physics, Mechanics & Astronomy, 60*, 084612.
37. Cao, B., Cheng, S., Li, A., Deng, Y., & Fang, Z. (2022). Fatigue crack propagation study of bridge steel Q345qD based on XFEM considering the influence of the stress ratio. *Applied Sciences, 12*, 12782.
38. OuladBrahim, A., Belaidi, I., Fahem, N., Khatir, S., Mirjalili, S., & Abdel Wahab, M. (2022). Prediction of the peak load and crack initiation energy of dynamic brittle fracture in X70 steel pipes using an improved artificial neural network and extended finite element method. *Theoretical and Applied Fracture Mechanics, 122*, 103627.
39. Prodromou, M., & Dow, R. S. (2022). Flexural response of composite coated steel components using the extended finite element method. *Marine Structures, 85*, 103266.
40. Chang, Y., Song, Q., Kuang, Z., Zhang, K., & Zheng, Z. (2019). Fracture analysis of cast iron materials with cracks based on elastoplastic extended finite element method. *Acta Mechanica Solida Sinica, 32*, 201–214.
41. Tsuda, T., Ohnishi, Y., Ohtagaki, R., Cho, K., & Fujimoto, T. (2015). Three-point bending crack propagation analysis of beam subjected to eccentric impact loading by X-FEM. In *Proceedings of the 10th European LS-DYNA conference*, Würzburg, Germany, 15–17 June 2015.
42. Gairola, S., & Jayaganthan, R. (2021). XFEM simulation of tensile and fracture behavior of ultrafine-grained Al 6061 alloy. *Metals, 11*, 1761.
43. Xie, X., Feng, Q., Zhang, M., & Liao, N. (2023). Evaluating fracture properties and interfacial strengths of silicon oxycarbide thin-films: XFEM simulations and experimental validation. *Engineering Fracture Mechanics, 292*, 109648.
44. Wang, Y., Cui, H., Zhao, Q., & Du, X. (2019). Chameleon-inspired structural-color actuators. *Matter, 1*, 626–638.
45. Kim, D.-H., Park, M. J., Chang, Y.-S., & Baek, U. B. (1813). Evaluation of fracture properties of two metallic materials under hydrogen gas conditions by using XFEM. *Metals, 2022*, 12.
46. Marques, I. J., Lupi, E., & Mercader, N. (2019). Model systems for regeneration: Zebrafish. *Development, 146*, dev167692.
47. Vellwock, A. E., Su, P., Zhang, Z., Feng, D., & Yao, H. (2022). Reconciling the conflict between optical transparency and fouling resistance with a nanowrinkled surface inspired by Zebrafish's cornea. *ACS Applied Materials & Interfaces, 14*, 7617–7625.

48. Yao, H., Xie, Z., He, C., & Dao, M. (2015). Fracture mode control: a bio-inspired strategy to combat catastrophic damage. *Science and Reports, 5*, 8011.
49. Tadayon, M., Amini, S., Masic, A., & Miserez, A. (2015). The mantis shrimp saddle: A biological spring combining stiffness and flexibility. *Advanced Functional Materials, 25*, 6437–6447.
50. Behera, R. P., & Le Ferrand, H. (2021). Impact-resistant materials inspired by the mantis shrimp's dactyl club. *Matter, 4*, 2831–2849.
51. Amini, S., Tadayon, M., Loke, J. J., Kumar, A., Kanagavel, D., Le Ferrand, H., Duchamp, M., Raida, M., Sobota, R. M., Chen, L., et al. (2019). A diecast mineralization process forms the tough mantis shrimp dactyl club. *Proceedings of the National academy of Sciences of the United States of America, 116*, 8685–8692.
52. Tadayon, M., Younes-Metzler, O., Shelef, Y., Zaslansky, P., Rechels, A., Berner, A., Zolotoyabko, E., Barth, F. G., Fratzl, P., Bar-On, B., et al. (2020). Adaptations for wear resistance and damage resilience: micromechanics of spider cuticular "tools." *Advanced Functional Materials, 30*, 2000400.
53. Libonati, F., & Vergani, L. (2016). Understanding the structure-property relationship in cortical bone to design a biomimetic composite. *Composite Structures, 139*, 188–198.
54. Casari, D., Michler, J., Zysset, P., & Schwiedrzik, J. (2021). Microtensile properties and failure mechanisms of cortical bone at the lamellar level. *Acta Biomaterialia, 120*, 135–145.
55. Chen, G., Lin, T., Guo, C., Richter, L., & Dai, N. (2022). Bending study of six biological models for design of high strength and tough structures. *Biomimetics, 7*, 176.
56. Gustafsson, A., Wallin, M., & Isaksson, H. (2020). The influence of microstructure on crack propagation in cortical bone at the mesoscale. *Journal of Biomechanics, 112*, 110020.
57. Gustafsson, A., Mathavan, N., Turunen, M. J., Engqvist, J., Khayyeri, H., Hall, S. A., & Isaksson, H. (2018). Linking multiscale deformation to microstructure in cortical bone using in situ loading, digital image correlation and synchrotron x-ray scattering. *Acta Biomaterialia, 69*, 323–331.
58. Yadav, R. N., Uniyal, P., Sihota, P., Kumar, S., Dhiman, V., Goni, V. G., Sahni, D., Bhadada, S. K., & Kumar, N. (2021). Effect of ageing on microstructure and fracture behavior of cortical bone as determined by experiment and Extended Finite Element Method (XFEM). *Medical Engineering & Physics, 93*, 100–112.
59. Chaturvedi, I., Jandyal, A., Wazir, I., Raina, A., & UlHaq, M. I. (2022). Biomimetics and 3D printing—Opportunities for design applications. *Sensors International, 3*, 100191.
60. Paar, M. J., & Petutschnigg, A. (2016). Biomimetic inspired, natural ventilated façade—A conceptual study. *Journal of Facade Design and Engineering, 4*, 131–142.
61. Zahra, O., Tolu, S., Zhou, P., Duan, A., & Navarro-Alarcon, D. (2022). A bio-inspired mechanism for learning robot motion from mirrored human demonstrations. *Frontiers in Neurorobotics, 16*, 826410.
62. Buccino, F., Aiazzi, I., Casto, A., Liu, B., Sbarra, M. C., Ziarelli, G., Vergani, L. M., & Bagherifard, S. (2021). Down to the bone: A novel bio-inspired design concept. *Materials, 14*, 4226.
63. Libonati, F., Colombo, C., & Vergani, L. (2014). Design and characterization of a biomimetic composite inspired to human bone. *Fatigue and Fracture of Engineering Materials and Structures, 37*, 772–781.
64. Vellwock, A. E., Vergani, L., & Libonati, F. (2018). A multiscale XFEM approach to investigate the fracture behavior of bio-inspired composite materials. *Composites. Part B, Engineering, 141*, 258–264.
65. Grezzana, G., Loh, H.-C., Qin, Z., Buehler, M. J., Masic, A., & Libonati, F. (2020). Probing the role of bone lamellar patterns through collagen microarchitecture mapping, numerical modeling, and 3D-printing. *Advanced Engineering Materials, 22*, 2000387.
66. Adamatzky, A. (2018). Generative complexity of Gray-Scott model. *Communications in Nonlinear Science and Numerical Simulation, 56*, 457–466.
67. Gray, P., & Scott, S. K. (1983). Autocatalytic reactions in the isothermal, continuous stirred tank reactor: Isolas and other forms of multistability. *Chemical Engineering Science, 38*(1), 29–43.

68. Gray, P., & Scott, S. K. (1984). Autocatalytic reactions in the isothermal, continuous stirred tank reactor: Oscillations and instabilities in the system A+ 2B→ 3B; B→ C. *Chemical Engineering Science, 39*(6), 1087–1097.
69. Gray, P., & Scott, S. K. (1985). Sustained oscillations and other exotic patterns of behavior in isothermal reactions. *The Journal of Physical Chemistry, 89*(1), 22–32.
70. Pearson, J. E. (1993). Complex patterns in a simple system. *Science, 261*(5118), 189–192.
71. Turing, A. M. (1990). The chemical basis of morphogenesis. *Bulletin of Mathematical Biology, 52*, 153–197.
72. Nishiura, Y., & Ueyama, D. (2001). Spatio-temporal chaos for the Gray-Scott model. *Physica D: Nonlinear Phenomena, 150*(3–4), 137–162.
73. Prasad, A., Varshney, V., Nepal, D., & Frank, G. J. (2023). Bioinspired design rules from highly mineralized natural composites for two-dimensional composite design. *Biomimetics, 8*(6), 500.
74. Roy, M., & Prasad, A. (2020). Biomechanics and computational analysis of vascular plant stem for composite design. *SSRN Electronic Journal.*
75. Lipatov, A., Lu, H., Alhabeb, M., Anasori, B., Gruverman, A., Gogotsi, Y., & Sinitskii, A. (2018). Elastic properties of 2D Ti3C2Tx MXene monolayers and bilayers. *Science Advances, 4*, eaat0491.
76. Wei, C., & Wu, C. (2020). Nonlinear fracture of two-dimensional transition metal carbides (MXenes). *Engineering Fracture Mechanics, 230*, 106978.
77. Borysiuk, V. N., Mochalin, V. N., & Gogotsi, Y. (2018). Bending rigidity of two-dimensional titanium carbide (MXene) nanoribbons: A molecular dynamics study. *Computational Materials Science, 143*, 418–424.
78. Zeleniakiene, D., Monastyreckis, G., Aniskevich, A., & Griskevicius, P. (2020). Deformation and Failure of MXeneNanosheets. *Materials, 13*, 1253.
79. Kilikevicius, S., Kvietkaite, S., Zukiene, K., Omastova, M., Aniskevich, A., & Zeleniakiene, D. (2020). Numerical investigation of the mechanical properties of a novel hybrid polymer composite reinforced with graphene and MXenenanosheets. *Computational Materials Science, 17*, 109497.
80. Monastyreckis, G., Mishnaevsky, L., Jr., Hatter, C. B., Aniskevich, A., Gogotsi, Y., & Zeleniakiene, D. (2020). Micromechanical modeling of MXene-polymer composites. *Carbon, 162*, 402–409.
81. Bowie, N. G. (2017). Terrorism events data: An inventory of databases and data sets, 1968–2017. *Perspectives on Terrorism, 11*(4), 50–72.
82. Ding, C., Ngo, T., Ghazlan, A., Lumantarna, R., & Mendis, P. (2015). Numerical simulation of structural responses to a far-field explosion. *Australian Journal of Structural Engineering, 16*(3), 226–236.
83. Flores-Johnson, E. A., Shen, L., Guiamatsia, I., & Nguyen, G. D. (2015). A numerical study of bioinspired nacre-like composite plates under blast loading. *Composite Structures, 126*, 329–336.
84. Tran, P., Ngo, T. D., & Mendis, P. (2014). Bio-inspired composite structures subjected to underwater impulsive loading. *Computational Materials Science, 82*, 134–139.
85. Ghazlan, A., Ngo, T., Le, T., et al. (2020). Blast performance of a bio-mimetic panel based on the structure of nacre—A numerical study. *Composite Structures, 234*, 111691.
86. Grujicic, M., Snipes, J. S., & Ramaswami, S. (2016). Ballistic impact behavior of Nacre-Like laminated composites consisting of B4C tablets and polyurea matrix. *Journal of Materials Engineering and Performance, 25*, 977–994.
87. Gu, G. X., Takaffoli, M., & Buehler, M. J. (2017). Hierarchically enhanced impact resistance of bioinspired composites. *Advanced Materials, 29*, 1700060.
88. Liu, P., Zhu, D., Yao, Y., et al. (2016). Numerical simulation of ballistic impact behavior of bio-inspired scale-like protection system. *Materials and Design, 99*, 201–210.
89. Browning, A., Ortiz, C., & Boyce, M. C. (2013). Mechanics of composite elasmoid fish scale assemblies and their bioinspired analogues. *Journal of the Mechanical Behavior of Biomedical Materials, 19*, 75–86.

90. Martini, R., & Barthelat, F. (2016). Stability of hard plates on soft substrates and application to the design of bioinspired segmented armor. *Journal of the Mechanics and Physics of Solids, 92*, 195–209.
91. Chintapalli, R. K., Mirkhalaf, M., Dastjerdi, A. K., et al. (2014). Fabrication, testing and modeling of a new flexible armor inspired from natural fish scales and osteoderms. *Bioinspiration & Biomimetics, 9*, 036005.
92. Grunenfelder, L. K., Suksangpanya, N., Salinas, C., et al. (2014). Bio-inspired impact-resistant composites. *Acta Biomaterialia, 10*, 3997–4008.
93. Ginzburg, D., Pinto, F., Iervolino, O., et al. (2017). Damage tolerance of bio-inspired helicoidal composites under low velocity impact. *Composite Structures, 161*, 187–203.
94. Huang, J., Durden, H., & Chowdhury, M. (2011). Bio-inspired armor protective material systems for ballistic shock mitigation. *Materials and Design, 32*, 3702–3710.
95. Ghazlan, A., Ngo, T., & Le, T., et al. (2020). A bio-mimetic cellular structure for mitigating the effects of impulsive loadings—A numerical study. *Journal of Sandwich Structures & Materials*. Epub ahead of print 25 February 2020.
96. Imbalzano, G., Linforth, S., Ngo, T. D., et al. (2018). Blast resistance of auxetic and honeycomb sandwich panels: Comparisons and parametric designs. *Composite Structures, 183*, 242–261.
97. Remennikov, A., Kalubadanage, D., Ngo, T., et al. (2019). Development and performance evaluation of large-scale auxetic protective systems for localised impulsive loads. *International Journal of Protective Structures, 10*, 390–417.
98. Qi, C., Remennikov, A., Pei, L. Z., et al. (2017). Impact and close-in blast response of auxetic honeycomb-cored sandwich panels: Experimental tests and numerical simulations. *Composite Structures, 180*, 161–178.
99. Baranowski, P., Płatek, P., Antolak-Dudka, A., et al. (2019). Deformation of honeycomb cellular structures manufactured with Laser Engineered Net Shaping (LENS) technology under quasi-static loading: Experimental testing and simulation. *Additive Manufacturing, 25*, 307–316.
100. Kucewicz, M., Baranowski, P., Małachowski, J., et al. (2018). Modelling, and characterization of 3D printed cellular structures. *Materials and Design, 142*, 177–189.
101. Qiao, J. X., & Chen, C. Q. (2015). Impact resistance of uniform and functionally graded auxetic double arrowhead honeycombs. *International Journal of Impact Engineering, 83*, 47–58.
102. Wang, H., Lu, Z., Yang, Z., et al. (2019). A novel re-entrant auxetic honeycomb with enhanced in-plane impact resistance. *Composite Structures, 208*, 758–770.
103. Ghazlan, A., Ngo, T., Nguyen, T., et al. (2020). Uncovering a high-performance bio-mimetic cellular structure from trabecular bone. *Scientific Reports, 10*, 14247.
104. Chawla, K. K. (2012). *Composite materials: Science and engineering* (3rd ed., p. 542). New York, NY, USA: Springer Science+Business Media.
105. Canale, G., Andrews, S., Rubino, F., Maligno, A., Citarella, R., & Weaver, P. M. (2018). Realistic stacking sequence optimisation of an aero-engine fan blade-like structure subjected to frequency, deformation and manufacturing constraints. *The Open Mechanical Engineering Journal, 12*, 151–163.
106. Herencia, J. E., Weaver, P. M., & Friswell, M. I. (2007). Optimization of long anisotropic laminated fiber composite panels with T-shaped stiffeners. *The American Institute of Aeronautics and Astronautics, 45*, 2497–2509.
107. Barbero, E. J. (2018). *Introduction to composite materials design* (3rd ed., p. 535). Boca Raton, FL, USA: CRC Press, Taylor & Francis Group.
108. Gibson, I., Rosen, D. W., & Stucker, B. (2015). *Additive manufacturing technologies: 3d printing, rapid prototyping and direct digital manufacturing* (2nd ed., p. 498). New York, NY, USA; London, UK: Springer.
109. Ghimire, A., Tsai, Y.-Y., Chen, P.-Y., & Chang, S.-W. (2021). Tunable interface hardening: designing tough bio-inspired composites through 3d printing, testing, and computational validation. *Composites. Part B, Engineering, 215*, 108754.

110. Studart, A. R. (2016). Additive manufacturing of biologically-inspired materials. *Chemical Society Reviews, 45*, 359–376.
111. Libonati, F., Gu, G. X., Qin, Z., Vergani, L., & Buehler, M. J. (2016). Bone-inspired materials by design: toughness amplification observed using 3d printing and testing. *Advanced Engineering Materials, 18*, 1354–1363.
112. Gu, G. X., Dimas, L., Qin, Z., & Buehler, M. J. (2016). Optimization of composite fracture properties: method, validation, and applications. *Journal of Applied Mechanics, 83*, 071006.
113. Holland, J. H. (1992). Genetic algorithm. *Scientific American, 267*, 66–72.
114. Mathias, J.-D., Balandraud, X., & Grediac, M. (2006). Applying a genetic algorithm to the optimization of composite patches. *Computers & Structures, 84*, 823–834.
115. Jenkins, W. M. (1991). Towards structural optimization via the genetic algorithm. *Computers & Structures, 40*, 1321–1327.
116. Vellwock, A. E., Vergani, L., & Libonati, F. (2018). A multiscale XFEM approach to investigate the fracture behavior of bio-inspired composite materials. *Composites Part B: Engineering, 141*, 258–264.
117. Yu, C. H., Tseng, B. Y., Yang, Z., Tung, C. C., Zhao, E., Ren, Z. F., Yu, S. S., Chen, P. Y., Chen, C. S., & Buehler, M. J. (2022). Hierarchical multiresolution design of bioinspired structural composites using progressive reinforcement learning. *Advanced Theory and Simulations, 5*(11), 2200459.
118. van Hasselt, H. (2010). *Advances in Neural Information Processing Systems* 23.
119. Gu, S., Lillicrap, T., Sutskever, I., & Levine, S. (2016). *Proceedings of 33rd International Conference on Machine Learning PMLR, 48*, 2829.
120. Silver, D., Schrittwieser, J., Simonyan, K., Antonoglou, I., Huang, A., Guez, A., Hubert T., Baker, L., Lai, M., Bolton, A., Chen, Y., Lillicrap, T., Hui, F., Sifre, L., Van Den Driessche, G., Graepel, T., & Hassabis, D. (2017). *Nature, 550*, 354.
121. Shamshirband, S., Patel, A., Anuar, N. B., Kiah, M. L. M., & Abraham, A. (2014). *Engineering Applications of Artificial Intelligence, 32*, 228.
122. Low, E. S., Ong, P., & Cheah, K. C. (2019). *Robotics and Autonomous Systems, 115*, 143.
123. Shah, T., Zhuo, L., Lai, P., LaRosa-Moreno, A. D., Amirkulova, F., & Gerstoft, P. (2021). *Journal of the Acoustical Society of America, 150*, 32.
124. Spielberg, S. P. K., Gopaluni, R. B., & Loewen, P. D. (2017). *6th international symposium on advanced control of industrial process. AdCONIP 2017* (Vol. 201).
125. Baker, B., Gupta, O., Naik, N., & Raskar, R. (2016). *Designing neural network architectures using reinforcement learning*. CoRR abs/1611.02167 (2016). arXiv:1611.02167.
126. Shahrabi, J., Adibi, M. A., & Mahootchi, M. (2017). A reinforcement learning approach to parameter estimation in dynamic job shop scheduling. *Computers & Industrial Engineering, 110*, 75–82.
127. Karras, T., Aila, T., Laine, S., & Lehtinen, J. (2017). *Progressive growing of GANs for improved quality, stability, and variation* (Vol 2). CoRR abs/1710.10196 (2017). arXiv:1710.10196.
128. Buehler, M. J., & Gao, H. (2006). Modeling dynamic fracture using large-scale atomistic simulations. In *Dynamic fracture mechanics* (pp. 1–68).
129. Jang, B., Kim, M., Harerimana, G., & Kim, J. W. (2019). Q-learning algorithms: A comprehensive classification and applications. *IEEE Access, 7*, 133653–133667.

Exploring the Functionality and Diversity of Biological Materials

Nisha Rani, Neeraj Bisht, Sonika Chauhan, and Deepa Singh

Abstract Nature has produced materials with characteristics and functioning well beyond the engineered materials. Its hierarchical structure has been primarily associated with the incredible efficiency of biological materials, including their exceptional qualities that rely on weak constituents, high performance per unit mass, and numerous functionalities. Because of the unique properties and functions, biological materials are absolutely necessary to study life sciences. Tissues, biomolecules, and cells are some of these materials. Materials with mechanical qualities and biocompatibility such as collagen, keratin, and cellulose are very essential to many biological processes. The hierarchical structure allows multifunctionality and situational adaptation at molecular and macroscopic levels. Recent advances in biotechnology and materials science enabled development of biomimetic materials, which imitate the properties of true biological materials that can be used in tissue engineering, drug delivery systems, and medical implants. The relationships between structures of biological materials and properties can help in understanding the mechanisms behind biocompatibility, resilience, and adaptability. Furthermore, biological materials play a vital role in the development of gentle, sustainable, and environmentally sound materials. A wide range of research on biological and bioinspired materials can be analyzed from mechanical and functional perspective. The development of bioinspired materials utilizes principles of biological system to enhance material performance for practical applications including, civil, marine, aerospace, and medical engineering. Bioinspired materials are now playing a significant role in encouraging advancements and development in the contemporary materials industry despite the obstacles.

Keywords Biological materials · Constituent of biological materials · Bioinspired materials

N. Rani · N. Bisht (✉) · S. Chauhan · D. Singh
Department of Mechanical Engineering, Govind Ballabh Pant University of Agriculture and Technology, Pantnagar, India
e-mail: neerajbisht.me@gbpuat-tech.ac.in

K. K. Dwivedi et al. (eds.), *Mechanics and Materials Science of Biological Materials*, Materials Horizons: From Nature to Nanomaterials,
https://doi.org/10.1007/978-981-96-5366-9_11

1 Introduction

Since the beginning of medicine in ancient civilizations, natural or artificial materials have been used to replace or integrate human functions or organs, protect aging tissue from traumatic or pathologic occurrences, promote tissue healing, or rectify defects. Innovations in therapy and the creation and application of sophisticated medical equipment enable greater patient survival, greatly enhance life quality, and prolong life expectancy. Biological examples are scaffolds, collagen utilized in bone repair, and coral implants, and so forth. The two classifications of materials are structural and functional. Mechanical properties are vital in structural materials; among their most crucial attributes are toughness, hardness, ductility, density, and strength. On the other hand, functional materials are designed for a variety of more significant properties, some of which are superconducting, magnetic, optical and electrical and energy storage. These distinctions are less obvious in biological materials since many of them have several functions. However, some serve primarily to maintain the structure, while others serve different purposes as their main characteristics.

Since the early twentieth century, biological systems have been studied as structures. Arcy W. Thompson's work [1], which was released in 1917, was regarded as the first significant contribution in this discipline. Considering biological systems as structural engineer, he reduced correlations that characterized their shapes and structure. Currey conducted extensive research on wide range of mineralized biological materials in 1970s and wrote a book named "Bone" [2]. "Structural Biological Materials" by Vincent is another important work [3]. During this time, biology has developed and flourished, but biologists have frequently rejected the engineering and materials techniques. A new and dynamic field, material science and engineering has developed into three areas since the 1950s: metals, polymers, and ceramics (as well as their mixtures and composites made from these materials). Its new future does undoubtedly include biological materials, which began to be added to its interests in the 1990s. It is possible to attain mechanical properties in many biological systems that are significantly higher than those of synthetic materials [3, 4]. With the relatively low strength of the fundamental minerals and polymers found in natural systems, this is an astonishing finding. Biological creatures develop composites including both organic and inorganic elements in an intricate structure. They are arranged at meso-, micro-, and nano-levels in a hierarchical manner. The emerging domain of biological materials presents an enormous number of novel prospects for materials scientists to excel in their specialization: resolving intricate, multidisciplinary scientific problems. A relatively new and quickly expanding branch of engineering and materials science is called biological materials science as illustrated in Fig. 1. The intricate limitations that biological systems face have certain features that are depicted in Fig. 1b. There are five parts to the modified pentahedron that Arzt [5] proposed: ambient temperature and pressure processing, hierarchy of self-assembly, evolution or environmental effects, structure, and functioning. Particularly, biological materials are different and unique. Among many characteristics that set biological systems apart

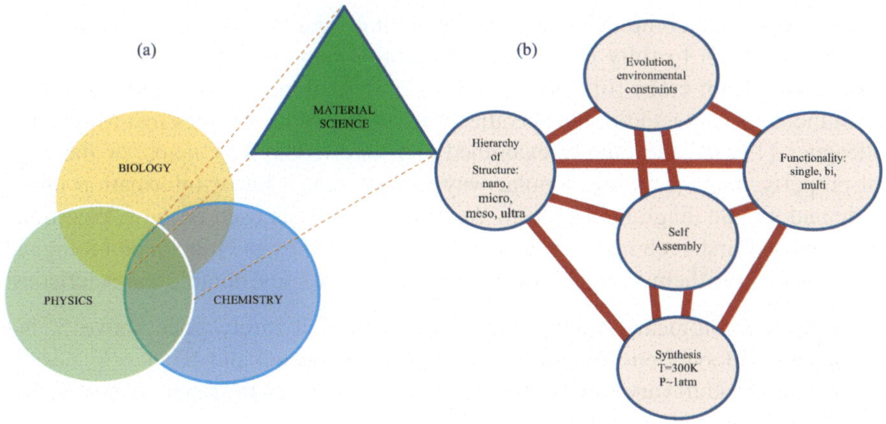

Fig. 1 **a** Contributing scientific fields in biological system and **b** components involved in the study of biological system

are their multifunctionality and evolution; multifunctionality still requires investigation, and evolution is not taken into account in synthetic materials. The diagrams in Fig. 1b illustrate the intricate relationships and contributions required to completely comprehend and utilize biological system. Several primary domains of investigation and endeavor within this field include:

- Natural materials: Biological material is the study of natural materials such as bone, skin, and feathers.
- Biomaterials: Artificial materials used in biological contexts, i.e., use of materials in the field of biomedical, e.g., implants.
- Biomimetics or bioinspired material design: Materials and design influenced by biology (is the creation of new materials, tools, and structures that draw inspiration from biology).

These three elements overlap with one another. For example, biological materials have effectively employed in several uses in the biomedical field.

Despite being an established field of study, materials scientists and engineers are just starting to examine biological materials and systems. In the final phase, the goal is to:

(a) Supply the resources needed to create materials with biological inspiration. It is the new field in materials research and is gaining more and more attention. This is also known as biomimetics.
(b) To improve our knowledge of how artificial materials and biological structures interact to facilitate the insertion of novel and intricate systems into human body, which could ultimately result in the addition or replacement of organs. They are referred as biomaterials.

The scope and complexity are overwhelming, and it will take several years of global research to fully understand. Presently, many people investigated upon exoskeletons from crabs [6], toucan beaks [7, 8], abalone [9, 10], and conch [11, 12]. It includes some additional specific systems and silica spicules that Mayer and colleagues [13, 14] thoroughly examined and reported. Ashby maps, or mechanical property maps [15], are popular way to represent a lot of information into a single and simple diagram. The term Weertman–Ashby is used because Weertman [16] proposed first such map for metals. Wegst and Ashby [17] included biological materials, making them a valuable design tool. There are some notable characteristics:

(a) Natural (biological) systems have low density. It rarely goes above 3, but densities of synthetic structural materials lie between 4 and 10.
(b) Young's moduli vary widely; their range varies between the ranges of 0.001–100 GPa. There are five orders of magnitude.
(c) With a variation encompassing four orders of magnitude from 0.1 to 1000 MPa, the range of strengths is nearly as wide as Young's moduli.
(d) Metals that often require high temperatures to process or high electric cuts are absent. Nature does not have these variables.

Biological materials (natural) are categorized in four types [17]:

(a) Ceramic composites or ceramic: Biological materials including a significant amount of mineral content, like sponge spicules, bones, teeth, diatoms, and shells.
(b) Polymer composites or polymers: These are exoskeletons of arthropods, silk, ligaments and tendons, and the hooves of mammals.
(c) Elastomers: These are biological materials with the unique ability to stretch or strain greatly. This group includes the body's soft tissues, muscles, blood vessels, skin, and individual cells.
(d) Cellular materials: Feathers, the inside of beaks, cancellous bone, and wood are examples of lightweight materials.

2 Constitute of Biological Material

There are few components that are composed of biological materials. Merely, 10% of 100 stable elements present in the periodic table serve as the fundamental constituents of both inorganic and organic tissues. Main components are silicon, phosphorus, carbon, sulfur, nitrogen, oxygen, and hydrogen. Furthermore, it has been discovered that in certain systems, trace elements—like iron—are essential for determining phase. For instance, chiton's radicular teeth contain iron oxide [18]. There are some significant exceptions, like catechol (in DOPA protein) and imidazole from histidine, when specific amino acid residues are present, metal ions act as cross-linkers to create metal complexes. These bonds are reversible, and self-healing materials can be developed using these kind of physio-chemical design [19–21]. The most widely used synthetic metals include copper, aluminum, and iron which are essentially negligible and are present only in limited amounts for specific purposes. A reason behind this is that these elements are processed at high temperatures, which are absent in the organisms.

2.1 Biopolymers

Polysaccharides (chitin, cellulose) or protein-forming polypeptides (primarily present in the biological materials: keratin, collagen, elastin, resilin, abductin, fibroin) are structural organic components of living tissues. These structural proteins are further categorized in two groups: first is primarily composed of amorphous, random-coil domains (found in elasticin, resilin, and abductin), while second is primarily composed of stiff secondary structure areas (such as the b-sheet strands in silk or b-keratin, or triple helix in collagen). Crystalline fibers are developed when molecular aggregates align in a specific direction [22]. The failure stresses of crystalline biopolymers are lower [22], but they are usually orders of magnitude stiffer than amorphous ones. Elastic energy is stored as an entropy change in elastomeric biopolymers, but it is stored in atomic bonds of crystalline biopolymers. Collagen and keratin resemble cross-linked polymers, but elastin, resilin, and abductin are comparable to rubber.

2.2 Proteins

A few well-known secondary structures seen in crystalline protein molecules are b-turn, a-helix, coiled-coil (supercoil bundle consisting of two or more a-helices), and b-sheet. Proteins are made up of long chains of amino acids with side chains that vary in size, which results in steric constraints, charge, and hydrophobic or hydrophilic properties. Each of the residues plays a distinct role in maintaining general stability of protein as well as giving them unique foldings and secondary structures. Post-translational modifications are chemically alter the side chains; proline hydroxylation, for example, crucial alteration in structural proteins like collagen. Collagen, the most prevalent protein in mammals, is present in blood vessels, cartilage, teeth, and bones, among other places. Collagen is also present in vertebrate muscle and fish scales. Polysaccharides are structural organic component found in plants, mollusks, and crustaceans.

90% of collagen in human bodies are composed of five types, despite the fact that there are 29 varieties. Type I is present in skin and bones, type II is mostly found in the cartilage, type III is typically seen in conjunction with type I, type IV is located in cell basement membrane, and type V is present in the placenta and hair. Amino acid sequence GlyPro-X or Gly-X-Hyp, that occurs frequently in collagen (Gly = glycine, Pro = proline, Hyp = hydroxyproline, X = various amino acids), is one of distinguishing feature. Smallest amino acid is glycine and the major residue in most protein-based fibers, with the exception of keratin. This residue is sufficiently small to accommodate twists along the triple helix. Tropocollagen is the result of twisting together of three collagen a-helix strands. Hydrogen bonding together stabilizes and holds the molecule. The tropocollagen forms approximately 300 nm-long strands that are spaced ~40 nm apart. Collagen fibrils is made from these strands arranged in a staggered pattern, with a diameter of ~1 μm and length ranging from micrometers to tens of micrometers [22–24].

3 Biological Materials and Bioinspired Materials

Nature uses biological materials, which are cleverly created and tuned tools, to help organisms survive and flourish in harsh conditions [25–28]. These techniques are considered elegant and serve a range of mechanical as well as functional purposes [26, 29, 30]. Their composition is typically simple, but their performance is efficient [31–33]. This is not like most engineering materials, which typically require costly manufacturing processes or complex chemical compositions and hence frequently require a trade-off between qualities (e.g., increasing weight to improve strength). Therefore, in recent decades, the development of innovative materials and structures has found constant inspiration in biological materials. The first essential step in putting this inspiration into action is to uncover mechanism of structure–property and develop systematic theories—a process called "Biological Materials Science" [28, 34–36]. It provides crucial insights with hitherto untapped tactics from natural designs, paving the way for next phase of developing a new sophisticated material. The growing comprehension and expertise of biological materials themselves goes hand in hand with this study, which is quickly moving to the forefront. Biological materials have numerous qualities and functions that are superior to those of manufactured materials because they use exquisite structures rather than chemical complexity. Their hierarchical structure holds the key to effectively securing these exceptional features [17, 25–27, 37]. This greatly increases the weak constituents' properties for mechanical performance. For example, shell nacre has fracture toughness (4–10 MPa $m^{1/2}$), Young's modulus (70–80 GPa) and high tensile strength (70–100 MPa) [38–40]. However, it is primarily made up of brittle mineral (atleast 95% by volume [41]), which exhibit fracture that is roughly 3000 times less compared to shell [40, 42]. However, because of their hierarchical structures, biological materials can perform significantly better per unit mass. For example, spider silk having a tensile strength (1.1 GPa) is similar to tensile strength of a high-strength steel (1.5 GPa) [43]; however, when taking into account, their density is 1.3 g/cm^3 compared to 7.8 g/cm^3 [44, 45], and spider silk is four times stronger. As a result, there are number of increasing high-performance bioinspired structural materials available, such as fracture-resistant composites inspired by crustaceans [46] and strong and durable materials inspired by nacres [47]. Biological structures undertake a variety of fascinating tasks. For instance, leaves of lotus have unique surface topographies enabling self-cleaning [48], through dry adhesion gecko feet's hierarchical structure enables wall scaling. Arzt [5], butterflies' microstructure interacts with light to produce beautiful colors [49], and many plants' fibrous structures deform themselves in response to humidity changes [50]. Biological materials offer some properties such as durability, nontoxicity, and reliability. Functional materials served as an inspiration for wide range of practical uses including bioinspired self-shaping composites [51], gecko-inspired high adhesion pads [52], nature-inspired reversible underwater adhesives [53], and high-performance bioinspired anticorrosion coatings [54]. Biological or bioinspired designs concentrate on a particular issue. Typical functions such as bioactivity, superwettability, stimuli-responsiveness, and mechanical properties, i.e., lightweight and high-toughness and lightweight and high strength, are provided by biological and bioinspired material. The structure-property mechanisms obtained from biological materials and bioinspired materials have amazing properties and functions for relevant applications. Further development of cutting-edge functional and structural materials offers difficulties and potential of biological and bioinspired designs.

4 Important Strategies for Using Biological Systems as Bioinspired Materials

The remarkable mechanical and functional qualities of biological materials, in spite of their weak constituents and moderate synthesis conditions, seem to be crucial for numerous engineering businesses in human society. The fundamental aspect of the hierarchical structure is the way the structural parts or building blocks are grouped and structured according to various length of scales. It has simple composition but effective in execution. These materials offer creative inspiration for creating contemporary advanced materials due to their excellent durability, dependability, and nontoxicity. It is true that exploring naturally occurring refined biological materials for innovation in the development of advanced materials has been an intriguing field, particularly in the last several years with the quick advancement of research into bioinspired materials. In general, the process of creating bioinspired materials entails actions that disclose the design principles found in nature. These principles serve as the foundation for creating innovative structured materials that are manufactured to solve specific technical challenges. The first step in this process is to identify the distinctive and fascinating properties of biological materials and systems. Then, using scientific analysis, the underlying mechanisms are systematized and theoreticalized to meet the requirements of pertinent engineering applications. The following phase uses innate principles to design and develop the materials with desired mechanical and functional qualities and improve their performance. Although there are obstacles such as the need for extremely advanced fabrication tools to create these structures and translation of small-scale laboratory synthesis to large-scale production, biological and bioinspired materials have yielded a wealth of insightful information that has aided in the advancement of innovation in a number of fields, including housing, transportation, medical, marine, and aerospace, as this work illustrates. Additionally, this process helps us to establish a more harmonious relationship between human civilization and the natural world by allowing us to reflect on our understanding of biology.

5 Development with Bioinspired Materials

Bioinspired structural materials will feature designs that offer high strength with low weight and high toughness with low weight; bioinspired functional materials will concentrate on designs that offer superwettability, bioactivity, and stimuli-responsiveness.

5.1 Superwettability

Superwettable structure and surfaces are obtained in many parts of nature such as the exoskeletons of the bettles [55, 56], spines of cactus [57, 58], shark skin [59, 60], lotus leaves [61, 62], rose petals [63, 64], and shark skin [59, 60]. These constructions use superwettability for a variety of vital purposes, such as directed liquid movement and the provision of an external barrier. Because functional surfaces with particular wettability are particularly useful in real-world applications, research on these surfaces has advanced remarkably in the recent few decades in terms of man-made materials. From a basic perspective, a surface's potential to be superwettable can be attributed to either its physical structure or its chemical composition. Because natural materials have a limited supply of certain constituents, nature typically favors the former over the latter. This offers an inspiration for creating textured, bioinspired structures with particular wettabilities. Super-hydrophobicity uses surface features to trap air pockets beneath liquid droplets in order to reject droplets [65]. Numerous plant leaves have these surfaces, which have a variety of nano/micro-topographies. The Salvinia effect refers to the fact that, despite hydrophilic components on the leaf surface, leaves of the Salvinia molesta plant exhibits behavior of super-hydrophobic surface. This is because hierarchical structures on a surface of leaf [66] effectively stabilize the air–water interface and retain air for an extended period of time. Using a three-dimensional (3D) printing approach, Chen et al. [67] effectively created bioinspired ultra-hydrophobic eggbeater like structure that was inspired by Salvinia effect. The technology for developing a bioinspired material is a 3D printing technology to reproduce intricate structures of a natural material at a much finer scale. The hydrophilic, photocurable substance is used to create the prepared eggbeater shape exhibits exceptional water adhesion and super-hydrophobic qualities. The structure which was modeled after Salvinia was then used to extract oil from water, effectively doing so in 6 s. Superwettable surfaces can provide many useful characteristics, such as self-cleaning, antireflection, and modification of hydrodynamic drag [58, 68–72].

Sharks, for instance, have many riblets the size of microns on their skin, which helps them swim efficiently. These ribs, which run the length of the shark's body, are able to significantly resist the flow of water across surface of skin surface. Through chemical modification and replication, Zhang et al. [72] created bioinspired polydimethylsiloxane (PDMS) sheets with super-hydrophobicity that resembled structure of shark skin. Compared to the non-grooved structures, the manufactured films of PDMS exhibit larger contact angles of water. Consequently, these PDMS films offer both self-cleaning and super-hydrophobicity. These PDMS films interact with surroundings fluid through a liquid–gas interface rather than a liquid–solid interface as they trap air. As a result, there is less flow resistance and effective slippage due to the fluid's ability to flow easily across the surface. Since shark skin reduces drag to a maximum of 5.4%, these super-hydrophobic PDMS films provide greater impact on drag reduction (maximum of around 21.7%). Directed liquid transfer is another use for superwettabile surfaces, and it may be seen in cactus spines. These cacti's spines use a hierarchical surface structure to gather water because of their arid surroundings. Cactis conical spines are divided into three distinct structural components, the directed gradient grooves, barbs and belt-structured trichomes that are arranged in a staggered fashion from tip toward base [68]. The three components work together to produce a surface-free energy gradient and Laplace pressure, which allows the droplets to travel in the direction of their collecting and feed the cactus. In particular, the gradient of the Laplace pressure causes the cactus spine's surface droplets to be directed toward the base of spine with larger radius. At same time, the base of spine has fewer micro-grooves than the tip, which results in a rough gradient. Roughness of gradient creates surface-free energy gradient acting as a driving force, directing water droplets in the direction of base of spine. Jiang et al. [71] created a bioinspired water collector having dual gradients by using chemical modification gradient and electrochemical corrosion gradient, inspired by mechanism of water in cactus. The synthetic spine of the cactus gathers water droplets at a comparatively rapid rate of approximately 2.08 $\mu L\ s^{-1}$ and transfers it at a speed of approximately 20.05 $\mu m\ s^{-1}$ [71].

5.2 *Bioactivity*

Because biological creatures are made up of materials which are biocompatible and bioactive by nature. Also, they are capable of carrying out the essential processes of life. This offers crucial direction for biomaterials to use them in biological applications. Essentially, unique composition and structure of biological materials interact with cells and tissues give rise to bioactive function. A research has been done to fabricate innovative bioinspired structures with improved biocompatibility and bioactivity. Typical methods includes, electrospinning, self-assembly, chemical manipulation, 3D printing, and templating. Every manufacturing method has special benefits of its own that can be used to achieve various morphologies. For instance, biomimetic mineralization and 3D printing have been combined to create scaffolds that are bioactive for bone tissue engineering [73]. On the surface of scaffolds, calcium and phosphate ions react to create hydroxyapatite (HAp) in-situ as part of the biomimetic mineralization process. Compared to materials that combines directly with HAp particles, they enables HAp to be distributed more evenly throughout the matrix. Furthermore, the chemical structure and material characteristics of this structure are comparable to those of real bone minerals. As a result, it possesses more biological activity and encourages cell division and growth more effectively [74–77].

Without biomimetic mineralization, scaffolds are more likely to collapse; these scaffolds have improved mechanical qualities after biomimetic mineralization to satisfy the demands of bone tissue restoration. In addition to 3D printed single-material used in bones, skin, muscles, and other tissues [78–80], the intricate structure and function of central nervous system make it difficult to imitate [81, 82]. Numerous cell types can be found in spinal cord tissue, and their arrangements are carefully regulated in terms of spatial distribution. The ability to replicate the architecture of spinal cord, which is crucial for regulating cell differentiation, depends on this spatial distribution.

Through multi-material 3D bioprinting, McAlpine et al. [83] created a spinal cord that is heterogeneously bioinspired in which spinal neuronal progenitor cells (sNPCs) and clusters of oligodendrocyte progenitor cells (OPCs) were precisely positioned in various scaffold segments during assembly. Bioinspired spinal cord mimics the native tissues as precisely as possible to exact placement of different cell types, which increases the scaffold's bioactivity. These biologically inspired spinal cords effectively demonstrated the in vitro differentiation of functionally mature neurons, paving the way for additional differentiation of stem cells and osteoproliferating cells into neurons that project oligodendrocytes and axons that myelinate the axons. This method of bioinspired offers fresh way in making scaffolds that are more bioactive so they can imitate central nervous systems, which opens up new treatment options for neurological disorders. Electrospun nanofibrous scaffolds have characteristics that mimic the extracellular matrix's (ECM) structure in addition to printing, opening up new possibilities for tissue regeneration and repair [84–86]. However, electrospinng often creates fibrous membranes of two-dimensional that needs to be stretched to match the collagen fibrils surface morphology in extracellular matrix (ECM) and three-dimensional geometry of the organs. In order to overcome this, Mi et al. [87] produced biomimetic scaffolds with notable biological activity, including a 3D nanofibrous foam (3D-PCL), by using wet electrospinning, carbon dioxide (CO_2) foaming, and controlled crystallization. More so than a 2D framework, the 3D-PCL enables cell infiltration and growth through the scaffolds. This nanofibrous foam was treated in a PCL/pentyl acetate solution to produce a shish-kebab nanostructure (3D-SK-PCL) on the nanofiber surface via controlled crystallization. These nanostructures on the surface of the nanofibers promote adhesion of cell and migration [88, 89] and can enhance scaffold's bioactive properties even more. The 3D-SK-PCL scaffolds cultivated HEF1 human fibroblast cells elongate and spread well, demonstrating the healthy state. Additionally, the population of cell population in 3D bioinspired scaffold is much larger than other scaffolds. This was explained by structural characteristics of a 3D-SK-PCL that imitate the native collagen fibril nanotopography and the true 3D geometry needed by living cells to promote cell adhesion, migration, and proliferation. Developing highly effective targeted drug delivery carriers by imitating the complex architecture of viruses is another noteworthy bioinspired design that offers increased bioactivity. Nucleic acid molecules comprise viruses, which are encased in a protein coat for protection. To accomplish cell invasion, some materials on coat surface, like glycoproteins, attach selectively to the host cell surface receptors.

An artificial system of virus with a structure of core–shell was developed by Pramod et al. [90] to target tissues of tumor using a simple and dependable self-assembly technique. Core of the synthetic virus is made up of DNA and doxorubicin adduct. Since doxorubicin is a particular type of anti-cancer medication that is harmful for healthy cells, it is added to artificial virus to prevent toxicity. Albumin linked to folate makes use of outer layer. Folate receptor expression is limited on normal cells and often overexpressed on cancer cells [91, 92], which enables the synthetic viruses to bind up and detect tumor tissue. Polypeptide complex develops outermost layer of fake virus and dissolves in an acidic pH [93]. Targeted drug delivery is thereby achieved when artificial virus enters tumor tissue, dissolves complex, and releases drug.

5.3 Stimuli-Responsiveness

Many naturally occurring materials have been cleverly constructed by nature to perform efficient actuating and sensing activities based upon passive architectures including non-living tissues. In the biological systems, an external stimulus causes varying degrees of interaction with the structure, resulting in a full cycle of detecting and responding activity. Depending on specific process, this typically converts thermal energy into strain energy for different color light and detection (by altered periodic structure). A process like this gets rid of the difficulties associated with managing complex chemical compounds or living cell activity by allowing a wide range of sensing and responding behaviors like converting force, light, heat, humidity, and other signal conversions into mechanical, electrical, or other types of responses [49, 94–99]. The natural sensors chameleons have are the capability to change their color immediately during fighting or courtship from a hidden (camouflaged) state to a highly visible one [100]. According to studies, the iridophore cells' guanine nanocrystal structure is actively tuned to produce this color shift. A loosely packed array of silica particles was included in the mechanochromic elastomers that were created by imitating the photonic structure of cell of chameleon iridophore [97]. This bioinspired sensor creates non-close-packed crystals by embedding stiff silica nanocrystal particles in an elastomer matrix.

Rapid volume changes in the sensor causes lattice parameters to alter the retaining of lattice structure. This alters light refraction and produces striking color shifts across visible spectrum. Sensor displays a shift of color from red to blue when stretched, and red to green color shift when compressed. Remarkably, this shift is reversible, much like their biological model. Promising uses for these mechanochromic sensors includes signboard displays, large-scale wallpaper, and optical recording, among other things. Spider hair is another fascinating sensing and responding mechanism; it functions as a wind sensor, enabling it to detect changes in the surrounding airflow brought on by predators or prey [101]. Su et al. [98] proposed wind sensors that are self-powered based upon flexible magnetoelectric material systems, which was inspired by animal hair sensors. Bioinspired sensor was made up of magnetic and electrical parts. Electrical component is thin sheet of polyethylene terephthalate (PET) screen-printed with silver nanoparticles in a triangular pattern like the fine hairs of a spider. As a result, this component is conductive for signal output and flexible for wind sensing. NdFeB, the magnetic component, is joined with the electrical component with the use of 3D printing. Magnetic flux is continuously supplied by this magnetic component to the electrical component. As a result, when the wind passes by the sensor, it will generate a distinct electrical output, or voltage. Sensors quickly distort and produce negative value when air flow blows. After that, inertia causes the sensor to bounce a little, producing a little reversal peak. The magnetic flux remains constant due to continuous blowing of wind, preventing any variation in voltage. The PET film recovers its initial state when air flow is stopped, and the relative positions of electrical and magnetic components shifts, resulting in a positive voltage of 20.3 μV. These are useful for the applications that need to monitor air flows in a challenging situations. Many immobile plants exhibit intriguing behavior of responding and sensing such as photoinduced, hygroscopic, and force-driven traits that are particularly used in practical applications with animal kingdom [51, 102–107]. The hygroscopic deformation of carrot umbel [83] and pinecone scales [51] is one well-known example. The corresponding structural elements are reversibly close to protect seeds from drying out upon hydration and open for dispersal of seed upon dehydration. The direction of the cellulose micro-fibrils, distribution of lignin and composition of tissues typically assessed using a bilayered structure linked to hygroscopic deformation. The two layers have different swelling characteristics. For example, when water enters the matrix, active layer at bottom swells and increases longitudinally, but passive layer at upper side does not swells. This causes overall bending. Based upon this, reinforcements within the matrix were meticulously aligned to generate bilayered structure, which was used to build false pinecone scales. As a result, the synthetic pinecone can display comparable humidity-induced deformation. This served as inspiration for Reichert et al. to create bioinspired building skins, can automatically sense changes in ambient humidity, and adapt movement of "skin" to open and close building [106]. This offers a novel solution of the current requirement for sophisticated mechanical or electronic systems since the design permits reversal movement without the need for motors. Controlling the hierarchical structure, it is essential to achieve the intended activities, and this can be done by dissecting the mechanisms of biological materials that exhibit bioactivity, superwettability, and stimuli-responsiveness. By modulating their structures, wide variety of bioinspired materials having enhanced and targeted functions can be developed. Advanced fabrication technologies such as chemical modification, high-resolution and multi-material 3D printing, replication, self-assembly, magnetic-assisted, and eletrospinning can be used for making composites.

6 Structural Materials with a Bioinspired Aesthetic

6.1 High-Strength Materials and Lightweight Materials

Despite having few, weak chemical components, biological materials exhibit outstanding mechanical qualities of strong and lightweight (i.e., minerals and polymers). The many, hierarchical structures of biological materials hold the secret to achieving this great strength and low weight without the need for a complicated system of available chemical ingredients [108]. Development of cutting-edge composite materials for a variety of contemporary industries including aviation, aerospace, marine, and land transportation offers tremendous inspiration. Commonly, cellular structural materials with good strength and low weight include cuttlebone [109–111], honeycomb [112], and cancellous bone [26]. For example, cuttlebone exhibits mechanical qualities and a high porosity that are balanced by hierarchical structures spanning the nanoscale to microscale. A high porous complex structure can withstand the pressure of its deep-sea environment in the cuttlebone by aragonite nanofibers with an organic phase forming micrometer-thick lamellae [110]. Using vanadia (V2O5) nanofibers, Burghard et al. created porous hierarchical scaffolds that mimicked this complex architecture through ice-templating. These bioinspired scaffolds have a porosity of around 99.8%, which is higher than cuttlebone. Compressive strength of cuttlebone-like scaffolds (V2O5-1) has superior mechanical properties about twice that of V2O5-0, when compared at random microstructures and same ultra-high porosity shown by the compressive stress–strain curves. This is due to the fact that the V2O5-1's regularly ordered rectangular pores are better at distributing applied stress than the V2O5-0's randomly built pores. Furthermore, elevating the concentration of vanadia nanofiber in the cell wall results in markedly improved mechanical attributes. For instance, Young's moduli of cuttlebone-shaped scaffolds, varying in nanofiber concentration, lies between 10 and 55 times greater than that of V2O5-0 (1.73 ± 0.71 kPa) specimens. This is because lamellae's lateral motion is reduced by rectangular pores, which increases Young's modulus. Many plants have structural traits in their stems that minimize weight while conferring strength against environmental stresses (wind, rain, ocean currents, etc.) found in their habitats [113–115]. Thalia dealbata, a perennial plant having height to diameter ratio nearly 200–350. The stem of this plant needs to be robust enough to withstand regular wild winds. According to Bai et al., porous stems of Thalia dealbata [115] showed an arrangement of layers of oriented lamellar with connected bridges along growth direction. Bidirectional freezing techniques are successfully used in graphene aerogels to simulate this hierarchical structure. Furthermore, the bidirectional freezing is a general technique that may be used by many other material systems to accomplish control of multiscale structure in the scalable way. With about 50% strain, prepared by graphene aerogel can hold more than 6000 times of its own weight. Furthermore, following unloading, it can fully recover without any permanent deformation. Since, at 50% strain, after compressive cycles of 1000 s, it retains around 85% of its initial compressive strength. These findings demonstrate that the Thalia dealbata's structural design may concurrently produce remarkable strength and low weight.

Numerous investigations had been conducted in improving absolute mechanical strength of wood inspired by biotechnology for applications of advanced engineering. Natural wood is inexpensive and sustainable [116–119]. One amorphous matrix substance found in natural wood, lignin, is responsible for binding cellulose fibrils and greatly influences total strength [120]. Yu et al. [121] provided a method to develop polymeric woods on a large scale using similar amorphous polyphenol matrix materials (phenol formaldehyde resin and melamine–formaldehyde resin) by thermocuring and self-assembly of conventional resins. The polymeric woods exhibit comparable compressive strength as real timbers about 45 MPa. Furthermore, bioinspired woods perform better in terms of axial compressive strength than other materials that resemble wood, such as foam materials made of ceramic. Density of polymeric woods is more as compared with other materials of engineering, suggesting good tunability in strength and weight. Spider silk is another well-known biological material that is more strong and light in weight. Lately, it has become evident that the structure—which consists of densely packed, highly ordered hydrogen-bonded β-sheet crystals inside semi-amorphous protein matrix—is primarily responsible for exceptional mechanical capabilities. Chen et al. created a composite film of bioinspired material [122] by inserting a graphene quantum dots (GOD, which resemble the β-sheet crystals) into polyvinyl alcohol (PVA, as the protein matrix). It was influenced by hierarchical nanostructure of spider silk. Analyzing the bioinspired GOD-reinforced PVA composite sheets' mechanical performance reveals that the mechanical characteristics can be significantly enhanced by adjusting the GOD content. In particular, when GOD content approaches 5.0 wt%, PVA's yield strength and elastic modulus rise by almost 66% (152.5 MPa) and 88% (up to 4.35 GPa), respectively, in comparison to 0.0 wt%.

6.2 *Lightweighted and Tough Materials*

The majority of biological materials are expertly engineered to demonstrate high toughness by retaining enough strength, in contrast to standard materials of engineering by obtaining two exclusive qualities of toughness and strength offering a substantial problem. Natural examples of strong, lightweight materials are highly mineralized (teeth and nacre from seashells), non-mineralized (human skin and wood) and intermediately mineralized (bone and fish scales). These materials' pertinent mechanical characteristics and structure have been thoroughly studied in an effort to isolate the key ideas for creating new, cutting-edge composites that are lightweight, highly durable, and favorably strong. Toughness has been mostly linked to the soft, polymeric properties of materials, is essentially the amount of energy that a material can withstand a load without failure while absorbing or dissipating through deformation. This focuses primarily on the main structural toughening methods found in biological and bioinspired materials that are typically strongly, intermediately, and non-mineralized. Highly mineralized biological materials, such mollusk shells and teeth, have attracted a lot of attention because to their extraordinary toughness and strength [40, 123, 124]. This is not surprising, given that the majority of the brittle component in calcium carbonate (>95 weight percent) is the same as that found in classroom chalk [42].

It is well known that the general stiffness and hardness are provided by the mineral components; on the other hand, the toughness is greatly enhanced by the organic components, which, despite their small amount, show significant deformability. The mineral building blocks are arranged into complex and weak interfaces that drive crack propagation into more convoluted configurations by dissipating more energy, hence enhancing the overall toughness [31]. The bulk materials with weak interface of enamel are made up of proteins and mineral rods having a diameter of around 4–6 μm. In order to prevent a catastrophic collapse, the fractures are directed to spread through the weak interface [125, 126]. Similar to this, the proteins and polysaccharides in nacre bind the minuscule calcium carbonate tablets together to create "strong–weak" microscopic characteristics.

To limit the development of cracks by dissipating energy, organic phase is crucial [28, 127–129]. Based on this, Ritchie et al. [130] created dental bioinspired materials by densifying with a methacrylate resin after freeze-casting zirconia polycrystals in suspension doped with yttria (3Y-TZP). This produced composites with brick-and-mortar structure and lamellar resembling nacre. Nacre is tough and malleable due to its sliding mechanism, which can release a significant amount of mechanical energy in addition to crack bridging and deflection [131–133]. Barthelat et al. produced toughened bioinspired glass using lamination and laser engraving production techniques [134]. The preparation procedure involves use of laser beam to carve the tablet's shapes into sheets of borosilicate glass. The engraved glass is then stacked using EVA and laminated to create clear glass with a bioinspired design. According to puncture tests, nacre (like glass) exhibits a more ductility with substantial deformations and high puncture energy when compared with laminated and regular glass. This is due to the fact that tablets in the nacre-like glass absorb lots of mechanical energy and move over one another across huge volumes. The structural organization and polymeric components of an intermediate and non-mineralized biological materials contribute toward significant deformations while sustaining stress makes them remarkable resistant. Biological collagenous materials include somewhat mineralized materials like fish scales and bone as well as non-mineralized tissues like skin, arteries, and corneas of the eyes. Various studies have been conducted on their mechanical properties, which span from the molecular to macroscales [31, 135–139]. The skin tissue's collagen fibers are bent and organized in various orientations. With the stretching of these fibers, they become straight and start spinning toward the direction of tensile stress. During the process, collagen develops mechanism of toughening that includes stretching at the molecular scale, fibril straightening, delaminating at microscale, and re-orienting so as to improve the toughness of skin and tear resistance. All of these mechanisms absorb a significant amount of energy. Human bone is an exceptional biological material with high toughness and strength for biological materials with an intermediate mineralization. It is made up of collagen molecules and nanoscale hydroxyapatite crystals, which pack into fibers and subsequently arrange into mesoscale osteons, microscale lamellae, and mature cortical bone. Strength [140, 141], elasticity [142–145], and creep characteristics [146, 147] are primarily determined by hydroxyapatite components of bone, whereas deformability and energy absorption are attributed to the collagen components. Depending upon the processes behind and front of crack tip, the toughening mechanisms in bone is typically categorized as either intrinsic or extrinsic [148–150]. Intrinsic toughening, which forms plastic zones around crack-like flaws by stretching and sliding collagen fibrils, inhibits degradation at nanoscales. In order to protect the local loads and strains, extrinsic toughening mechanisms primarily work on the crack wake at the micrometer scale. Through techniques like deflection of crack, micro-cracking restriction and bridging of crack successfully prevents growth of crack. Because of these two methods, larger driving force is required for propagation of cracks, hence, increasing materials' total toughness. Biological keratinous materials are among the toughest biological materials, along with collagenous materials. One prominent example is the keratin-based baleen structure seen in the mouths of baleen whales, which serves as a filter for food throughout the whale's life. This means that in order to facilitate ease of usage, baleen must be lightweight enough to mechanically withstand a range of stresses from both water and prey.

Key components of this hierarchical structure are nanoscale filament-matrix with mineral structures, microscale tubular lamellae, and macroscale sandwich-tubules [151]. Baleen is one of the toughest biological keratinous materials due to its substantial fracture toughness. Wang et al. employed multi-material 3D printing to create a range of baleen-like models [150, 152]. Their findings indicate the baleen model exhibiting overall superior qualities in comparison to the similar structures that lack some of the toughening mechanisms of baleen. In addition to viscoelasticity, the structure significantly affects behavior of strain-rate of these materials. Enormous biological materials are examples of lightweight, durable, and adequately strong structural materials. Nearly all contemporary engineering domains, including aircraft, navigational transit, and automotive engineering, have traditionally revolved around these fundamental characteristics. The use of hierarchical structures and cellular patterns confer greater resistance against deformation and failure by introducing a variety of toughening mechanisms. The development of fabrication techniques, including 3D printing, laser engraving, thermocuring, self-assembly, and freeze casting, is underway in preparation for the subsequent stage of advanced structural material mass production and scaling.

7 Conclusion

With millions of years to perfect materials and structures, nature is the ultimate designer. It offers an abundance of fascinating biological materials with creatively inspiring qualities that can be used to create a wide range of new materials. This leads to a remarkable advancement in the fundamental knowledge of biological materials and developing bioinspired materials with a wide range of characteristics. Although the field has expanded rapidly, there are some unsolved difficulties that must be addressed if bioinspired and biological materials are promising future. Recent developments in computational simulation and nanotechnology can partially support this, but still it is challenging to conduct experiments at small scales like the molecular mechanisms that protein chains and fibrils use to toughen themselves. The fact is that these techniques have to account for polymer's brittleness which can be destroyed by vacuum, desiccation, and electron beams making it worse. However, current theories mostly relate to a uniform, single-scaled structure. It is challenging to deduce relevant mechanical theories that integrate the operating mechanism at various scales. Also, research has been done for characterization of various biological materials on a broad scale. On the other hand, practical applications and a wide range of engineering applications will benefit from conserved or unifying principles that underlie these diverse natural phenomena. It must be acknowledged that in the process of generating bioinspired materials, efficiencies of biological materials are hard to replicate since hierarchical structure requires the fabrication of intricate architectural features that are beyond existing capabilities of most of nanotechnologies. Furthermore, majority of high-performance artificial materials that are bioinspired by nature were created in laboratories using a small amount of raw materials, and their characteristics are assessed on a case-by-case basis without following set criteria. Furthermore, the manufacturing techniques that can ensure structural characteristics across various length scales are typically costly and time-consuming, which presents a significant obstacle to industrial mass production. Therefore, mentioned difficulties along with several more will be resolved as this study moves quickly forward, which will help convert our basic knowledge of biological materials into useful engineering applications through the use of bioinspired materials.

Future materials will need to be environmentally benign, have excellent mechanical characteristics, and several functions in order to meet the ever-increasing needs from a variety of businesses. Coincidentally, many biological materials have these built-in benefits. Therefore, bioinspired materials—those derived from biological systems—have a great chance of taking center stage in the next wave of sophisticated material design. In the present study, we explore various materials in terms of structural and functional categories and outline important techniques from biological to bioinspired materials.

Data Availability Not applicable.

Declaration of Competing Interest The authors have no conflicts to disclose.

References

1. Thompson, S. D. A. W. (1917). *On growth and form. [With illustrations.]*. Cambridge.
2. Currey, J. D. (2002). *Bones: Structure and mechanics*. Princeton University Press.
3. Vincent, J. (2012). *Structural biomaterials*. Princeton University Press.
4. Srinivasan, A. V., Haritos, G. K., & Hedberg, F. L. (1991). Biomimetics: Advancing man-made materials through guidance from nature 463–482.
5. Arzt, E. (2006). Biological and artificial attachment devices: Lessons for materials scientists from flies and geckos. *Materials Science and Engineering: C, 26*(8), 1245–1250.
6. Chen, P. Y., Lin, A. Y. M., Mckittrick, J., & Meyers, M. A. (2008). Structure and mechanical properties of crab exoskeletons. *Acta Biomaterialia, 4*(3), 587–596.
7. Seki, Y., Schneider, M. S., & Meyers, M. A. (2005). Structure and mechanical behavior of a toucan beak. *Acta materialia, 53*(20), 5281–5296.
8. Seki, Y., Kad, B., Benson, D., & Meyers, M. A. (2006). The toucan beak: Structure and mechanical response. *Materials Science and Engineering: C, 26*(8), 1412–1420.
9. Menig, R., Meyers, M. H., Meyers, M. A., & Vecchio, K. S. (2000). Quasi-static and dynamic mechanical response of Haliotis rufescens (abalone) shells. *Acta Materialia, 48*(9), 2383–2398.
10. Lin, A., & Meyers, M. A. (2005). Growth and structure in abalone shell. *Materials Science and Engineering: A, 390*(1–2), 27–41.
11. Lin, A. Y. M., Meyers, M. A., & Vecchio, K. S. (2006). Mechanical properties and structure of Strombus gigas, Tridacna gigas, and Haliotis rufescens sea shells: A comparative study. *Materials Science and Engineering: C, 26*(8), 1380–1389.
12. Menig, R., Meyers, M. H., Meyers, M. A., & Vecchio, K. S. (2001). Quasi-static and dynamic mechanical response of Strombus gigas (conch) shells. *Materials Science and Engineering: A, 297*(1–2), 203–211.
13. Sarikaya, M. (1994). An introduction to biomimetics: A structural viewpoint. *Microscopy Research and Technique, 27*(5), 360–375.
14. Mann, S. (2001). *Biomineralization: Principles and concepts in bioinorganic materials chemistry*.
15. Ashby, M. F. (1994). Materials selection in mechanical design. *Metallurgia Italiana, 86*, 475–475.
16. Weertman, J., & Weertman, J. R. (1970). *Physical metallurgy* (Cahn, R. W. (Ed.)) (Vol. 983) Amsterdam: North-Holland.
17. Wegst, U. G. K., & Ashby, M. F. (2004). The mechanical efficiency of natural materials. *Philosophical Magazine, 21*, 2167–2186. https://doi.org/10.1080/14786430410001680935
18. Saunders, M., Kong, C., Shaw, J. A., Macey, D. J., & Clode, P. L. (2009). Characterization of biominerals in the radula teeth of the chiton, Acanthopleura hirtosa. *Journal of Structural Biology, 167*(1), 55–61.
19. Harrington, M. J., & Waite, J. H. (2007). Holdfast heroics: Comparing the molecular and mechanical properties of Mytilus californianus byssal threads. *Journal of Experimental Biology, 210*(24), 4307–4318.
20. Harrington, M. J., Masic, A., Holten-Andersen, N., Waite, J. H., & Fratzl, P. (2010). Iron-clad fibers: A metal-based biological strategy for hard flexible coatings. *Science, 328*(5975), 216–220.
21. Broomell, C. C., Zok, F. W., & Waite, J. H. (2008). Role of transition metals in sclerotization of biological tissue. *Acta Biomaterialia, 4*(6), 2045–2051.
22. Chen, P. Y., McKittrick, J., & Meyers, M. A. (2012). Biological materials: Functional adaptations and bioinspired designs. *Progress in Materials Science, 57*(8), 1492–1704.
23. Giraud-Guille, M. M. (1988). Twisted plywood architecture of collagen fibrils in human compact bone osteons. *Calcified Tissue International, 42*, 167–180.
24. Giraud-Guille, M. M. (1998). Plywood structures in nature. *Current Opinion in Solid State and Materials Science, 3*(3), 221–227.

25. Meyers, M. A., McKittrick, J., & Chen, P. Y. (2013). Structural biological materials: critical mechanics-materials connections. *Science, 339*(6121), 773–779.
26. Wegst, U. G., Bai, H., Saiz, E., Tomsia, A. P., & Ritchie, R. O. (2015). Bioinspired structural materials. *Nature Materials, 14*(1), 23–36.
27. Eder, M., Amini, S., & Fratzl, P. (2018). Biological composites—Complex structures for functional diversity. *Science, 362*(6414), 543–547.
28. Meyers, M. A., Chen, P. Y., Lin, A. Y. M., & Seki, Y. (2008). Biological materials: Structure and mechanical properties. *Progress in Materials Science, 53*(1), 1–206.
29. Yan, X., Jin, Y., Chen, X., Zhang, C., Hao, C., & Wang, Z. (2020). Nature-inspired surface topography: Design and function. *Science China Physics, Mechanics & Astronomy, 63*(2), 224601.
30. Cui, M., Wang, B., & Wang, Z. (2019). Nature-inspired strategy for anticorrosion. *Advanced Engineering Materials, 21*(7), 1801379.
31. Ritchie, R. O. (2011). The conflicts between strength and toughness. *Nature Materials, 10*(11), 817–822.
32. Zioupos, P., & Currey, J. D. (1998). Changes in the stiffness, strength, and toughness of human cortical bone with age. *Bone, 22*(1), 57–66.
33. Wang, B., Yang, W., McKittrick, J., & Meyers, M. A. (2016). Keratin: Structure, mechanical properties, occurrence in biological organisms, and efforts at bioinspiration. *Progress in Materials Science, 76*, 229–318.
34. Wang, B. (2016). *Structural and functional design strategies of biological keratinous materials*. University of California.
35. Naleway, S. E., Porter, M. M., McKittrick, J., & Meyers, M. A. (2015). Structural design elements in biological materials: Application to bioinspiration. *Advanced Materials, 27*(37), 5455–5476.
36. Antonietti, M., & Fratzl, P. (2010). Biomimetic principles in polymer and material science. *Macromolecular Chemistry and Physics, 211*(2), 166–170.
37. Baer, E., Hiltner, A., & Morgan, R. J. (1992). Biological and synthetic hierarchical composites. *Physics Today, 45*(10), 60–67.
38. Wang, R. Z., Suo, Z., Evans, A. G., Yao, N., & Aksay, I. A. (2001). Deformation mechanisms in nacre. *Journal of Materials Research, 16*(9), 2485–2493.
39. Currey, J. D. (1977). Mechanical properties of mother of pearl in tension. *Proceedings of the Royal society of London. Series B. Biological sciences, 196*(1125): 443–463.
40. Jackson, A. P., Vincent, J. F., & Turner, R. M. (1988). The mechanical design of nacre. *Proceedings of the Royal Society of London. Series B. Biological Sciences, 234*(1277), 415–440.
41. Currey, J. D., & Taylor, J. D. (1974). The mechanical behaviour of some molluscan hard tissues. *Journal of Zoology, 173*(3), 395–406.
42. Ji, B., & Gao, H. (2004). Mechanical properties of nanostructure of biological materials. *Journal of the Mechanics and Physics of Solids, 52*(9), 1963–1990.
43. Gosline, J. M., Guerette, P. A., Ortlepp, C. S., & Savage, K. N. (1999). The mechanical design of spider silks: From fibroin sequence to mechanical function. *Journal of Experimental Biology, 202*(23), 3295–3303.
44. Saravanan, D. (2006). Spider silk-structure, properties and spinning. *Journal of Textile and Apparel, Technology and Management, 5*(1), 1–20.
45. Capelli, F. (2005). Stainless steel: A new structural automotive material. In *9th international conference & exhibition*, Florence ATA.
46. Zhang, M., Jiao, D., Tan, G., Zhang, J., Wang, S., Wang, J., Liu, Z., Zhang, Z., & Ritchie, R. O. (2019). Strong, fracture-resistant biomimetic silicon carbide composites with laminated interwoven nanoarchitectures inspired by the crustacean exoskeleton. *ACS Applied Nano Materials, 2*(2), 1111–1119.
47. Gao, H. L., Chen, S. M., Mao, L., Song, Z., Yao, H., Colfen, H., Luo, X., Zhang, F., Pan, Z., Meng, Y., Ni, Y., & Yu, S. (2017). Mass production of bulk artificial nacre with excellent mechanical properties. *Nature Communications, 8*(1), 287.

48. Zhang, L., Zhou, Z., Cheng, B., DeSimone, J. M., & Samulski, E. T. (2006). Superhydrophobic behavior of a perfluoropolyether lotus-leaf-like topography. *Langmuir, 22*(20), 8576–8580.
49. Xu, D., Yu, H., Xu, Q., Xu, G., & Wang, K. (2015). Thermoresponsive photonic crystal: Synergistic effect of poly (N-isopropylacrylamide)-co-acrylic acid and morpho butterfly wing. *ACS Applied Materials & Interfaces, 7*(16), 8750–8756.
50. Dawson, C., Vincent, J. F., & Rocca, A. M. (1997). How pine cones open. *Nature, 390*(6661), 668–668.
51. Erb, R. M., Sander, J. S., Grisch, R., & Studart, A. R. (2013). Self-shaping composites with programmable bioinspired microstructures. *Nature Communications, 4*(1), 1712.
52. Gould, J. (2015). Learning from nature's best. *Nature, 519*(7544 S1), S2.
53. Zhao, Y., Wu, Y., Wang, L., Zhang, M., Chen, X., Liu, M., Fan, J., Liu, J., Zhou, F., & Wang, Z. (2017). Bio-inspired reversible underwater adhesive. *Nature Communications, 8*(1), 2218.
54. Cui, M., Wang, P. Y., Wang, Z., & Wang, B. (2019). Mangrove inspired anti-corrosion coatings. *Coatings, 9*(11), 725.
55. Zhang, S., Huang, J., Chen, Z., & Lai, Y. (2017). Bioinspired special wettability surfaces: From fundamental research to water harvesting applications. *Small (Weinheim an der Bergstrasse, Germany), 13*(3), 1602992.
56. Zhu, H., Duan, R., Wang, X., Yang, J., Wang, J., Huang, Y., & Xia, F. (2018). Prewetting dichloromethane induced aqueous solution adhered on Cassie superhydrophobic substrates to fabricate efficient fog-harvesting materials inspired by Namib Desert beetles and mussels. *Nanoscale, 10*(27), 13045–13054.
57. Ju, J., Yao, X., Yang, S., Wang, L., Sun, R., He, Y., & Jiang, L. (2014). Cactus stem inspired cone-arrayed surfaces for efficient fog collection. *Advanced Functional Materials, 24*(44), 6933–6938.
58. Malik, F. T., Clement, R. M., Gethin, D. T., Kiernan, M., Goral, T., Griffiths, P., Beynon, D., & Parker, A. R. (2016). Hierarchical structures of cactus spines that aid in the directional movement of dew droplets. *Philosophical Transactions of the Royal Society A: Mathematical, Physical and Engineering Sciences, 374*(2073), 20160110. https://doi.org/10.1098/rsta.2016.0110
59. Dundar Arisoy, F., Kolewe, K. W., Homyak, B., Kurtz, I. S., Schiffman, J. D., & Watkins, J. J. (2018). Bioinspired photocatalytic shark-skin surfaces with antibacterial and antifouling activity via nanoimprint lithography. *ACS Applied Materials & Interfaces, 10*(23), 20055–20063.
60. Bixler, G. D., & Bhushan, B. (2012). Bioinspired rice leaf and butterfly wing surface structures combining shark skin and lotus effects. *Soft Matter, 8*(44), 11271–11284.
61. Zhang, P., Lin, L., Zang, D., Guo, X., & Liu, M. (2017). Designing bioinspired anti-biofouling surfaces based on a superwettability strategy. *Small (Weinheim an der Bergstrasse, Germany), 13*(4), 1503334.
62. Zhao, Y., Yu, C., Lan, H., Cao, M., & Jiang, L. (2017). Improved interfacial floatability of superhydrophobic/superhydrophilic Janus sheet inspired by lotus leaf. *Advanced Functional Materials, 27*(27), 1701466.
63. Zong, C., Hu, M., Azhar, U., Chen, X., Zhang, Y., Zhang, S., & Lu, C. (2019). Smart copolymer-functionalized flexible surfaces with photoswitchable wettability: From superhydrophobicity with "rose petal" effect to superhydrophilicity. *ACS Applied Materials & Interfaces, 11*(28), 25436–25444.
64. Tian, Y., & Jiang, L. (2015). Design of bioinspired, smart, multiscale interfacial materials with superwettability. *MRS Bulletin, 40*(2), 155–165.
65. Darmanin, T., & Guittard, F. (2015). Superhydrophobic and superoleophobic properties in nature. *Materials today, 18*(5), 273–285.
66. Barthlott, W., Schimmel, T., Wiersch, S., Koch, K., Brede, M., Barczewski, M., Wallheim, S., Weis, A., Kaltenmaier, A., Leder, A., & Bohn, H. F. (2010). The Salvinia paradox: Superhydrophobic surfaces with hydrophilic pins for air retention under water. *Advanced Materials, 22*(21), 2325–2328.

67. Yang, Y., Li, X., Zheng, X., Chen, Z., Zhou, Q., & Chen, Y. (2018). 3D-printed biomimetic super-hydrophobic structure for microdroplet manipulation and oil/water separation. *Advanced materials, 30*(9), 1704912.
68. Ju, J., Bai, H., Zheng, Y., Zhao, T., Fang, R., & Jiang, L. (2012). A multi-structural and multi-functional integrated fog collection system in cactus. *Nature communications, 3*(1), 1247.
69. Han, Z., Jiao, Z., Niu, S., & Ren, L. (2019). Ascendant bioinspired antireflective materials: Opportunities and challenges coexist. *Progress in Materials Science, 103*, 1–68.
70. Fang, Y., Yong, J., Chen, F., Huo, J., Yang, Q., Bian, H., Du, G., & Hou, X. (2016). Durability of the tunable adhesive superhydrophobic PTFE surfaces for harsh environment applications. *Applied Physics A, 122*, 1–7.
71. Ju, J., Xiao, K., Yao, X., Bai, H., & Jiang, L. (2013). Bioinspired conical copper wire with gradient wettability for continuous and efficient fog collection. *Advanced Materials, 25*(41), 5937–5942.
72. Liu, Y., Gu, H., Jia, Y., Liu, J., Zhang, H., Wang, R., Zhang, B., Zhang, H., & Zhang, Q. (2019). Design and preparation of biomimetic polydimethylsiloxane (PDMS) films with superhydrophobic, self-healing and drag reduction properties via replication of shark skin and SI-ATRP. *Chemical Engineering Journal, 356*, 318–328.
73. Abouzeid, R. E., Khiari, R., Beneventi, D., & Dufresne, A. (2018). Biomimetic mineralization of three-dimensional printed alginate/TEMPO-oxidized cellulose nanofibril scaffolds for bone tissue engineering. *Biomacromolecules, 19*(11), 4442–4452.
74. Rodriguez, K., Renneckar, S., & Gatenholm, P. (2011). Biomimetic calcium phosphate crystal mineralization on electrospun cellulose-based scaffolds. *ACS Applied Materials & Interfaces, 3*(3), 681–689.
75. Ethirajan, A., Ziener, U., & Landfester, K. (2009). Surface-functionalized polymeric nanoparticles as templates for biomimetic mineralization of hydroxyapatite. *Chemistry of Materials, 21*(11), 2218–2225.
76. Abou-Zeid, R. E., Hassan, E. A., Bettaieb, F., Khiari, R., & Hassan, M. L. (2015). Use of cellulose and oxidized cellulose nanocrystals from olive stones in chitosan bionanocomposites. *Journal of Nanomaterials, 2015*(1), 687490.
77. Lopes, D., Martins-Cruz, C., Oliveira, M. B., & Mano, J. F. (2018). Bone physiology as inspiration for tissue regenerative therapies. *Biomaterials, 185*, 240–275.
78. Mironov, V., Boland, T., Trusk, T., Forgacs, G., & Markwald, R. R. (2003). Organ printing: Computer-aided jet-based 3D tissue engineering. *TRENDS in Biotechnology, 21*(4), 157–161.
79. Kang, H. W., Lee, S. J., Ko, I. K., Kengla, C., Yoo, J. J., & Atala, A. (2016). A 3D bioprinting system to produce human-scale tissue constructs with structural integrity. *Nature Biotechnology, 34*(3), 312–319.
80. Kolesky, D. B., Homan, K. A., Skylar-Scott, M. A., & Lewis, J. A. (2016). Three-dimensional bioprinting of thick vascularized tissues. *Proceedings of the National Academy of Sciences, 113*(12), 3179–3184.
81. Hopkins, A. M., DeSimone, E., Chwalek, K., & Kaplan, D. L. (2015). 3D in vitro modeling of the central nervous system. *Progress in Neurobiology, 125*, 1–25.
82. Giger, R. J., Hollis, E. R., & Tuszynski, M. H. (2010). Guidance molecules in axon regeneration. *Cold Spring Harbor Perspectives in Biology, 2*(7), a001867.
83. Joung, D., Truong, V., Neitzke, C. C., Guo, S. Z., Walsh, P. J., Monat, J. R., Meng, F., Dutton, J. R., Parr, A. M., & McAlpine, M. C. (2018). 3D printed stem-cell derived neural progenitors generate spinal cord scaffolds. *Advanced Functional Materials, 28*(39), 1801850.
84. Xu, Y., Bao, J., Zhang, X., Li, W., Xie, Y., Sun, S., Zhao, W., & Zhao, C. (2019). Functionalized polyethersulfone nanofibrous membranes with ultra-high adsorption capacity for organic dyes by one-step electrospinning. *Journal of Colloid and Interface Science, 533*, 526–538.
85. Zhang, L., & Webster, T. J. (2009). Nanotechnology and nanomaterials: Promises for improved tissue regeneration. *Nano Today, 4*(1), 66–80.
86. Miller, R. J., Chan, C. Y., Rastogi, A., Grant, A. M., White, C. M., Bette, N., Schaub, N. J., & Corey, J. M. (2018). Combining electrospun nanofibers with cell-encapsulating hydrogel

fibers for neural tissue engineering. *Journal of Biomaterials Science, Polymer Edition, 29*(13), 1625–1642.
87. Jing, X., Li, H., Mi, H. Y., Liu, Y. J., & Tan, Y. M. (2019). Fabrication of fluffy shish-kebab structured nanofibers by electrospinning, CO2 escaping foaming and controlled crystallization for biomimetic tissue engineering scaffolds. *Chemical Engineering Journal, 372*, 785–795. https://doi.org/10.1016/j.cej.2019.04.194
88. Rouede, D., Schaub, E., Bellanger, J. J., Ezan, F., Scimeca, J. C., Baffet, G., & Tiaho, F. (2017). Determination of extracellular matrix collagen fibril architectures and pathological remodeling by polarization dependent second harmonic microscopy. *Scientific Reports, 7*(1), 12197.
89. Mouw, J. K., Ou, G., & Weaver, V. M. (2014). Extracellular matrix assembly: A multiscale deconstruction. *Nature reviews Molecular cell biology, 15*(12), 771–785.
90. Ajithkumar, K. C., & Pramod, K. (2018). Doxorubicin-DNA adduct entrenched and motif tethered artificial virus encased in pH-responsive polypeptide complex for targeted cancer therapy. *Materials Science and Engineering: C, 89*, 387–400.
91. Franklin, W. A., Waintrub, M., Edwards, D., Christensen, K., Prendegrast, P., Woods, J., Bunn, P. A., & Kolhouse, S. F. (1994). New anti-lung-cancer antibody cluster 12 reacts with human folate receptors present on adenocarcinoma. *International Journal of Cancer, 57*(S8), 89–95.
92. Lu, Y., & Low, P. S. (2002). Folate-mediated delivery of macromolecular anticancer therapeutic agents. *Advanced Drug Delivery Reviews, 54*(5), 675–693.
93. Black, K. A., Priftis, D., Perry, S. L., Yip, J., Byun, W. Y., & Tirrell, M. (2014). Protein encapsulation via polypeptide complex coacervation. *ACS Macro Letters, 3*(10), 1088–1091. https://doi.org/10.1021/mz500529v
94. Dong-il, C., & Tae-Jae, L. (2015). A review of bioinspired vision sensors and their applications. *Sensors and Materials, 27*, 447–463.
95. Pan, H., Jing, X., Sun, W., & Li, Z. (2018). Analysis and design of a bioinspired vibration sensor system in noisy environment. *IEEE/ASME Transactions on Mechatronics, 23*(2), 845–855. https://doi.org/10.1109/Tmech.2018.2803284
96. Zhang, M., Wu, X., Cui, N., Engheta, N., & Van der Spiegel, J. (2014). Bioinspired focal-plane polarization image sensor design: From application to implementation. *Proceedings of the IEEE, 102*(10), 1435–1449. https://doi.org/10.1109/Jproc.2014.2347351
97. Lee, G. H., Choi, T. M., Kim, B., Han, S. H., Lee, J. M., & Kim, S. H. (2017). Chameleon-inspired mechanochromic photonic films composed of non-close-packed colloidal arrays. *ACS Nano, 11*(11), 11350–11357. https://doi.org/10.1021/acsnano.7b05885
98. Wu, Z., Ai, J., Ma, Z., Zhang, X., Du, Z., Liu, Z., ... & Su, B. (2019). Flexible out-of-plane wind sensors with a self-powered feature inspired by fine hairs of the spider. *ACS Applied Materials & Interfaces, 11*(47), 44865–44873. https://doi.org/10.1021/acsami.9b15382
99. Ye, R., Zhu, C., Song, Y., Lu, Q., Ge, X., Yang, X., ... & Lin, Y. (2016). Bioinspired synthesis of all-in-one organic-inorganic hybrid nanoflowers combined with a handheld ph meter for on-site detection of food pathogen. *Small (Weinheim an Der Bergstrasse, Germany), 12*(23), 3094–3100.
100. Teyssier, J., Saenko, S. V., Van Der Marel, D., & Milinkovitch, M. C. (2015). Photonic crystals cause active colour change in chameleons. *Nature Communications, 6*(1), 6368. https://doi.org/10.1038/ncomms7368
101. Barth, F. G., & Holler, A. (1999). Dynamics of arthropod filiform hairs. V. The response of spider trichobothria to natural stimuli. *Philosophical Transactions of the Royal Society of London. Series B: Biological Sciences, 354*(1380), 183–192. https://doi.org/10.1098/rstb.1999.0370
102. Correa, D., Papadopoulou, A., Guberan, C., Jhaveri, N., Reichert, S., Menges, A., & Tibbits, S. (2015). 3D-printed wood: programming hygroscopic material transformations. *3D Printing and Additive Manufacturing, 2*(3), 106–116. https://doi.org/10.1089/3dp.2015.0022
103. Korner, A., Born, L., Mader, A., Sachse, R., Saffarian, S., Westermeier, A. S., Poppinga, S., Bischoff, M., Gresser, G. T., Milwich, M., Speck, T., & Knippers, J. (2017). Flectofold—A biomimetic compliant shading device for complex free form facades. *Smart Materials and Structures, 27*(1), 017001.

104. Zhang, X., Yu, Z., Wang, C., Zarrouk, D., Seo, J. W. T., Cheng, J. C., Buchan, A. D., Takei, K., Zhao, Y., Ager, J. W., Zhang, J., Hettick, M., Hersam, M. C., Pisano, A. P., Fearing, R. S., & Javey, A. (2014). Photoactuators and motors based on carbon nanotubes with selective chirality distributions. *Nature Communications, 5*(1), 2983. https://doi.org/10.1038/ncomms 3983
105. Lienhard, J., Schleicher, S., Poppinga, S., Masselter, T., Milwich, M., Speck, T., & Knippers, J. (2011). Flectofin: A hingeless flapping mechanism inspired by nature. *Bioinspiration & Biomimetics, 6*(4), 045001. https://doi.org/10.1088/1748-3182/6/4/045001
106. Reichert, S., Menges, A., & Correa, D. (2015). Meteorosensitive architecture: Biomimetic building skins based on materially embedded and hygroscopically enabled responsiveness. *Computer-Aided Design, 60*, 50–69.
107. Zhang, P., Chen, P. Y., Wang, B., Yu, R., Pan, H., & Wang, B. (2019). Evaluating the hierarchical, hygroscopic deformation of the Daucus carota umbel through structural characterization and mechanical analysis. *Acta Biomaterialia, 99*, 457–468.
108. Gantenbein, S., Masania, K., Woigk, W., Sesseg, J. P., Tervoort, T. A., & Studart, A. R. (2018). Three-dimensional printing of hierarchical liquid-crystal-polymer structures. *Nature, 561*(7722), 226–230.
109. Knoller, A., Runcevski, T., Dinnebier, R. E., Bill, J., & Burghard, Z. (2017). Cuttlebone-like V2O5 nanofibre scaffolds–advances in structuring cellular solids. *Scientific Reports, 7*(1), 42951.
110. Birchall, J. D., & Thomas, N. L. (1983). On the architecture and function of cuttlefish bone. *Journal of Materials Science, 18*, 2081–2086.
111. Jia, X., Ma, X., Wei, D., Dong, J., & Qian, W. (2008). Direct formation of silver nanoparticles in cuttlebone-derived organic matrix for catalytic applications. *Colloids and Surfaces A: Physicochemical and Engineering Aspects, 330*(2–3), 234–240.
112. Xu, X., Heng, L., Zhao, X., Ma, J., Lin, L., & Jiang, L. (2012). Multiscale bio-inspired honeycomb structure material with high mechanical strength and low density. *Journal of Materials Chemistry, 22*(21), 10883–10888.
113. Naik, D. L., & Kiran, R. (2018). Naïve Bayes classifier, multivariate linear regression and experimental testing for classification and characterization of wheat straw based on mechanical properties. *Industrial Crops and Products, 112*, 434–448.
114. Tan, T., Rahbar, N., Allameh, S. M., Kwofie, S., Dissmore, D., Ghavami, K., & Soboyejo, W. O. (2011). Mechanical properties of functionally graded hierarchical bamboo structures. *Acta Biomaterialia, 7*(10), 3796–3803.
115. Yang, M., Zhao, N., Cui, Y., Gao, W., Zhao, Q., Gao, C., Bai, H., & Xie, T. (2017). Biomimetic architectured graphene aerogel with exceptional strength and resilience. *ACS Nano, 11*(7), 6817–6824.
116. Fang, C. H., Mariotti, N., Cloutier, A., Koubaa, A., & Blanchet, P. (2012). Densification of wood veneers by compression combined with heat and steam. *European Journal of Wood and Wood Products, 70*(1), 155.
117. Paril, P., Brabec, M., Maňák, O., Rousek, R., Rademacher, P., Cermak, P., & Dejmal, A. (2014). Comparison of selected physical and mechanical properties of densified beech wood plasticized by ammonia and saturated steam. *European Journal of Wood and Wood Products, 72*, 583–591.
118. Laine, K., Segerholm, K., Wålinder, M., Rautkari, L., & Hughes, M. (2016). Wood densification and thermal modification: Hardness, set-recovery and micromorphology. *Wood Science and Technology, 50*, 883–894.
119. Kutnar, A., & Kamke, F. A. (2012). Compression of wood under saturated steam, superheated steam, and transient conditions at 150 C, 160 C, and 170 C. *Wood Science and technology, 46*, 73–88.
120. Hofstetter, K., Hellmich, C., Eberhardsteiner, J., & Mang, H. A. (2008). Micromechanical estimates for elastic limit states in wood materials, revealing nanostructural failure mechanisms. *Mechanics of Advanced Materials and Structures, 15*(6–7), 474–484.

121. Yu, Z. L., Yang, N., Zhou, L. C., Ma, Z. Y., Zhu, Y. B., Lu, Y. Y., Qin, B., Xing, W. Y., Ma, T., Li, S. C., Gao, H. L., Wu, H. A., & Yu, S. H. (2018). Bioinspired polymeric woods. *Science Advances, 4*(8), eaat7223.
122. Song, P., Dai, J., Chen, G., Yu, Y., Fang, Z., Lei, W., Fu, S., Wang, H., & Chen, Z. G. (2018). Bioinspired design of strong, tough, and thermally stable polymeric materials via nanoconfinement. *ACS Nano, 12*(9), 9266–9278.
123. Barthelat, F., Tang, H., Zavattieri, P. D., Li, C. M., & Espinosa, H. D. (2007). On the mechanics of mother-of-pearl: A key feature in the material hierarchical structure. *Journal of the Mechanics and Physics of Solids, 55*(2), 306–337.
124. Espinosa, H. D., Rim, J. E., Barthelat, F., & Buehler, M. J. (2009). Merger of structure and material in nacre and bone–Perspectives on de novo biomimetic materials. *Progress in Materials Science, 54*(8), 1059–1100.
125. Yahyazadehfar, M., Bajaj, D., & Arola, D. D. (2013). Hidden contributions of the enamel rods on the fracture resistance of human teeth. *Acta Biomaterialia, 9*(1), 4806–4814.
126. Mirkhalaf, M., Dastjerdi, A. K., & Barthelat, F. J. N. C. (2014). Overcoming the brittleness of glass through bio-inspiration and micro-architecture. *Nature Communications, 5*(1), 3166.
127. Rousseau, M., Lopez, E., Stempflé, P., Brendlé, M., Franke, L., Guette, A., Naslain, R., & Bourrat, X. (2005). Multiscale structure of sheet nacre. *Biomaterials, 26*(31), 6254–6262.
128. Smith, B. L., Schäffer, T. E., Viani, M., Thompson, J. B., Frederick, N. A., Kindt, J., Belcher, A., Stucky, G. D., Morse, D. E., & Hansma, P. K. (1999). Molecular mechanistic origin of the toughness of natural adhesives, fibres and composites. *Nature, 399*(6738), 761–763.
129. Barthelat, F., Rabiei, R., & Dastjerdi, A. K. (2012). Multiscale toughness amplification in natural composites. *MRS Online Proceedings Library (OPL), 1420*, mrsf11-1420.
130. Tan, G., Zhang, J., Zheng, L., Jiao, D., Liu, Z., Zhang, Z., & Ritchie, R. O. (2019). Nature-inspired nacre-like composites combining human tooth-matching elasticity and hardness with exceptional damage tolerance. *Advanced Materials, 31*(52), 1904603.
131. Barthelat, F., & Rabiei, R. (2011). Toughness amplification in natural composites. *Journal of the Mechanics and Physics of Solids, 59*(4), 829–840.
132. Barthelat, F. (2014). Designing nacre-like materials for simultaneous stiffness, strength and toughness: Optimum materials, composition, microstructure and size. *Journal of the Mechanics and Physics of Solids, 73*, 22–37.
133. Yin, Z., Hannard, F., & Barthelat, F. (2019). Impact-resistant nacre-like transparent materials. *Science, 364*(6447), 1260–1263.
134. Yang, W., Meyers, M. A., & Ritchie, R. O. (2019). Structural architectures with toughening mechanisms in nature: A review of the materials science of type-I collagenous materials. *Progress in Materials Science, 103*, 425–483.
135. Nair, A. K., Gautieri, A., Chang, S. W., & Buehler, M. J. (2013). Molecular mechanics of mineralized collagen fibrils in bone. *Nature Communications, 4*(1), 1724.
136. Buehler, M. J. (2007). Molecular nanomechanics of nascent bone: Fibrillar toughening by mineralization. *Nanotechnology, 18*(29), 295102.
137. Gautieri, A., Buehler, M. J., & Redaelli, A. (2009). Deformation rate controls elasticity and unfolding pathway of single tropocollagen molecules. *Journal of the Mechanical Behavior of Biomedical Materials, 2*(2), 130–137.
138. Ritchie, R. O., Koester, K. J., Ionova, S., Yao, W., Lane, N. E., & Ager Iii, J. W. (2008). Measurement of the toughness of bone: A tutorial with special reference to small animal studies. *Bone, 43*(5), 798–812.
139. Kennedy, J. C., Hawkins, R. J., Willis, R., & Danylchuck, K. D. (1976). Tension studies of human knee ligaments. Yield point, ultimate failure, and disruption of the cruciate and tibial collateral ligaments. *JBJS, 58*(3), 350–355.
140. Fritsch, A., Hellmich, C., & Dormieux, L. (2009). Ductile sliding between mineral crystals followed by rupture of collagen crosslinks: Experimentally supported micromechanical explanation of bone strength. *Journal of Theoretical Biology, 260*(2), 230–252.
141. Morin, C., Vass, V., & Hellmich, C. (2017). Micromechanics of elastoplastic porous polycrystals: Theory, algorithm, and application to osteonal bone. *International Journal of Plasticity, 91*, 238–267.

142. Hellmich, C., & Ulm, F. J. (2002). Are mineralized tissues open crystal foams reinforced by crosslinked collagen?—Some energy arguments. *Journal of Biomechanics, 35*(9), 1199–1212.
143. Hellmich, C., Barthélémy, J. F., & Dormieux, L. (2004). Mineral–collagen interactions in elasticity of bone ultrastructure–a continuum micromechanics approach. *European Journal of Mechanics-A/Solids, 23*(5), 783–810.
144. Fritsch, A., & Hellmich, C. (2007). 'Universal' microstructural patterns in cortical and trabecular, extracellular and extravascular bone materials: Micromechanics-based prediction of anisotropic elasticity. *Journal of Theoretical Biology, 244*(4), 597–620.
145. Eberhardsteiner, L., Hellmich, C., & Scheiner, S. (2014). Layered water in crystal interfaces as source for bone viscoelasticity: Arguments from a multiscale approach. *Computer Methods in Biomechanics and Biomedical Engineering, 17*(1), 48–63.
146. Shahidi, M., Pichler, B., & Hellmich, C. (2014). Viscous interfaces as source for material creep: A continuum micromechanics approach. *European Journal of Mechanics-A/Solids, 45*, 41–58.
147. Ritchie, R. O., Buehler, M. J., & Hansma, P. (2009). Plasticity and toughness in bone. *Physics today, 62*(6), 41–47.
148. Nalla, R. K., Kruzic, J. J., & Ritchie, R. O. (2004). On the origin of the toughness of mineralized tissue: Microcracking or crack bridging? *Bone, 34*(5), 790–798.
149. Dastjerdi, A. K., & Barthelat, F. (2015). Teleost fish scales amongst the toughest collagenous materials. *Journal of the Mechanical Behavior of Biomedical Materials, 52*, 95–107.
150. Ritchie, R. O. (1988). Mechanisms of fatigue crack propagation in metals, ceramics and composites: Role of crack tip shielding. *Materials Science and Engineering, 103*, 15–28.
151. Wang, B., Sullivan, T. N., Pissarenko, A., Zaheri, A., Espinosa, H. D., & Meyers, M. A. (2019). Lessons from the ocean: Whale baleen fracture resistance. *Advanced Materials, 31*(3), 1804574. https://doi.org/10.1002/adma.201804574
152. Szewciw, L. J., De Kerckhove, D. G., Grime, G. W., & Fudge, D. S. (2010). Calcification provides mechanical reinforcement to whale baleen α-keratin. *Proceedings of the Royal Society B: Biological Sciences, 277*(1694), 2597–2605.

Mechanical Stimulation and Collagen Fiber Deformation

Ravinder Jhorar and Krashn Kumar Dwivedi

Abstract The molecular structure's reaction to a mechanical signal is soft tissue remodeling. It is unclear, therefore, how the molecular structure of soft tissues reacts to external mechanical stimuli. Although the origin of tissue remodeling is unknown, certain investigations have shown that it is dependent on the rate of mechanical stimulation. Developing physiotherapeutic criteria to enhance tissue health and wound healing requires an understanding of how molecular structure reacts to mechanical stimulation. In this work, we examined how the rate of mechanical stimulation affected the changes in the skin's molecular and nanostructure. Pig skin was cyclically stretched at three distinct frequencies (0.1, 2.0, and 5.0 Hz) in order to provide mechanical stimulation. In contrast, the molecular and nanostructure of mechanical tested and virgin tissue was investigated using atomic force microscope and Attenuated Total Reflectance Fourier Transform Infrared spectroscopy. On the other hand, attenuated total reflectance and an atomic force microscope (AFM) were used to evaluate the nano and molecular structure of both virgin and mechanically tested skin. respectively, Fourier transform infrared spectroscopy (FTIR–ATR). The change in collagen fibril d-periodicity and collagen molecular (tropocollagen, protein structure) was found to be much bigger at 0.1 Hz than at 2.0 and 5.0 Hz, indicating that the alteration in nano and molecular structure was sensitive to the rate of mechanical stimulation. The frequency of stretching and changes in the structure of the collagen protein were found to be significantly correlated negatively ($r = 0.74$, $p = 0.041$). These findings suggest that tissue remodeling may be enhanced by gradual mechanical stimulation during physical therapy.

R. Jhorar
Delhi Skill and Entrepreneurship University, New Delhi, India

K. K. Dwivedi (✉)
Department of Mechanical Engineering, Washington University in St. Louis, St. Louis, USA
e-mail: kkdwivedi159@gmail.com

K. K. Dwivedi et al. (eds.), *Mechanics and Materials Science of Biological Materials*, Materials Horizons: From Nature to Nanomaterials,
https://doi.org/10.1007/978-981-96-5366-9_12

1 Introduction

One of the extrinsic factors that regulate the remodeling of soft tissues is mechanical stimulation [1]. Through various micro and nano-level deformation mechanics, mechanical stimulations at the tissue level are transferred to the molecular level [2] and disrupt the tissue's molecular structural balance. Tissue remodeling results from the activation of a metabolic cascade brought on by this disruption in molecular structure. Collagen fibers facilitate the transfer of applied mechanical signals in the majority of soft tissues. Additionally, it has been discovered that changes in the protein structure of collagen molecules in connective soft tissues are what trigger the tissue remodeling cascade [3]. Collagen fibers have a major role in supporting mechanical stress in the majority of connective soft tissues [4] and exhibit complicated deformation response at various length scales [5]. Rotation, bending, stretching, and shearing of collagen fibers with intera and interfiber cross-linkers breaking are thought to be involved in the transmission of mechanical signals from tissue to the molecular level [5]. These intricate deformation mechanisms are caused by the hierarchical organization of collagen fibers within tissue [6], and they are unique to the type and rate of mechanical stimulation [2].

1.1 Structure of Collagen Fibers

The polypeptide strands of the protein, which are left-handed alpha helices, are the first to assemble into the collagen fiber [6]. Procollagen is a macromolecular unit of collagen fibers that is formed by the assembly of three sets of these left-hand alpha helices [7]. Procollagen matures into tropocollagen, the fundamental load-bearing unit of collagen fibers, through the cleavage of the C and N terminals [8]. It is thought that changes in tropocollagen brought on by mechanical stimulations (stretching-dominated deformation) could initiate the biochemical remodeling process. Tropocollagen typically measures 300 nm in length and 1.5 nm in diameter. Moreover, the collagen fibrils, a nanoscale unit of collagen fibers, are formed by the staggered arrangement of tropocollagen. Enzymatic cross-linking connects the tropocollagens within collagen fibrils. Shearing and stretching share the total mechanical stimulation delivered to collagen fibrils [2]. D-periodicity typically ranges between 64 and 67 nm; any deviation from this range brought on by mechanical means indicates a change in the structure of collagen fibrils. Collagen fibrils were organized into collagen fibers by interfibrillar cross-linking of glycosaminoglycan, and collagen fibers take the shape of bundles when combined with proteoglycan chains. Collagen fibers are primarily in charge of giving soft connective tissues strength and supporting mechanical loads. Collagen fiber bundles are assembled with water and other non-protein substances like proteoglycans to create the extracellular matrix of soft tissues.

1.2 The Mechanism Through Which Mechanical Stimulus Transferred from Tissue to Molecules

Mechanical stimulation, such as strain or load applied to tissues, is initially distributed between the extracellular matrix (ECM) and collagen fibers [4]. However, due to the matrix components' relatively compliant nature, they carry only a small portion of the load, with the majority being borne by the collagen fibers. When subjected to load-induced strain, collagen fibers undergo a two-step process of straightening and stretching. Initially, the fibers straighten, and as they continue to elongate, they stretch. Once this deformation occurs, the strain or load is transferred to the collagen fibrils. At this stage, strain is transmitted through interfibrillar shearing, or sliding, which stretches the interfibrillar cross-links. This stretching imposes tensile strain on the collagen fibrils, and their deformation can be assessed by tracking changes in d-periodicity [2].

The strain experienced by collagen fibrils is then passed on to tropocollagen molecules via intrafibrillar shearing and the stretching of intrafibrillar cross-links. This process may alter the molecular structure of tropocollagen, activating biochemical pathways that influence tissue growth and remodeling. The degree of tissue remodeling—whether normal or pathological—depends on the magnitude of the mechanical stimulation reaching the tropocollagen. The interplay between mechanical deformation and molecular responses is critical in determining the outcome of these remodeling processes.

Due to the complex structure of soft tissues and collagen fibers, the amount of strain or load transferred to the molecular level likely depends on the nature and rate of mechanical stimulation, which may influence tissue remodeling. While the effects of mechanical stimulation on tissue properties across various scales are well-documented, it is established that the mechanical properties of soft tissues—such as stiffness, failure behavior, and modulus—are rate dependent at tissue, nano, and molecular levels [9, 10]. However, the impact of loading rate on strain transfer at the molecular scale remains unclear. This study aims to investigate how the rate of mechanical stimulation alters tropocollagen structure and potentially affects tissue remodeling. To explore this, Attenuated Total Reflection Fourier Transform Infrared Spectroscopy (ATR-FTIR) was employed to analyze mechanically induced protein structure changes.

1.3 Attenuated Total Reflection Fourier Transform Infrared Spectroscopy (ATR-FTIR)

ATR-FTIR is frequently utilized to detect changes in collagen molecules brought on by mechanical pressure and a variety of pathological situations (such as aging and illnesses). Based on the research by Rabotyagova et al. [11], the amide I band, which is connected to the secondary structure of proteins, is shown by the peak in the ATR-FTIR spectrum that is between 1590 and 1710 cm^{-1}. Collagen molecule structural information is carried by this band.

The deconvolution of the amide I band into subpeaks provides detailed insights into the native and altered structures of collagen proteins [12, 13].

1.4 Summery of Study

In this research, pig skin was cyclically stretched at three distinct frequencies: 0.1 Hz (within the physiological range), 2 Hz (moderate), and 5 Hz (exceeding the physiological range). Each frequency was applied for 1000 cycles over a three-hour period, using different samples for each test. Afterward, AFM and ATR-FTIR were utilized to investigate the nano- and molecular structures of both untreated and mechanically tested specimens. AFM measured strain in collagen fibrils, while ATR-FTIR evaluated changes in collagen molecules. The selected frequencies allowed investigation of rate-dependent changes in collagen structure. Pig skin was chosen due to its relevance in tissue expansion for plastic surgery and the use of oscillatory stimulation to promote wound healing and tissue regeneration [14].

2 Material and Method

2.1 Sample Preparation

The experimental data was adopted from the previous published study [2]. The skin from the loin area was collected along the spine and stored in gauze soaked with phosphate-buffered saline (PBS) at 4 °C. Before preparing the specimens, the fat, hair, and epidermis were carefully removed using a surgical scalpel and razor. Dog-bone-shaped samples were cut from the dermis layer parallel to the spine using a punching machine (SCD, Kolkata, India) fitted with an aluminum die in accordance with ASTM D412 standards. To prevent dehydration during the preparation, PBS was continuously sprayed on the samples. All dermis samples were 30 mm in length (gauge length), 4 mm in width (gauge area), and had an average thickness of 2.75 ± 0.212 mm. Thickness measurements were taken at multiple points along the gauge area using a laser thickness measurement device (ILD1320-10, Micro-Epsilon, Germany). The samples were stored in PBS-soaked gauze and kept at 4 °C for preservation. The experimental procedure was based on methods described in previous studies [15–19].

2.2 Mechanical Testing

A total of six specimens per group (n = 6) were subjected to 1000 cycles of tensile testing at frequencies of 0.1, 2.0, and 5.0 Hz. During testing, each sample was stretched to 7.5 mm, equivalent to 25% strain, and then returned to its initial position, with the stretching and relaxation phases aligned to the respective test frequency.

2.3 Attenuated Total Reflection Fourier Transform Infrared Spectroscopy (ATR-FTIR)

The FTIR spectra of dermis samples were obtained using a Bruker Tensor 27 FTIR spectrometer in Attenuated Total Reflection (ATR) mode. Initially, scans were performed in transmission mode with a resolution of 2 cm^{-1} and later converted to absorption mode. Measurements were taken at four distinct locations along the specimen's gage length, with 32 scans recorded before and after testing, across a wavenumber range of 400–3600 cm^{-1}. The amide band was extracted and deconvoluted into Gaussian subpeaks using the second derivative method in Origin software (Version 2018).

2.4 Atomic Force Microscopy

Following ATR-FTIR analysis, both untested and mechanically tested skin samples were embedded in Optimal Cutting Temperature (OCT) compound and stored at −80 °C. The OCT-embedded specimens were subsequently placed in a cryostat, where 50 μm-thick skin sections were sliced. These sections were then floated in PBS and examined using atomic force microscopy (AFM) with a triangular tip. The scanning was carried out in water using tapping mode to prevent the tissue from dehydrating. Initially, a 10.0 μm^2 area was scanned to identify the region of interest, and then a 1 μm^2 ROI was chosen for all further imaging and analysis. The d-periodicity of each collagen fibril at each location was quantified using NanoScope Analysis software, and the average d-periodicity value was used for comparison. The change in d-periodicity between the tested and untreated specimens was used to assess strain accumulation in the collagen fibrils. This approach was adapted from previous research. [2].

2.5 *Statistical Analysis*

Statistical analyses were conducted using R-studio. Data normality was evaluated with the Kolmogorov–Smirnov test. Pairwise comparisons used the Student's t-test, while one-way ANOVA with Tukey's HSD test was applied for group comparisons. Statistical significance was set at $p < 0.05$. Results are reported as mean ± standard deviation (SD) with corresponding p-values.

3 Result

3.1 *Mechanical Experiments*

The dermis's J-shaped stress–strain curves, which demonstrate the skin's nonlinear mechanical response, were acquired for all three frequencies—0.1, 2.0, and 5.0 Hz. These findings demonstrate a distinct variation in the skin's tissue-level mechanical response. This mechanical reaction may be caused by a rate-specific, variable response of the skin's intricate structure at various length scales. These findings indicate that permanent or inelastic skin deformation was more associated with lower frequency values than with moderate and high-frequency values. Microscopic damage or structural changes are typically the cause of the material's accumulation of inelastic strain. After 1000 cycles, it was discovered that the skin stiffness decreased much more at 0.1 than at 2 and 5.0 Hz frequency. The stiffness was reduced by approximately 41 ± 7, 21 ± 3, and 16 ± 5%, correspond to frequencies of 0.1, 2.0, and 5.0 Hz, respectively. Since collagen is primarily responsible for the soft tissue's rigidity, these findings also provide insight into the rate-specific modification of collagen structure [4].

3.2 *Atomic Force Microscopy Results*

After 1000 cycles, the difference in d-periodicity between the tested and untested specimens for each of the three frequencies was used to calculate the fibrils' inelastic strain. For all three frequencies, there was a substantial difference in plastic strain among all frequencies. A significant portion of applied strain is transferred to fibril stretching, as indicated by the high fibril plastic strain value at 0.1 Hz. These findings suggest that mechanical stimulation transmission is rate dependent. The nano and molecular structures are significantly altered by high plastic strain values.

3.3 Molecular Structural Response of Collagen Depends on Loading Rate

The FTIR spectrum showed key amide bands, with amide I (~1645 cm^{-1}) analyzed for protein conformation changes. Peaks at ~1631 cm^{-1} (α-helix) and ~1657 cm^{-1} (triple helix) were used to calculate the intensity ratio (1631/1657 cm^{-1}), revealing increases after cyclic stretching: 0.1 Hz (2.03 vs 1.384, $p = 0.000179$), 2 Hz (1.503 vs 1.277, $p = 0.02044$), and 5 Hz (1.38 vs 1.283, $p = 0.3809$). Significant differences were found between 0.1 Hz versus 2 Hz and 0.1 Hz versus 5 Hz, but not between 2 and 5 Hz. A negative correlation ($r = -0.74$, $p = 0.041$) was observed between protein alterations and frequency.

4 Discussion

This research demonstrates a significant decrease in the plastic strain within collagen fibrils when subjected to the same number of cycles at increasing frequencies. The findings indicate that slower stretching likely leads to irreversible damage in the glycosaminoglycan (GAG) bridges connecting collagen fibrils and/or a progressive unwinding of tropocollagen due to inter/intra-fibril sliding. This process potentially widens the gaps within collagen fibrils, increasing the d-periodicity and causing substantial irreversible strain accumulation at low frequencies. Conversely, at higher frequencies, this deformation mechanism does not occur due to time constraints. Thus, permanent strain in the fibrils at high frequencies is primarily due to the progressive and irreversible straightening of the fibrils under repeated loading. Moreover, the macroscopic inelastic strain exceeds the fibril's permanent strain, implying that the remaining macroscopic inelastic strain results from the permanent straightening or realignment of the collagen fibers, as well as damage to the interfiber matrix caused by sliding between the matrix and the collagen fibers.

Attenuated Total Reflectance Fourier Transform Infrared spectroscopy (ATR-FTIR) is the gold standard for analyzing skin's molecular structure. This study concentrated on the amide I region, indicative of protein secondary structure, particularly collagen, where changes in relative intensity reflect conformational shifts. This protein band corresponds to the vibration of amide carbonyls along the polypeptide backbone. The triple helix band intensity decreases in denatured collagen, while the α-helix intensity increases. To our knowledge, this is the first instance where this phenomenon has been used to monitor alterations in the native collagen structure of porcine dermis under cyclic loading at varying frequencies. We used the intensity ratio (1632 cm^{-1}/1657 cm^{-1}) to observe collagen conformation, comparing this ratio between denatured and native collagen states. Our observations showed an increase in the intensity ratio (1632 cm^{-1}/1657 cm^{-1}) after 1000 stretching cycles, demonstrating collagen denaturation under cyclic tensile testing. Furthermore, for the same cycle count, collagen denaturation (indicated by a higher intensity ratio) was significantly greater at low frequencies, while high-frequency stretching did not result in significant denaturation. Based on these results, we propose the following mechanism: slow cyclic stretching prolongs the duration of dermal stretching, transferring applied stress to the molecular level. This stress unfolds and loosens collagen molecules, denaturing the collagen triple helix. This denaturation may activate biochemical pathways for tissue remodeling. Our study suggests that the rate and duration of mechanical stimulation influence mechanically stimulated tissue remodeling.

5 Conclusion

The acquired experiment's findings demonstrate that collagen denaturation is greater at low stretching frequencies and insignificant at high ones, indicating rate-dependent protein molecule stretching and denature. The collagen molecule's rate-dependent modification may have an impact on tissue remodeling. Overall, this study's findings will contribute to our understanding of how collagen remodeling in the dermis and other connective tissue is impacted by cyclic mechanical stimulation at varying frequencies.

Conflict of Interest The authors have no conflict of interest.

References

1. Sheng, R., Jiang, Y., Backman, L. J., Zhang, W., & Chen, J. (2020). The application of mechanical stimulations in tendon tissue engineering. In *Stem Cells International*.
2. Dwivedi, K. K., Lakhani, P., Kumar, S., & Kumar, N. (2020). Frequency dependent inelastic response of collagen architecture of pig dermis under cyclic tensile loading: An experimental study. *Journal of the Mechanical Behavior of Biomedical Materials, 112*, 104030.
3. Bonnans, C., Chou, J., & Werb, Z. (2014). Remodelling the extracellular matrix in development and disease. *Nature Reviews Molecular Cell Biology, 15*(12), 786–801.
4. Fung, Y. C. (2013). *Biomechanics: Mechanical properties of living tissues*. Springer Science & Business Media.
5. Yang, W., Sherman, V. R., Gludovatz, B., Schaible, E., Stewart, P., Ritchie, R. O., & Meyers, M. A. (2015). On the tear resistance of skin. *Nature Communications, 6*(1), 6649.
6. Sherman, V. R., Yang, W., & Meyers, M. A. (2015). The materials science of collagen. *Journal of the Mechanical Behavior of Biomedical Materials, 52*, 22–50.
7. Hollingsworth, S.A., & Karplus, P. A. (2010). A fresh look at the Ramachandran plot and the occurrence of standard structures in proteins.
8. Buehler, M. J. (2006). Atomistic and continuum modeling of mechanical properties of collagen: Elasticity, fracture, and self-assembly. *Journal of Materials Research, 21*(8), 1947–1961.
9. Bose, S., Li, S., Mele, E., & Silberschmidt, V. V. (2022). Exploring the mechanical properties and performance of Type-I collagen at various length scales: A progress report. *Materials, 15*(8), 2753.
10. Gautieri, A., Vesentini, S., Redaelli, A., & Buehler, M. J. (2012). Viscoelastic properties of model segments of collagen molecules. *Matrix Biology, 31*(2), 141–149.

11. Rabotyagova, O. S., Cebe, P., & Kaplan, D. L. (2008). Collagen structural hierarchy and susceptibility to degradation by ultraviolet radiation. *Materials Science and Engineering: C, 28*(8), 1420–1429.
12. Uniyal, P., Sihota, P., & Kumar, N. (2022). Effect of organic matrix alteration on strain rate dependent mechanical behaviour of cortical bone. *Journal of the Mechanical Behavior of Biomedical Materials, 125*, 104910.
13. Dwivedi, K. K., Lakhani, P., Sihota, P., Tikoo, K., Kumar, S., & Kumar, N. (2023). The multiscale characterization and constitutive modeling of healthy and type 2 diabetes mellitus Sprague Dawley rat skin. *Acta Biomaterialia, 158*, 324–346.
14. Kuehlmann, B., Bonham, C. A., Zucal, I., Prantl, L., & Gurtner, G. C. (2020). Mechanotransduction in wound healing and fibrosis. *Journal of Clinical Medicine, 9*(5), 1423.
15. Lakhani, P., Dwivedi, K. K., & Kumar, N. (2020). Directional dependent variation in mechanical properties of planar anisotropic porcine skin tissue. *Journal of the Mechanical Behavior of Biomedical Materials, 104*, 103693.
16. Dwivedi, K. K., Lakhani, P., Kumar, S., & Kumar, N. (2020). The effect of strain rate on the stress relaxation of the pig dermis: A hyper-viscoelastic approach. *Journal of Biomechanical Engineering, 142*(9), 091006.
17. Lakhani, P., Dwivedi, K. K., Parashar, A., & Kumar, N. (2021). Non-invasive in vivo quantification of directional dependent variation in mechanical properties for human skin. *Frontiers in Bioengineering and Biotechnology, 9*, 749492.
18. Dwivedi, K. K., Lakhani, P., Kumar, S., & Kumar, N. (2022). A hyperelastic model to capture the mechanical behaviour and histological aspects of the soft tissues. *Journal of the Mechanical Behavior of Biomedical Materials, 126*, 105013.
19. Dwivedi, K. K., Lakhani, P., Kumar, S., & Kumar, N. (2022). Effect of collagen fibre orientation on the Poisson's ratio and stress relaxation of skin: An ex vivo and in vivo study. *Royal Society Open Science, 9*(3), 211301.

Mechanical Characterization of Hydrated and Dehydrated Goat Bone

Ram Naresh Yadav and Navin Kumar

Abstract The ability to determine the mechanical characterizations of bone at microscopic level has a great importance in biological and clinical application. Fatigue and static fracture generally occur due to repeated load or low intensity load for long term rather than trauma, most of the cases of fatigue failure is found in military, athletes, ballet dancer and in young children and frequently increases. The purpose of present study is to examine the mechanical properties of hydrated and non-hydrated bovine compact bone using nanoindentation technique. Here we examine elastic modulus, hardness, compressive and tensile strength and their viscoelastic nature by measuring storage and loss modulus of both grouped bones (Hydrated and non-hydrated).

Keywords Nanoindentation · Bone · Hydrated · Bovine · Mechanical property

1 Introduction

Bone is a composite material whose mechanical performance is strongly dependent on the complex internal hierarchical structure shown in Fig. 1. At the micron length scale, the mechanical properties of bone are dominated by close interactions between its organic matrix and inorganic mineral component [1]. Mineral phase (mainly inorganic hydroxypatite) provides the strength and stiffness, and organic matrix (mainly type 1 collagen) provides the softness of bone. While mineral and collagen each contribute to the bone's competency, as do microarchitecture (e.g., porosity and trabecular connectivity), macrostructure (e.g., curvature of diaphysis and thickness of cortical shell), and in vivo microdamage (e.g., microcracks and diffuse cracks) their interaction with water is equally [2].

Nanoindentation is widely used in material science [3]. It is a relatively new tool for testing biological tissues. The obtained precise parameters, which have previously

R. N. Yadav (✉) · N. Kumar
Indian Institute of Technology Ropar, Punjab, India
e-mail: ramnaresh.yadav@iitrpr.ac.in

K. K. Dwivedi et al. (eds.), *Mechanics and Materials Science of Biological Materials*,
Materials Horizons: From Nature to Nanomaterials,
https://doi.org/10.1007/978-981-96-5366-9_13

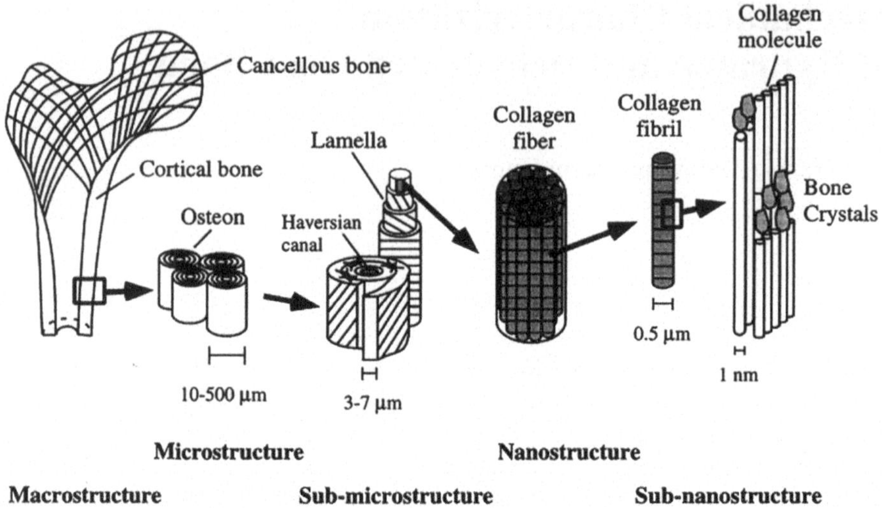

Fig. 1 Demonstrates the hierarchical structure of bone

been quantified in bone, include hardness and reduced elastic modulus. However In contrast, determination of the viscoelastic properties of bone or mineralized tissues by nanoindentation is less covered [4]. Several factors are affecting the mechanical properties of bone: species type, tissue type, age, sample orientation, anatomical location, collagen orientation and degree of mineralization, and these have been studied by nanoindentation as reviewed by Feng et al. [5].

Bone is an inhomogeneous, anisotropic and viscoelastic material, and interaction with water is equally important for the mechanical property of bone. Thus, bone is also a fluid imbibed material in which the distribution of water affects the mechanical properties of bone. The distribution of water in bone appears to change throughout life. It has been reported that water in bone tissues decreases with skeletal growth [2]. Hence, measurements on the interaction with water at nanoscale add new aspects to the inter-individual biological variation and require precise measurement protocols. To address this issue, the present study investigated the effect of water on the mechanical behavior of goat cortical bone. Specifically nanoindentation technique has been used for studying mechanical properties by using quasistatic and dynamic mechanical analysis.

2 Materials and Methods

2.1 Sample Preparation

Tibia bone of goat has been used in this study. The bone has been obtained from local slaughter house and placed in deep freezer at −80 °C in solution of ethanol and distilled water. Cross section of bone has been cut into two small pieces using low speed circular saw (Buehler IsoMet^{TM}). These pieces of samples were divided into two groups. These two sets of samples- hydrated (wet) and dehydrated (dry) were prepared for nanoindentation.

For the wet samples, care was taken to keep the tibia moist with ethanol solution at all times during their preparation and testing. Samples were frozen at −80 °C. Dry samples were stored at −5 °C after each step of sample preparation and testing. Prior to nanoindentation bone samples of both the group are embedded in epoxy solution (Mixture of EpoThin Resin & EpoThin Hardener) in longitudinal orientation and the mixture was then allowed to cure for 9 h. The samples were then grounded and polished using Buehler EcoMet 250 grinder and polisher. Force of 10N was used for this grinding and polishing process. Grinding was done by using sand papers of grit sizes 400, 600 and 1200. For polishing, diamond solutions of 9, 6, 3, 1 and 0.5 μm particle sizes were used.

2.2 Nanoindentation

Nano-indentation was performed using a TI-950 TriboIndenter (Hysitron Inc., Minneapolis, MN) equipped with Berkovich tip. Calibration of the instrument was done using standard fused Quartz and polycarbonate sample. A three-sided pyramidal tip with an included angle of 142.3° and a tip radius of ~150 nm was utilized. The testing temperature was 22 °C and relative humidity was 60%. Regions for testing were identified using an optical microscope integrated into the Nanoindentation system. The "tip to optics calibration" was done by performing 7 indents in "H Pattern". For area function calibration, a series of indents with different contact depths were performed on standard sample of known elastic modulus, and the contact area was calculated. A plot of the calculated area as a function of contact depth was created and fitted by the TriboScan software.

2.2.1 Quasi Static Indentation

For quasi static analysis of sample, 20 subsequent indents were performed along cross-section of both groups bone samples, specified locations with user-specified parameters as shown in Fig. 2. Indentations with contact depths of an order of magnitude larger than local surface roughness are thought to be sufficiently deep to avoid

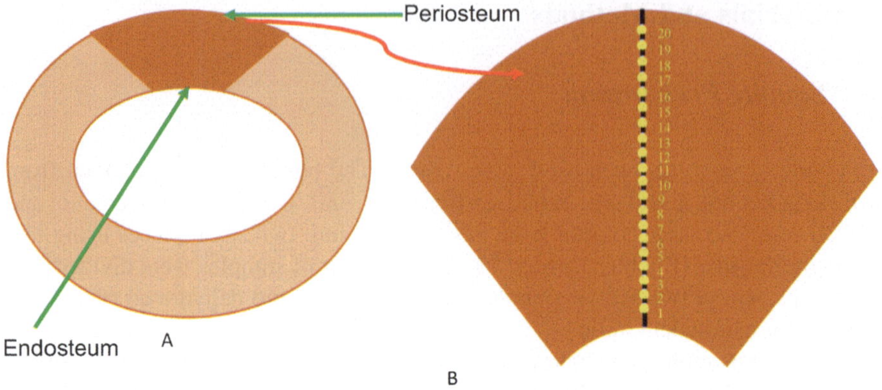

Fig. 2 Top view of cross section of bone sample (**a**), and site of indent location (**b**). Indentations have been done from endosteum to periosteum

a strong effect of roughness on the measured properties. Hence, peak load 5000 μN was selected. A load function consisting of a 10s loading to peak force segment, followed by a 30s hold segment and a 10 s unloading segment were used, shown in Fig. 3. The hardness and the reduced modulus were automatically computed using the Oliver and Pharr method. The reduced modulus E_r is related to the young's modulus E_s of testing material through the following relationship [6]:

$$\frac{1}{E_r} = \frac{(1 - \upsilon_i^2)}{E_i} + \frac{(1 - \upsilon_s^2)}{E_s}, \tag{1}$$

where Ei and υ_i are the elastic modulus and Poisson's ratio of indenter material, respectively. Es and υ_s are the elastic modulus and Poisson's ratio for the substrate materials, respectively. Reduced modulus was calculated by taking the value of elastic modulus and Poisson's ratio to be 1141 GPa and 0.07, respectively, (for diamond indenter tip).

2.2.2 Dynamic Mechanical Analysis

The dynamic nanoindentation tests were accomplished using the frequency sweep method. Analysis was performed by using nano DMA software in the Histro Triboscope. In this method, the maximum applied load is specified to the nanoindentor, probe is swing through a range of frequency (5–200 Hz). By computing the resultant load amplitude, displacement amplitude and phase lag throughout the test, the DMA technique permit the computation of the storage modulus, loss modulus, and tan δ using following relation.

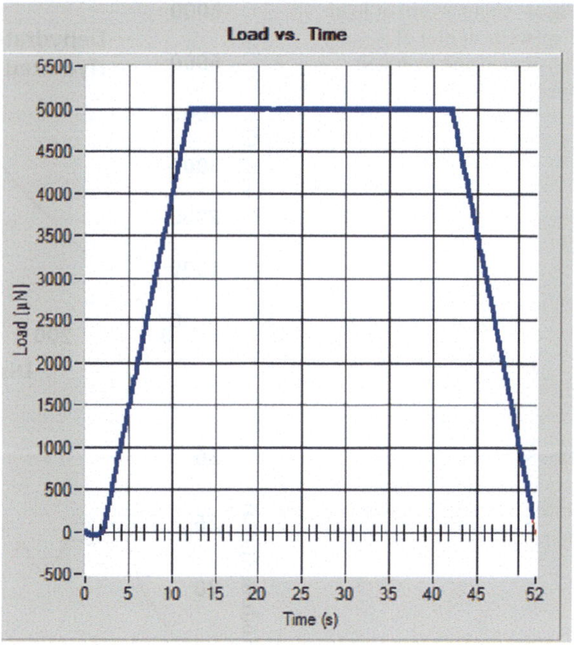

Fig. 3 Illustrates the force function which was used during nanoindentation of bone. This force function contains loading, holding, and unloading segments

$$E' = \frac{K_s\sqrt{\pi}}{2\sqrt{A_c}},\ E'' = \frac{\omega C_s}{2\sqrt{A_c}} \text{ and } \tan\delta = \frac{E'}{E''}$$

where E' and E'' are the storage and loss modulus, ω is the frequency of the applied load, K_s and C_s are the stiffness and damping coefficient of sample and A_C is the projected contact area of the indent on the surface of the specimen [7].

3 Result

3.1 Quasi Static Indentation

Comparisons of load displacement plot obtained from quasi static indentation have been shown in Fig. 4. The elastic moduli were 21.2 ± 4.7 GPa and 12 ± 4.5 GPa for dehydrated and hydrated cortical bone, respectively. Variations of elastic moduli from endosteum to periosteum have been shown in Fig. 5 for dehydrated and hydrated cortical bone. Hardness of dehydrated and hydrated cortical bone was 0.80 ± 0.30 and 0.54 ± 0.36 respectively. Variation of hardness from endosteum to periosteum has been shown in Fig. 6 for both groups of bone. From these all plots it was notice that modulus and hardness increases due to dehydration. Elastic modulus of hydrated

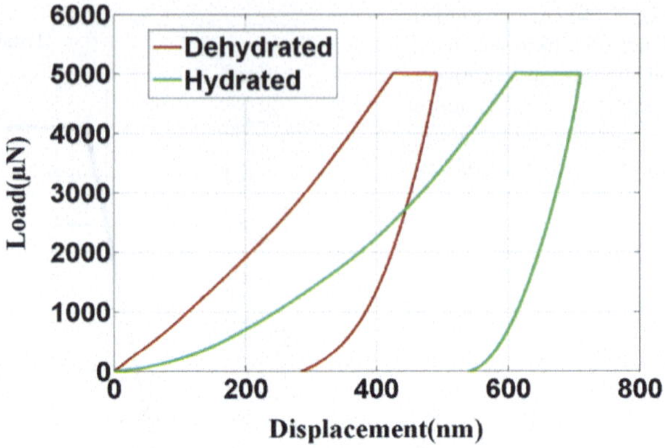

Fig. 4 Comparison of load displacement plot of dehydrated and hydrated bone

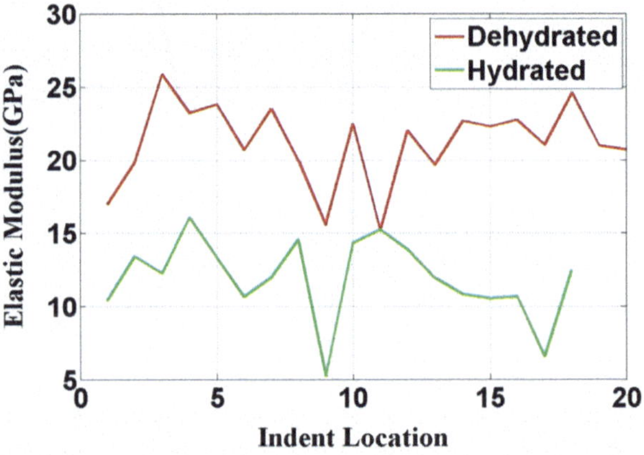

Fig. 5 Variation of elastic modulus with indent location for dehydrated and hydrated bone

bone is about 43% of dehydrated bone. Hardness of hydration bone is about 32% of dehydrated bone (Fig. 7).

3.2 *Dynamic Mechanical Analysis*

DMA is performed on center portion of cross-sectional cortical bone. Variation of hardness with frequency was shown in Fig. 8 for dehydrated and hydrated bone. For comparison of viscoelastic behavior of both group bones, we have done study, storage modulus, loss modulus and their variation with frequency is shown in Figs. 8 and 9, respectively.

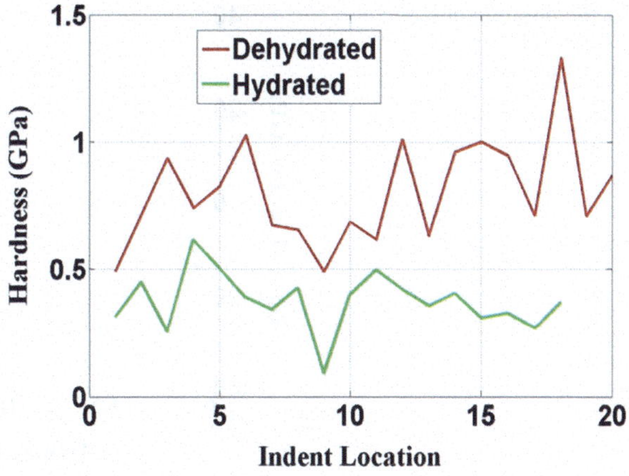

Fig. 6 Variation of hardness with indent location for dehydrated and hydrated bone

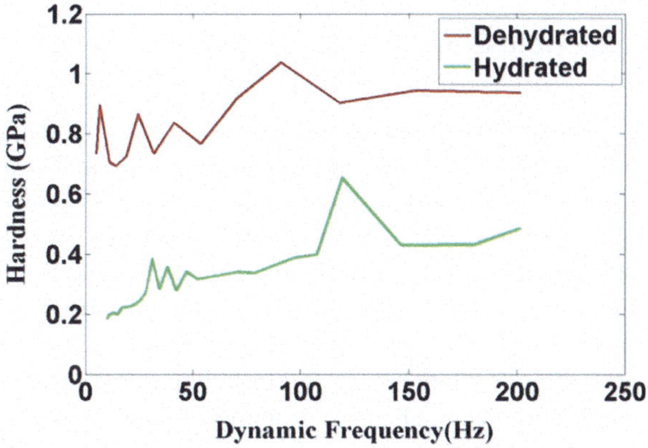

Fig. 7 Variation of hardness with dynamic frequency for dehydrated and hydrated bone

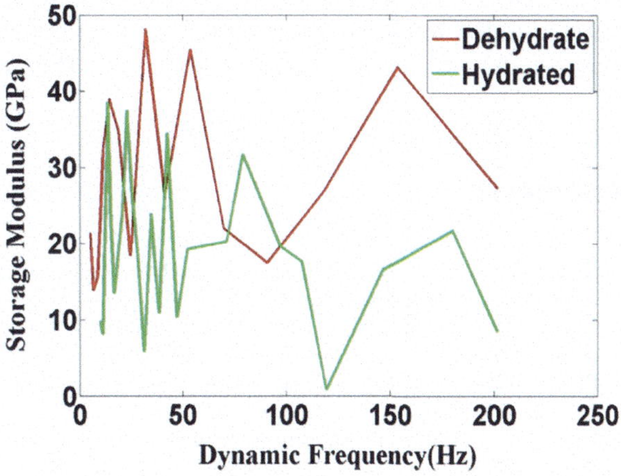

Fig. 8 Variation of storage modulus with frequency for both group bones

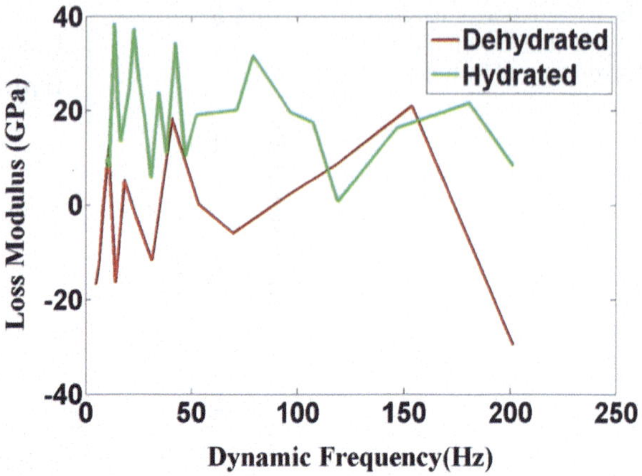

Fig. 9 Variation of Loss modulus with frequency of both group bones

4 Conclusion

This study investigated the influence of water on mechanical characterization of goat bone, under quasi static and dynamic mechanical loading. Due to hydration, bone becomes soft as compared to dehydrated bone as shown in Figs. 4, 5, and 6 (As results obtained from quasi static indentation test). For investigating the effect of hydration from dynamic mechanical analysis, it was found that storage modulus of hydrated bone decrease and loss modulus increase as compared to dehydrated bone, as shown in Figs. 8 and in 9, respectively. From the result of dynamic analysis it was clear that hydrated bone show more viscous behavior [4]. Thus water plays an important role for affecting the mechanical property of bone. How it affects the bone at tissue level was not clearly understood. Previous studies suggest that under dry condition no interaction of water with collagen and mineral were present. In the absence of water collagen fibrils might stiffen and contract longitudinally compressing the mineral phase, leading to a higher strength of bone [8]. This statement supports our result.

References

1. Rho, J.-Y., et al. (1998). Mechanical properties and the hierarchical structure of bone. *Medical Engineering & Physics, 20*(2), 92–102.
2. Nyman, J. S., et al. (2006). The influence of water removal on the strength and toughness of cortical bone. *Journal of Biomechanics, 39*(5), 931–938.
3. Ebenstein, D. M., & Pruitt, L. A. (2006). Nanoindentation of biological materials. *Nano Today, 1*(3), 26–33.
4. Pathak, S., et al. (2011). Measuring the dynamic mechanical response of hydrated mouse bone by nanoindentation. *Journal of the Mechanical Behavior of Biomedical Materials, 4*(1), 34–43.
5. Feng, L., et al. (2012). Mechanical properties of porcine femoral cortical bone measured by nanoindentation. *Journal of Biomechanics, 45*(10), 1775–1782.

6. Herbert, E., et al. (2008). Nanoindentation and the dynamic characterization of viscoelastic solids. *Journal of Physics D: Applied Physics, 41*(7), 074021.
7. Oyen, M. L., & Cook, R. F. (2009). A practical guide for analysis of nanoindentation data. *Journal of the Mechanical Behavior of Biomedical Materials,* 2(4), 396–407.
8. Lee, K.-L., et al. (2012). Nanomechanical characterization of canine femur bone for strain rate sensitivity in the quasistatic range under dry versus wet conditions. *International Journal of Biomaterials, 2012*(1), 415230.

Investigation of Friction and Dry Sliding Wear Behaviors of Al6063/SiC Metal Matrix Composite

Gaurang Deep, Naman Jain, Mohit Bhatt, and Akarsh Verma

Abstract In the present article, an attempt has been made to develop a conventional low-cost method of producing aluminum–silicon carbide (Al-SiC) metal matrix composite (MMC) and investigate the influence of percentage variation, applied load, and sliding distance on the dry sliding wear of Al-SiC MMC. Here, five different composition samples of Al-SiC composites with varying weight fractions of SiC (i.e., 5, 10, 15, 20, and 25%) were prepared using the stir-casting technique. Al6063 and SiC powder (400 grit size) were used to prepare the samples. Then the dry sliding wear tests on prepared samples were carried out using a Pin-on-Disk wear test rate with normal loads of 10, 20, and 30 N at a constant sliding velocity of 3.769 m/s and varying sliding distances of 1131, 2262, and 3393 m. Wear loss, wear rate, and the coefficient of friction (CoF) were measured in the wear test. Taguchi method was used to optimize the process parameters leading to minimum wear rate and CoF. It was observed that the addition of SiC up to 15% improves the wear behavior, but the addition of SiC above 15% increases the wear rate. Furthermore, it was observed from the ANOVA analysis that the applied load (41.93%) has the greatest influence on the wear rate of samples, while the percentage of SiC (72.91%) has the greatest influence on the CoF of samples.

Keywords Metal matrix composite · Silicon carbide · Wear · Aluminum matrix composite · Taguchi method

G. Deep
Department of Mechanical Engineering, Meerut Institute of Engineering and Technology, Meerut, India

N. Jain
Department of Mechanical Engineering, ABES Engineering College, Ghaziabad, India

M. Bhatt
Department of Physics, School of Advanced Engineering, UPES, Dehradun, India

A. Verma (✉)
Department of Mechanical Engineering, UPES (University of Petroleum and Energy Studies), Dehradun 248007, India
e-mail: akarshverma007@gmail.com

K. K. Dwivedi et al. (eds.), *Mechanics and Materials Science of Biological Materials*, Materials Horizons: From Nature to Nanomaterials,
https://doi.org/10.1007/978-981-96-5366-9_14

1 Introduction

One of the most frequent industrial issues that necessitate the replacement of engineering assemblies and components is wear [1, 2]. Many attempts have been made in the past few years to develop more robust materials and methods to lessen tool and engineering component wear. Compared to other metal alloys, aluminum (Al) and its alloys are used in the industry more often for machine components [3, 4]. Al-Si alloys, among other Al alloys, are widely used in cylinder heads and blocks, cylinder liners, axles and wheels, automobile transmission casings, internal combustion engine pistons, pump components, high-speed rotating parts, and impellers, among other uses.

As an alternative to metal alloys for possible applications, several metal matrix composites (MMCs) are now being researched [5–9]. Adding the required qualities of metals and ceramics is the goal of MMC design. Because of its intrinsic qualities, such as hardness and wear resistance, good creep behavior, lightweight, high strength-to-weight ratio, design flexibility, and low wear rate, Al-SiC MMC is being used in the aerospace, automotive, and other sectors. In addition to different metals, alloys, and nonmetals, a lot of effort has been done recently to identify composite materials with wear behavior and friction. The purpose of the analysis of investigating the friction and wear properties is to minimize the coefficient of friction (CoF) and wear rate. Literature also suggests that the wear and friction properties of a MMC can be enhanced by changing the composition of reinforced material and by optimizing the wear parameters.

Sun et al. [10] investigated the impact of particle size on the mechanical characteristics and microstructures of SiC particle-reinforced pure Al composites made using the powder metallurgy process earlier in 2011. It was determined that while the gaps are found to coexist with the clustered and large-sized SiC particles, the smaller-sized SiC particles have a tendency to cluster readily. Tribological test results for Al-Si alloy A356 (EN-Al Si7Mg0.3), with and without 3 weight percent Al2O3 reinforcement, were reported by Vencl et al. [11]. The friction coefficient of the matrix material was found to be independent of the applied load, but the friction coefficient of the composite material decreased as the applied stress increased. **Singh et al.** [12] focused on studying the change in behavior of Al by adding different % amounts of "SiC" and "Al_2O_3." It was concluded from their investigation that as the weight % of reinforcement goes on increasing, the mechanical properties such as hardness, yield strength, and ultimate strength also increase. But at the same time, elongation decreased, and the behavior of material changed from ductile to brittle. **Singh et al.** [13] adopted the mechanical alloying method of powder metallurgy to develop Al -based silicon carbide (SiC) particulate MMC. The results showed that this method is moderately successful in attaining uniform dispersion of reinforcement in the matrix. An overview of the stir-casting procedure, process parameters, and the creation of aluminum matrix composite (AMC) material utilizing Al as the matrix form and SiC, Al2O3, and graphite as reinforcement in different proportions was provided by **Gangaram et al.** [14]. They discovered throughout the AMC's manufacture that such process variables are crucial to the consistent distribution of reinforcement. The tribological behavior of semi-metallic lining material (brake pad), stir cast Al-10% SiC composite, and LM13 cast Al alloy sliding against a gray cast iron disk was reported by **Londhe et al.** [15]. At room temperature, the composite had a high frictional force and low wear. **Saravanan et al.** [16] fabricated Al-SiC with mica in the stir-casting method to obtain a hybrid MMC. Al-SiC MMCs and their characteristics were also investigated by **Nidhish and Sijo** [17]. Previous research showed that characteristics improved up to a certain point and fracture toughness decreased after that as the percentage (%) of SiC rose. The inclusion of 25% SiC produced the best fracture toughness results. Additional data revealed a decrease in the stress concentration factor; the causes of this decrease are yet up for investigation. To create the MMC, **Satyanarayen et al.** [18] mixed SiC to LM25 alloy at weight percentages of 3, 6, and 9%. It was discovered that as the percentage of SiC reinforcement grew, so did the mechanical characteristics like hardness and tensile strength. The impact of wear factors such as sliding distance, applied load, and sliding speed on the dry sliding wear of Al silicon titanium alloys was examined by **Jambukar et al.** [19]. The experiment's design strategy was used to use the Taguchi technique to collect data in a controlled manner. It has been discovered that adding titanium to Al silicon alloys alters their microstructure and enhances their tribological properties. Using a stir-casting technique, **Marwaha et al.** [20] created an Al alloy base matrix supplemented with SiC (10%) and graphite (5%) particles. For data processing, a L9 orthogonal array was chosen. According to their findings, weight and sliding speed have the greatest effects, followed by track diameter.

2 Materials and Methods

2.1 Materials

A coal-fired furnace was constructed in the Mechanical Engineering Department workshop of the Meerut Institute of Engineering Technology in Meerut, Uttar Pradesh, India, in order to fabricate the Al-SiC composite. The furnace is equipped with a stirrer system that uses an electric motor and a steel rod. The control panel, which comes separately with the furnace, has an ON/OFF type digital temperature indicator/controller with a temperature precision of ± 5 °C. The prepared coal-fired stir-casting furnace is seen in Fig. 1. A plate made of Al6063 aluminum alloy was bought from Kamal Steel, located in the Hasti Industrial Estate in Navi Mumbai, Maharashtra, India. Si (0.513%) and Mg (0.521%) are the main alloying elements in Al6063. Lastly, the SiC particles having grit size 400 were purchased from Sigma-Aldrich [21].

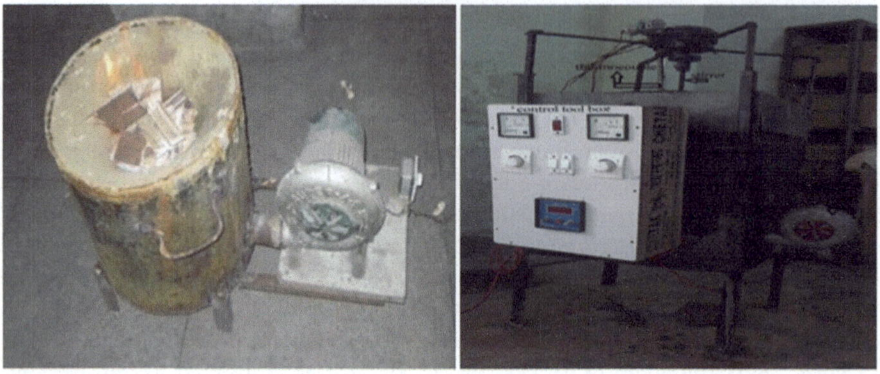

Fig. 1 Coal fired stir-casting furnace

2.2 Method

In the present work, Al-SiC composites were prepared by varying weight fractions of SiC (5, 10, 15, 20, 25, and 30%). The SiC particles were initially preheated up to 750 °C for 3 h in a muffle furnace. Al alloy was then melted in the graphite crucible of the furnace at a temperature of about 700–750 °C. The composite slurry (Al and SiC powder) was heated in a furnace crucible to approximately 900 °C. The slurry was then stirred by a stirrer for 40 min at 100 rpm in clockwise and anticlockwise directions so as to create a vortex. Sand casting method was employed to prepare the wear test specimen.

2.3 Wear Test Equipment and Procedure

A Pin-on-Disk machine made by Contech Microsystems was used for wear tests. The wear test was carried out in the mechanical engineering department's metallurgy lab at M.J.P. Rohilkhand University in Bareilly, India. The trials were carried out utilizing the counter face of a steel disk 316L/EN-32 hardened to 62–65 HRC in accordance with the ASTM G99-90 standard. Figure 2 depicts the Pin-on-Disk device that was utilized in this experiment. As seen in Fig. 2c, the track radius was maintained at 60 mm while the pin, which is normal to the sliding contact for various loads, was subjected to load during the wear test. In the current investigation, the sliding velocity was maintained at 3.769 m/s throughout all trials. Normal load and sliding distance are the factors used in the wear test. All samples in the current study were subjected to trials at loads of 1, 2, and 3 kg, respectively. As the studies lasted five, ten, and fifteen minutes, respectively, the sliding distance was changed. During sliding, worn debris was removed from the track using a camel hair brush. A weighing system that detected weight loss (in grams) at predetermined intervals of time was used to

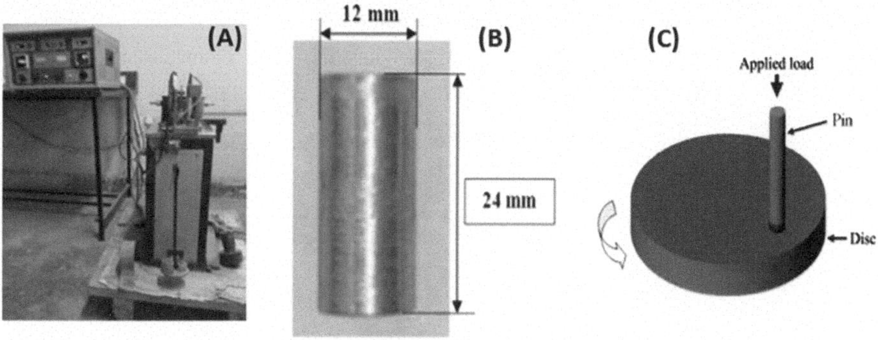

Fig. 2 **a** Experimental setup of Pin-on-Disk apparatus, **b** wear test specimen, and **c** Pin-on-Disk mechanisms

assess the amount of material lost during Pin-on-Disk wear testing, while the CoF was continually monitored. Figure 2b displays the test specimen's dimensions.

2.4 Process Optimization and Taguchi Method

The applied stress, sliding distance, and percentage of SiC addition were the variables that required to be considered in the wear test. The wear test's objective is to identify the critical elements and combinations of elements affecting the wear process in order to minimize wear rate and coefficient of friction. The process parameters were optimized using the Taguchi technique, which resulted in the lowest wear rate and CoF [22, 23]. To show how process factors affected wear rate and CoF, a statistical analysis of variance (ANOVA) was used in addition to the S/N ratio. The experimental details of the control components and their amounts are displayed in Table 1. For the design of experiments, or the creation and analysis of Taguchi design, MINITAB 17 software was utilized. Because not every control element was at the same level, an orthogonal array design of mixed-level type L18 (6^1 3^2) was employed. Signal-to-noise (S/N) ratios were derived from the experimental findings. Based on the kind of reaction, three different S/N ratios are possible. The S/N ratio is in the "smaller is better" range for minimal specified wear rate and CoF.

3 Results and Discussion

To begin with, the Taguchi method was employed for the wear test analysis in which % addition of SiC in Al6063, applied load, and sliding distance were taken as the control factors. For every experiment set up on a Pin-on-Disk machine, the wear loss, wear rate, and CoF were measured. Hence, response data in the Taguchi design was

Table 1 Experimental details of control factors and their level

Control factors	% addition of SiC	Applied load (N)	Sliding distance (m)
Level 1	0	10	1131
Level 2	5	20	2262
Level 3	10	30	3393
Level 4	15	–	–
Level 5	20	–	–
Level 6	25	–	–

wear loss, wear rate, and CoF, respectively. Taguchi analysis was carried out for each response data, and the S/N ratio and mean are the two commonly known approaches in the Taguchi analysis. In the present work, the influence of the control factor on wear loss, wear rate, and CoF was evaluated using both the S/N ratio response and mean analysis by the software MINITAB 17.

3.1 Analysis of the Wear Loss

The Taguchi L18 orthogonal array employed in this investigation is displayed in Table 2. 18 tests make up the orthogonal array, as seen in Table 2's first column. The percentage of SiC added to Al6063, the applied load, and the sliding distance are displayed in the second, third, and fourth columns of this table, respectively. Each test run's wear loss value in grams is shown in column five. The "smaller the better" wear loss characteristic was used for this investigation. The program produced the primary effect charts. The main impact graphs for the mean and S/N ratio are displayed in Figs. 3 and 4, respectively. The best quality with the least amount of variance is always produced by process parameter choices with the highest S/N ratio and lowest mean. The results show that the % addition of SiC reduces the wear losses up to 15%, and further addition increases the wear loss, as shown in Fig. 3. This may be due to the non-uniform dispersion and coagulation of SiC particles into the Al matrix above 15%. Minimum loss was obtained at 15% SiC content, 10 N applied load and a siding distance of 1131 m.

Analysis of variance (ANOVA) was performed on the wear loss to ascertain the statistical significance of several parameters, such as the percentage addition of SiC, applied load, and sliding distance. For the current investigation, a significance threshold of 5% is taken into account. The ANOVA analysis shows that the percentage of SiC added, the applied force, and the sliding distance all affect the samples' wear loss. In comparison to the sliding distance (27.01), Table 3 shows that the percentage addition of SiC (36.78%) and load (35.08%) have a significant impact on the wear loss.

Table 2 Results of L_{18} orthogonal array for wear loss

Test run	% of SiC	Applied load (N)	Sliding distance (m)	Wear loss (g)	Signal-to-noise (S/N) ratio (db)	Mean
1	0	10	1131	0.06	24.43697	0.06
2	0	20	2262	0.14	17.07744	0.14
3	0	30	3393	0.25	12.0412	0.25
4	5	10	1131	0.05	26.0206	0.05
5	5	20	2262	0.09	20.91515	0.09
6	5	30	3393	0.18	14.89455	0.18
7	10	10	2262	0.05	26.0206	0.05
8	10	20	3393	0.09	20.91515	0.09
9	10	30	1131	0.07	23.09804	0.07
10	15	10	3393	0.05	26.0206	0.05
11	15	20	1131	0.04	27.9588	0.04
12	15	30	2262	0.08	21.9382	0.08
13	20	10	2262	0.07	23.09804	0.07
14	20	20	3393	0.1	20	0.1
15	20	30	1131	0.09	20.91515	0.09
16	25	10	3393	0.11	19.17215	0.11
17	25	20	1131	0.08	21.9382	0.08
18	25	30	2262	0.16	15.9176	0.16

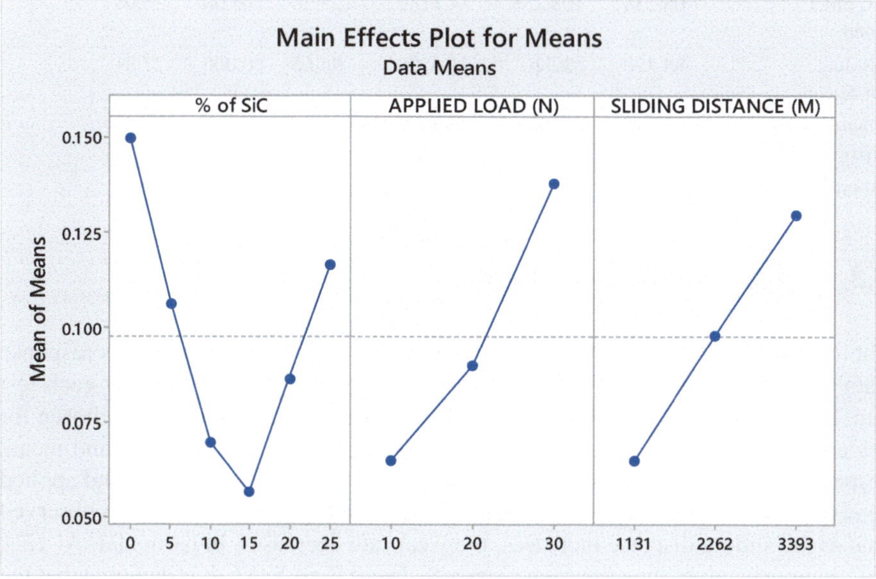

Fig. 3 Means versus input parameters (wear loss)

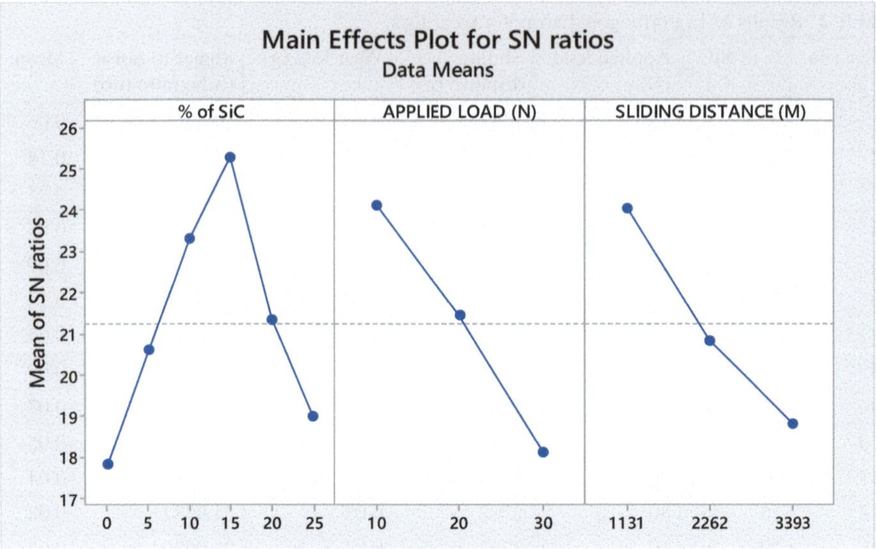

Fig. 4 Signal-to-noise (S/N) ratio versus input parameter (wear loss)

Table 3 Analysis of the variance for wear loss (signal-to-noise (S/N) ratios)

Source	DF	Seq. SS	Adj. SS	Adj. MS	F	P	% contribution
% SiC	5	113.467	113.467	22.6933	52.65	0.00	36.78
Applied load	2	108.238	108.238	54.1189	125.56	0.000	35.08
Sliding distance	2	83.320	83.320	41.6599	96.66	0.000	27.01
Residual error	8	3.448	0.448	30.4310	–	–	1.13
Total	17	308.472	–	–	–	–	100

3.2 Analysis of the Wear Rate

Table 4 shows the Taguchi L_{18} orthogonal array used for the wear rate as response data, and column five of this table depicts the value of the wear rate of each test run. In this study, the "smaller the better" characteristic of wear rate was chosen for the analysis. Figures 5 and 6 are the main effect plots for the S/N ratio and mean, respectively. The trend of wear rate with respect to the % addition of SiC and applied load was found to be similar to wear loss, as shown in Fig. 5. However, it was observed that as the sliding distance increases, the wear rate decreases. In particular, Al-15% SiC composite gives the minimum wear rate. From Figs. 5 to 6, it is depicted that the factor combination of SiC addition of 15%, applied load 10 N, and sliding distance

Table 4 Results of L_{18} orthogonal array for wear rate

Test run	% of SiC	Applied load (N)	Sliding distance (m)	Wear rate (gm/m)	Signal-to-noise (S/N) ratio (db)	Mean
1	0	10	1131	4.42	−12.9084	4.42
2	0	20	2262	6.189	−15.8324	6.189
3	0	30	3393	7.368	−17.347	7.368
4	5	10	1131	4.42	−12.9084	4.42
5	5	20	2262	3.978	−11.9933	3.978
6	5	30	3393	5.305	−14.4937	5.305
7	10	10	2262	2.21	−6.88785	2.21
8	10	20	3393	2.652	−8.47147	2.652
9	10	30	1131	6.189	−15.8324	6.189
10	15	10	3393	1.473	−3.36405	1.473
11	15	20	1131	3.536	−10.9702	3.536
12	15	30	2262	3.536	−10.9702	3.536
13	20	10	2262	3.095	−9.81321	3.095
14	20	20	3393	2.947	−9.3876	2.947
15	20	30	1131	7.597	−17.6128	7.597
16	25	10	3393	3.242	−10.2163	3.242
17	25	20	1131	7.073	−16.9921	7.073
8	25	30	2262	7.073	−16.9921	7.073

of 3393 m gives the minimum rate of wear. Unlike wear loss, it can be observed from the ANOVA Table 5 that the applied load (41.93%) has a major influence on the wear rate as compared to the SiC content (37.70%) and sliding distance (17.71%).

3.3 *Analysis of the Coefficient of Friction*

Table 6 showcases the Taguchi L_{18} orthogonal array used for the CoF as response data, and column five of this table depicts the value of the CoF obtained from each test run. In this study, the "smaller the better" characteristic of the CoF was chosen for the analysis. Table 6 also reveals the values of the S/N ratio and mean for each test run. Figures 7 and 8 are the main effect plots for the S/N ratio and mean, respectively. It is observed that the factor combination of SiC addition of 5%, applied load 30 N, and sliding distance of 1131 m gives the minimum CoF. The results show that the CoF is minimum for 5% addition of SiC and maximum for 15% as well as 20% addition of SiC. Moreover, the CoF decreases as the applied load increases, and it increases as the sliding distance increases. It can be seen from the ANOVA Table 7 that the % addition of SiC (72.91%) is the most important control parameter that

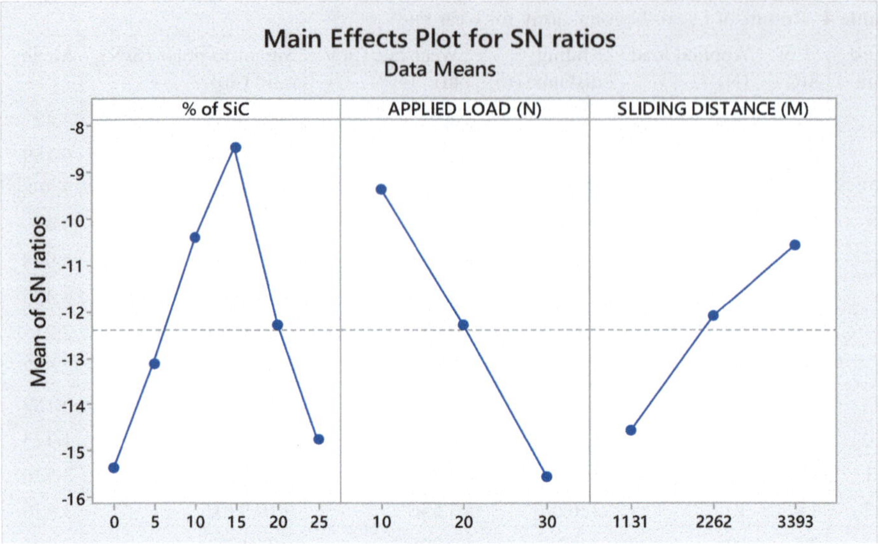

Fig. 5 Signal-to-noise (S/N) ratios versus input parameters (wear rate)

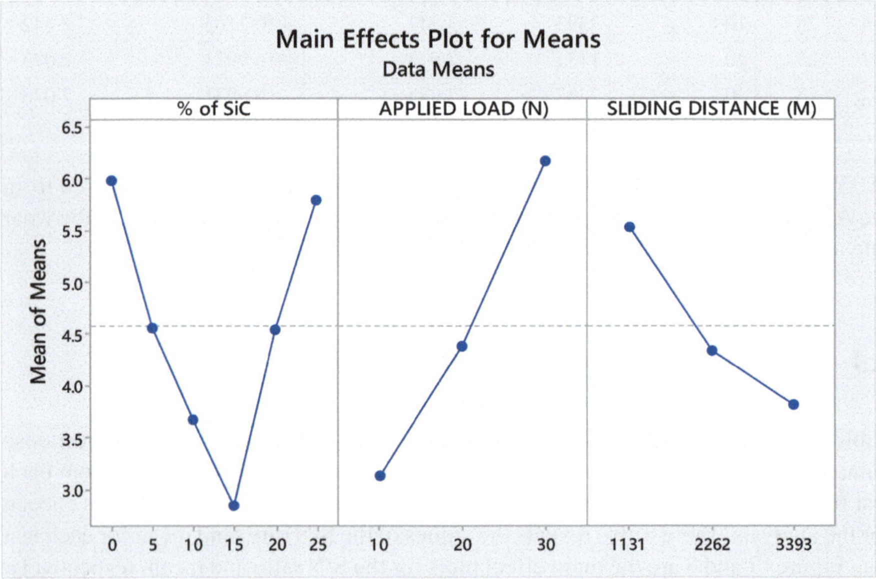

Fig. 6 Means versus input parameters (wear rate)

Table 5 Analysis of the variance for wear rate (signal-to-noise (S/N) ratios)

Source	DF	Seq. SS	Adj. SS	Adj. MS	F	P	% contribution
% SiC	5	103.521	103.521	20.7042	22.81	0.00	37.70
Applied load	2	115.127	115.127	57.5636	63.4	0.000	41.93
Sliding distance	2	48.626	48.626	24.3131	26.79	0.000	17.71
Residual error	8	7.262	7.262	0.907	–	–	2.66
Total	17	274.536	–	–	–	–	100

affects the CoF of samples. The consequence of the other two input parameters (i.e., applied load and sliding distance) was found to be less. It can be perceived from this ANOVA analysis that the % contribution of SiC is much more as compared to wear loss and wear rate.

Table 6 Results of L_{18} orthogonal array for CoF

Test run	% of SiC	Applied load (N)	Sliding distance (m)	Coefficient of friction	Signal-to-noise (S/N) ratio (db)	Mean
1	0	10	1131	0.56	5.036239	0.56
2	0	20	2262	0.52	5.679933	0.52
3	0	30	3393	0.52	5.679933	0.52
4	5	10	1131	0.45	6.93575	0.45
5	5	20	2262	0.44	7.130946	0.44
6	5	30	3393	0.42	7.535014	0.42
7	10	10	2262	0.56	5.036239	0.56
8	10	20	3393	0.54	5.352125	0.54
9	10	30	1131	0.47	6.558043	0.47
10	15	10	3393	0.61	4.293403	0.61
11	15	20	1131	0.55	5.192746	0.55
12	15	30	2262	0.53	5.514483	0.53
13	20	10	2262	0.59	4.58296	0.59
14	20	20	3393	0.57	4.882503	0.57
15	20	30	1131	0.53	5.514483	0.53
16	25	10	3393	0.53	5.514483	0.53
17	25	20	1131	0.47	6.558043	0.47
18	25	30	2262	0.45	6.93575	0.45

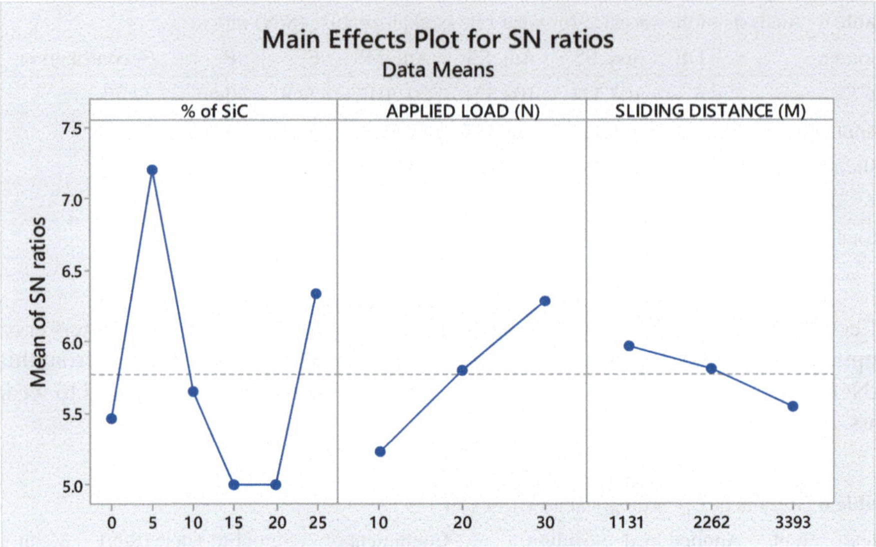

Fig. 7 Signal-to-noise (S/N) ratios versus input parameters (CoF)

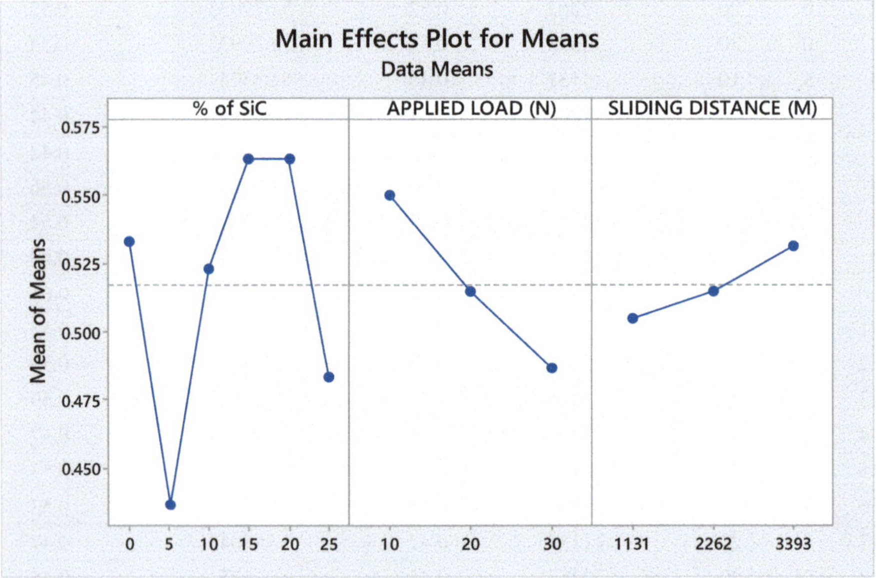

Fig. 8 Means versus input parameters (CoF)

Table 7 Analysis of the variance for CoF (signal-to-noise (S/N) ratios)

Source	DF	Seq. SS	Adj. SS	Adj. MS	F	P	% contribution
% SiC	5	11.0106	11.0106	2.20212	94.85	0.00	72.91
Applied load	2	3.3540	3.3540	1.67698	72.23	0.000	22.21
Sliding distance	2	0.5506	0.5506	0.27532	11.86	0.004	3.64
Residual error	8	0.1857	0.1857	0.02322	–	–	1.24
Total	17	15.1009	–	–	–	–	100

4 Conclusions

This experimental investigation of friction and wear properties has led to the following specific conclusions:

i. The presence of SiC fraction in commercial Al up to 15% weight fraction improved their wear characteristics. On further addition of SiC over 15%, wear loss increased. The results showed that the Al-15% SiC composite was found to be the best one as far as the improvement in wear performance is concerned.
ii. From the Taguchi design and analysis, it was observed that the factor combination of SiC addition of 15% and an applied load of 10 N gives the minimum wear loss as well as wear rate. However, the sliding distances of 1131 m and 3393 m gave the minimum wear loss and minimum wear rate, respectively.
iii. The CoF decreased as the applied load increased, and it increased as the sliding distance increased. However, the trend of the CoF with respect to the % addition of SiC was found to be unpredictable. From the Taguchi design and analysis, it can be concluded that the factor combination of SiC addition of 5%, applied load of 30 N, and sliding distance of 1131 m gave the minimum CoF.
iv. Significant control factors affecting the wear loss, wear rate, and CoF have been identified through the ANOVA analysis. The input parameters, i.e., % addition of SiC (36.78%), applied load (41.95%), and again % addition of SiC (72.91%) greatly influence the output parameters that are wear loss, wear rate, and CoF, respectively. However, the effect of the sliding distance was found to be less on output parameters as compared to other control parameters.

Acknowledgements Author "Akarsh Verma" acknowledges the academic support from the University of Petroleum and Energy Studies (UPES), Dehradun 248007, India.

Data Availability Not applicable.

Declaration of Competing Interest The authors have no conflicts to disclose.

References

1. Sharath, B. N., Madhu, P., & Verma, A. (2024). Wear behaviour of aluminium-based hybrid composites processed by equal channel angular pressing. *Proceedings of the Institution of Mechanical Engineers, Part J: Journal of Engineering Tribology, 238*(9), 1079–1090. https://doi.org/10.1177/13506501241247616
2. Thimmaiah, S. H., Narayanappa, K., Thyavihalli Girijappa, Y., Gulihonenahali Rajakumara, A., Hemath, M., Thiagamani, S. M. K., & Verma, A. (2023). An artificial neural network and Taguchi prediction on wear characteristics of Kenaf-Kevlar fabric reinforced hybrid polyester composites. *Polymer Composites, 44*(1), 261–273. https://doi.org/10.1002/pc.27043
3. Sharath, B. N., Madhu, P., & Verma, A. (2023). Enhancing tribological performance: A review of ceramic reinforced aluminium hybrid composites for high-temperature engineering applications. *Hybrid Advances, 4*, 100094. https://doi.org/10.1016/j.hybadv.2023.100094
4. Chaturvedi, S., Verma, A., Sethi, S. K., & Ogata, S. (2022). Defect energy calculations of nickel, copper and aluminium (and their alloys): molecular dynamics approach. In *Forcefields for atomistic-scale simulations: materials and applications* (pp. 157–186). Singapore: Springer Nature Singapore. https://doi.org/10.1007/978-981-19-3092-8_8
5. Kafaltiya, S., Sharma, P., Belwal, N., Singh, V. K., Chauhan, S., & Verma, A. (2024). Evolutionary techniques in optimization of machining parameters on hybrid metal matrix composites—A review. *Unraveling New Frontiers and Advances in Bioinformatics* 365–385. https://doi.org/10.1007/978-981-97-7123-3_16
6. Pathak, A., Singh, V. K., Chauhan, S., & Verma, A. (2024). Reinforcement materials improve the characteristics of metal matrix composites—A review. *Unraveling New Frontiers and Advances in Bioinformatics*, 267–292. https://doi.org/10.1007/978-981-97-7123-3_12
7. Gupta, H. S., Verma, D., & Verma, A. (2024). Combination of different kinds of materials for hybrid composites: A brief review. *Unraveling New Frontiers and Advances in Bioinformatics* 331–363. https://doi.org/10.1007/978-981-97-7123-3_15
8. Rawat, N. K., Jain, N., Mishra, A. K., & Verma, A. (2024). Atomistic-scale simulations on grain boundary migration mechanisms involved in metals and alloys: A critical review. *Archives of Computational Methods in Engineering* 1–38. https://doi.org/10.1007/s11831-024-10201-8
9. Sharath, B. N., Madhu, P., Pradeep, D. G., Thanush, H. G., Kumar, K. M., Manoj, S., & Verma, A. (2024). Unveiling the potential of age hardened aluminum alloys: Strengthening solutions for engineering challenges. In *Hybrid composite materials: experimental and theoretical analysis* (pp. 349–369). Singapore: Springer Nature Singapore. https://doi.org/10.1007/978-981-97-2104-7_15
10. Sun, C., Song, M., Wang, Z., & He, Y. (2011). Effect of particle size on the microstructures and mechanical properties of SiC-reinforced pure aluminum composites. *Journal of Materials Engineering and Performance, 20*(9), 1606–1612. https://doi.org/10.1007/s11665-010-9801-3
11. Vencl, A., Bobić, I., & Mišković, Z. (n.d.). Tribological properties of Al-Si alloy A356 reinforced With Al2O3 particles. *Tribology in Industry, 28*(1).
12. Singh, D., Singh, H., Kumar, S., & Singh, G. (n.d.). An experimental investigation of mechanical behavior of aluminum by adding SiC and alumina.
13. Singh, R., & Singla, E. (2012). Tribological characterization of aluminium-silicon carbide composite prepared by mechanical alloying. *International Journal of Applied Engineering Research, 7*(11), 1420–1423.
14. Rajeshkumar Gangaram Bhandare, P., & Sonawane, M. (2013). Preparation of aluminium matrix composite by using stir casting method. *International Journal of Engineering and Advanced Technology, 3*, 2249–8958.
15. Londhe, V. D., Mhaske, M. S., & Kapgate, R. A. (2014). Evaluation of tribological behaviour of LM13—Silicon carbide composite under ambient & elevated temperature. *International Journal of Scientific Research Engineering & Technology, 3*(3), 334–340.
16. Saravanan, P., Shanmuga Priyan, M., Raghuram, R., Sharath Kumar, S., & Anand, T. Investigation of mechanical properties in aluminium silicon carbide mica hybrid metal matrix composite. *International Journal of Engineering Science and Advanced Technology, 5*(2), 112–117.

17. Nidhish, B. N., & Sijo, M. T. (2014). Effect of silicon carbide percentage on fracture toughness of aluminum silicon carbide metal matrix composites. *International Journal of Research in Engineering and Technology, 3*(11), 412–415.
18. Satyanarayen, D. R., Shreesaravanan, M. B., & Devanathan, C. (2015). A study on mechanical properties of aluminium LM25-SiC composites fabricated using stir casting technique. *International Journal of Advanced Research Trends in Engineering and Technology, II*(Special Issue XXII).
19. Jambukar, G. B., & Kharde, Y. R. (2014) Dry sliding wear behaviour of Al-Si-Ti alloys using Taguchi method. *International Journal of Science and Research, 3.*
20. Marwaha, R., Gupta, M. D., Jain, V., & Sharma, K. K. (2013). Experimental investigation and analysis of wear parameters on Al/SiC/Gr metal matrix hybrid composite by Taguchi method. *Global Journal of Researches in Engineering, 13*(9), 15–22. Version 1.0.
21. Bisht, N., Verma, A., Chauhan, S., & Singh, V. K. (2021). Effect of functionalized silicon carbide nano-particles as additive in cross-linked PVA based composites for vibration damping application. *Journal of Vinyl and Additive Technology, 27*(4), 920–932.
22. Thimmaiah, S. H., Narayanappa, K., Thyavihalli Girijappa, Y., Gulihonenahali Rajakumara, A., Hemath, M., Thiagamani, S. M. K., & Verma, A. (2023). An artificial neural network and Taguchi prediction on wear characteristics of Kenaf-Kevlar fabric reinforced hybrid polyester composites. *Polymer Composites, 44*(1), 261–273.
23. Raja, S., Verma, A., Rangappa, S. M., & Siengchin, S. (2022). Development and experimental analysis of polymer based composite bipolar plate using Aquila Taguchi optimization: Design of experiments. *Polymer Composites, 43*(8), 5522–5533.

Fin Design and Aerodynamics in High-Speed Missile and Rocket Systems

Vidhu Dixit, Gautam Goplani, Aman Aggarwal, and Harshit Shukla

Abstract Fins are crucial for flight stability and control, particularly at supersonic and hypersonic speeds. Shock waves, boundary layer behaviour, and flow separation are examples of complex aerodynamic phenomena that must be taken into account during the design phase since they significantly affect structural integrity and performance. This chapter covers the physics involved in the fin design for rocket systems and high-speed missiles and offers a literature analysis on the design and applications of fins in rockets and supersonic missiles, with a focus on optimizing the aerodynamic performance characteristics of fins. Fins are essential for flight stability and control because of their greater impact at high speeds, when aerodynamic forces significantly affect flight performance. This study examines the effects of various fin configurations on critical parameters like lift, drag, and overall stability. Fin placement, size, and other geometric changes are also taken into account to minimize wave drag. This article reviews the experimental and computational approaches to fin behaviour analysis. Special emphasis is made on the challenges of supersonic flight regimes, where phenomena like shock wave interactions, boundary layer effects, and flow separation come into effect. This paper summarizes earlier studies and provides useful information on the trade-offs necessary to strike a balance between structural durability and aerodynamic efficiency, serving as a guide for future advancements in missile and rocket fin design.

Keywords Aerodynamics · Fin design · Missile aerodynamics · Supersonic missiles · Defence technology

V. Dixit · G. Goplani · A. Aggarwal · H. Shukla (✉)
Department of Aerospace Engineering, School of Advanced Engineering, UPES, Dehradun, India
e-mail: hshukla21@gmail.com

K. K. Dwivedi et al. (eds.), *Mechanics and Materials Science of Biological Materials*, Materials Horizons: From Nature to Nanomaterials,
https://doi.org/10.1007/978-981-96-5366-9_15

1 Introduction

Fins are an essential component of missile and rocket design because they provide vital stability and control, particularly during high-speed flight. They ensure precision when hitting objects since they are aerodynamic surfaces that increase agility, withstand destabilizing forces, and help maintain the intended flight path. If fins are not constructed properly, projectiles may deviate significantly from turbulence, airflow disturbances, or changes in their centre of pressure, which could lead to mission failure.

Fins govern two fundamental functions: control (the intentional alteration of direction) and stability (the capacity to revert to an initial orientation following disturbances). These surfaces maintain stable flight from launch to target contact by fending off destabilizing forces. Lift, drag, and stability are three important aerodynamic performance measures that are impacted by fin design. Drag prevents forward motion, lift makes stabilization easier, and stability guarantees that the aerodynamic centre stays behind the centre of gravity. In the case that the missile rotates, this placement creates a restoring moment that keeps it on course (Fig. 1).

The tremendous aerodynamic forces experienced in the supersonic (Mach 1–5) and hypersonic (Mach 5+) flight regimes increase the complexity of fin design. Shock waves change the pressure distribution during supersonic flight, which affects fin performance. Aerodynamic performance and structural integrity are put to the test

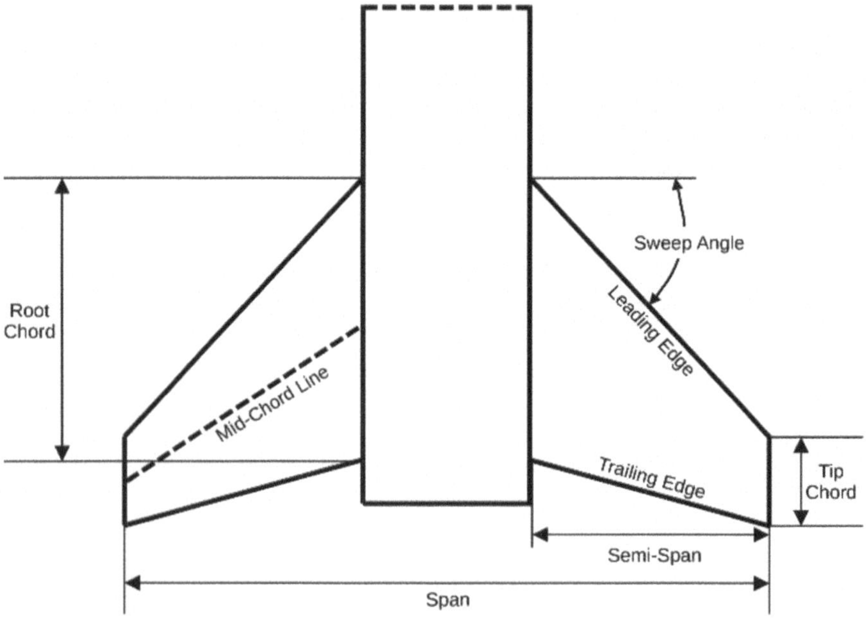

Fig. 1 Fin nomenclature

in hypersonic environments with fluctuating air densities and intense heat loads. Heat resistance, shockwave interaction, and preserving aerodynamic stability are crucial factors in fin design.

Fin types, fin placement, and their effects on lift and drag are just a few of the variables covered by basic fin design principles. In various speed ranges, lift generation, directional control, and induced drag must all be considered by designers. By improving camber, aspect ratio, and leading edge design, wave drag can be decreased. Methods like as wind tunnel testing and computational fluid dynamics (CFD) simulations offer vital information about fin performance.

A missile's fin design has a major impact on its structural integrity, stability, and aerodynamic performance while in flight. The choice of fin configuration is influenced by the missile's operating speeds, mission objectives, and environmental conditions. Each fin design has unique advantages and disadvantages that impact fabrication complexity, flutter resistance, drag, and stability. To guide the design process, Table 1 provides a comprehensive analysis of common fin configurations, highlighting both their benefits and drawbacks [1]. Designers can choose fins that best balance performance, durability, and manufacturing viability by taking these characteristics into consideration (Fig. 2).

This study highlights the literature on fin design, paying special attention to aerodynamic principles, jet interactions, and state-of-the-art design optimization techniques. By examining significant research, we intend to highlight significant advancements and identify knowledge gaps that will direct future investigations into enhancing the aerodynamic performance of high-speed rocket and missile systems.

2 Literature Survey

This survey of the literature looks at the significant contributions made to the field of fin design and aerodynamics in high-speed rocket and missile systems. The primary subjects are the optimization of aerodynamic performance and the various factors influencing stability and control at supersonic speeds. From computer modelling to empirical research, the reviewed publications provide a comprehensive understanding of the challenges and advancements in missile fin design.

In their study, Vitale et al. [2] investigated the aerodynamics of wrap-around fin (WAF) missile configurations at supersonic Mach numbers ranging from 2.75 to 5.15. They analysed the loss of static stability as velocity increased and examined the effects of fin geometry on induced roll and side moments. The researchers used both empirical and computational methods, such as PRODAS and EAGLE, to report abnormalities in static stability that was associated with stress structures in the fin area. Traditional roll and side moment characteristics that are consistent with WAF designs were indicated by their findings, which showed a high degree of agreement between computational and empirical predictions.

Table 1 Advantages and disadvantages associated with various fin configurations

Fin configuration	Advantages	Disadvantages
Rectangular fin	• Simple and easy to fabricate • Structurally robust, offering good durability • Resistant to flutter	• Aesthetically less appealing, often considered unattractive • May be susceptible to damage upon landing, especially the corners
Swept fin	• Structurally robust, providing strength and durability • Resistant to flutter, which enhances aerodynamic stability at high speeds	• Susceptible to damage during landing, particularly if landing on uneven surfaces
Delta fin	• Resistant to flutter, making it ideal for high-speed rockets	• The sharp tip may be prone to damage and can pose a handling hazard (this can be mitigated by rounding the tip) • Has a lower force coefficient compared to other shapes, which can reduce aerodynamic efficiency
Tapered swept fin	• Ideal for high-speed flights, especially for upper stages of rockets • Can be clipped to move the CP forward, improving stability	• Prone to damage upon landing due to its geometry • May be susceptible to flutter, requiring careful design considerations
Clipped delta fin	• Structurally robust, offering durability • Clipping allows for the forward shift of CP, improving stability	• Susceptible to damage during landing, particularly the edges
Trapezoidal fin	• A reliable, robust design often used due to its structural integrity • Can be modified (clipped) to shift the centre of pressure (CP) forward, enhancing stability • Resistant to flutter	• Less aerodynamically efficient compared to swept planforms, which may affect flight performance
Ellipsoidal fin	Known for its aerodynamic efficiency, potentially reducing drag • Structurally robust and resistant to flutter	• Difficult to fabricate due to its complex shape • Not compatible with Barrowman equations, making CP calculations more complicated
Clipped taper swept fin	• An enhanced version of the tapered swept design, offering better aerodynamic performance • Clipping shifts the CP forward, improving stability	Prone to damage upon landing • May be susceptible to flutter under certain conditions

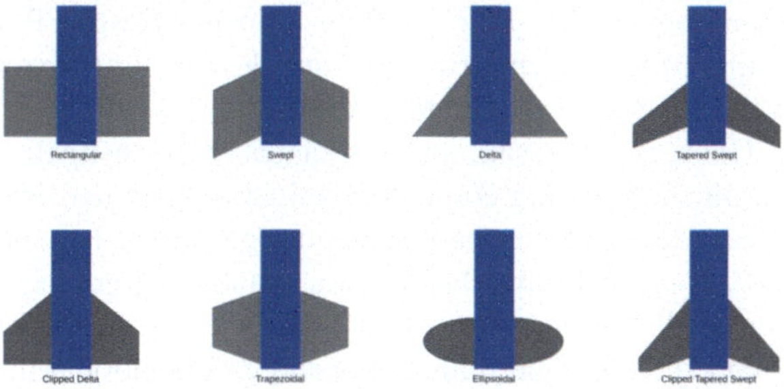

Fig. 2 Fin configurations

Brandeis and Gill [3] explored the impact of jet interaction on missile configurations featuring lifting surfaces. Their experimental investigation focused on an ogive-cylinder body with various surface combinations, including wings, tails, and strakes. The findings revealed a significant amplification of forces and moments resulting from the interaction between the jet-induced flow field and the planar surfaces, even at Mach numbers as low as 2. This amplification of forces varied inversely with jet pressure, while the interaction moments contributed to aerodynamic lifting forces, thereby enhancing control for certain configurations.

Tanıl and Platin et al. [4] developed a design tool in MATLAB® called EXCON, which focuses on aerodynamic configuration sizing for missiles. By integrating Missile DATCOM for aerodynamic coefficient prediction, the tool was applied to a subsonic cruise missile as a numerical example. Their study demonstrated that missile geometry optimization is mission-dependent and influenced by factors such as flight trajectory and motor type. EXCON proved effective in designing tailored configurations based on specific mission profiles.

In 2014, Nguyen et al. [5] validated body-wing-tail missile configurations using Missile DATCOM and compared the results with experimental data. They developed an aerodynamic database (Aero DB) that was integrated into the missile design process to optimize air intercept missile configurations. The optimized design achieved a 27.8% improvement in range compared to the baseline configuration. Their application of ANSYS Fluent for high-fidelity analysis yielded more reliable results than traditional spreadsheet analyses, particularly at angles of attack up to 10 degrees.

Sethunathan et al. [6] examined the aerodynamic characteristics of anti-aircraft missiles, specifically focusing on the body, wing, and tail sections. They employed linear and slender wing theories to predict the centre of pressure and size the tail for static neutral stability. According to their research, performance improves at supersonic speeds as drag decreases with height. Additionally, they emphasized that in order to achieve static stability, the tail area must be optimized according to the relative positions of the centre of gravity and centre of pressure.

Allen and Ghoreyshi [7] used time-accurate computational fluid dynamics (CFD) simulations to study forced movements and aerodynamic stability models for missile designs. To predict aerodynamic stability derivatives, they employed Mentor's SST turbulence model, Mach sweeps, and the Kestrel flow solver. According to their results, Mach sweeps provide a feasible substitute for discrete flying state simulations in the computation of static coefficients.

Ghoreyshi, Jirasek, and associates [8] investigated extended strake-tail missile modifications using computational aerodynamics. Their simulations included both steady-state and unsteady-state configurations and encompassed Mach numbers of 1.17, 2.49, and 4.39. There were notable differences at Mach 1.17 and at significant angles of attack at Mach 2.49, but overall, the force and moment coefficients obtained from wind tunnel data and computational fluid dynamics (CFD) matched rather well. They also examined the impact of fin deflections and changes in Mach numbers, coming to the conclusion that increases in moment coefficients become less significant as Mach numbers climb.

The literature study presented here highlights important advancements in the design and aerodynamic optimization of fins for high-speed rocket and missile systems. Numerous significant research areas are covered by the evaluated papers, including design optimization methods, jet interaction effects, missile system aerodynamics, and aerodynamic stability. Notably, time-accurate computational fluid dynamics (CFD) has been used to study forced movements and model stability, offering alternatives to conventional methods for assessing aerodynamic performance. Additionally, by streamlining mission-specific features, the combination of aerodynamic databases like Aero DB and design tools like EXCON has improved performance results. The combined findings show how crucial elements like tail optimization, fin shape, and jet interactions are to obtaining better stability and control at supersonic speeds.

3 Discussion and Key Takeaways

The literature review highlights significant advancements in fin design and aerodynamic optimization for high-speed missile and rocket designs, while also concentrating on other crucial research areas:

- ***Aerodynamic Stability***: The significance of fin geometry and centre of pressure placement in maintaining static stability is demonstrated by experiments conducted by Sethunathan et al. [3] and Brandeis and Gill [3]. Shock structure-specific anomalies underscore the importance of thorough evaluations during the design phase.
- ***Jet Interaction Effects***: Brandeis and Gill [3] demonstrated that jet-induced flow interactions can significantly amplify forces and moments, enhancing control and performance even at lower Mach numbers.
- ***Design Optimization Tools***: The development of tools like EXCON by Tanıl and Platin et al. [4] and the aerodynamic database (Aero DB) by Nguyen et al. [5] showcases the benefits of integrating computational methods for mission-specific configurations and improved aerodynamic coefficients.

- ***CFD Applications***: Research by Allen and Ghoreyshi [7] and Ghoreyshi et al. [8] highlights the effectiveness of time-accurate CFD simulations in predicting aerodynamic stability derivatives and moment coefficients, offering efficient alternatives to traditional simulations.
- ***High-Speed Flight Considerations***: As well as the importance of fin design, tail optimization, and jet interactions in achieving stability and control at supersonic speeds, the results underscore the challenges posed by shockwaves and heat resistance.

4 Conclusion

Fin design and optimization are essential for enhancing rocket and missile aerodynamic performance, particularly in the challenging supersonic and hypersonic flight regimes. This survey of the literature includes significant research findings that highlight the intricate relationships among aerodynamic stability, control, and design innovation. The studies under review highlight the significance of fin shape, jet interaction effects, and contemporary computational techniques in tackling the difficulties of high-speed flight.

As the field advances, more research into these elements will be necessary to design more effective and efficient rocket and missile systems. In addition to advancing our knowledge of aerodynamic principles, the data gathered from ongoing research opens the door for possible future developments that could improve the dependability and performance of high-speed aerospace vehicles.

References

1. 6. Fin, *Introduction to rocket design, Richard Nakka's experimental rocketry web site.*
2. Vitale, R. E., Abate, G. L., Winchenbach, G. L., & Riner, W. (1992). Aerodynamic test and analysis of a missile configuration with curved fins. In *Astrodynamics conference, 1992* (pp. 788–798). American Institute of Aeronautics and Astronautics Inc., AIAA. https://doi.org/10.2514/6.1992-4495.
3. Brandeis, J., & Gill, J. (1998). Experimental investigation of super- and hypersonic jet interaction on missile configurations. *Journal of Spacecraft and Rockets, 35*, 296–302.
4. Tanıl, Ç., Platin, B. E., & Mahmutyazıcıoğlu, G. (2009). *External configuration optimization of missiles in conceptual design.*
5. Nguyen, N.-V., Tyan, M., Lee, J.-W., & Byun, Y.-H. *Investigations on missile configuration aerodynamic characteristics for design optimization.*
6. Sethunathan, P., Professor, A., Sugendran, R. N., & Anbarasan, T. *Aerodynamic Configuration Design of a Missile.* www.ijert.org.
7. Allen, J., & Ghoreyshi, M. (2018). Forced motions design for aerodynamic identification and modeling of a generic missile configuration. *Aerospace Science and Technology, 77*, 742–754.
8. Ghoreyshi, M., Jirasek, A., Aref, P., & Seidel, J. (2022). Computational aerodynamic investigation of long strake-tail missile configurations. *Aerospace Science and Technology, 127*, 107704.

The manufacturer's authorised representative in the EU is Springer Nature Customer Service Centre GmbH, Europaplatz 3, 69115 Heidelberg, Germany. If you have any concerns regarding our products, please contact ProductSafety@springernature.com

Printed and bound by CPI Group (UK) Ltd, Croydon, CR0 4YY
01/10/2026
02225921-0006